英　汉
直升机术语辞典

English-Chinese Dictionary of Helicopter Terms

主　编 ◎ 申秋元　姚剑华　程新平

北京理工大学出版社
BEIJING INSTITUTE OF TECHNOLOGY PRESS

图书在版编目（CIP）数据

英汉直升机术语辞典 / 申秋元，姚剑华，程新平主编. —北京：北京理工大学出版社，2022.4

ISBN 978-7-5763-1221-8

Ⅰ.①英… Ⅱ.①申… ②姚… ③程… Ⅲ.①直升机－名词术语－对照词典－英、汉 Ⅳ.①V275-61

中国版本图书馆CIP数据核字（2022）第056200号

出版发行 / 北京理工大学出版社有限责任公司
社　　址 / 北京市海淀区中关村南大街 5 号
邮　　编 / 100081
电　　话 /（010）68914775（总编室）
（010）82562903（教材售后服务热线）
（010）68944723（其他图书服务热线）
网　　址 / http://www.bitpress.com.cn
经　　销 / 全国各地新华书店
印　　刷 / 保定市中画美凯印刷有限公司
开　　本 / 880 毫米 ×1230 毫米 1/32
印　　张 / 17.25
字　　数 / 407千字
版　　次 / 2022 年 4 月第 1 版 2022 年 4 月第 1 次印刷
定　　价 / 62.00 元

责任编辑 / 多海鹏
文案编辑 / 把明宇
责任校对 / 周瑞红
责任印制 / 李志强

《英汉直升机术语辞典》主创团队

Editors of English-Chinese Dictionary of Helicopter Terms

主　　编： 申秋元　姚剑华　程新平

副 主 编： 牟善成　鄂爱军　张滕龙　黄俊华　李　健

编纂人员：（按姓氏笔画顺序）

王　萌　王君慧　牛剑桥　田　英　闫春蕾

江　韵　孙明月　李　雷　李　鹏　李世忠

张　凯　张云龙　陈媛媛　范鹏程　罗　波

徐　希　高　巍　郭　艳　黄汉超　董　萍

前　言

直升机因其低空飞行、垂直起降、机动灵活等特点，广泛应用于军用和民用领域。我国直升机产业始终坚持“自主发展为主、积极开展国际合作”的方针，通过几十年的发展，在直升机机型数量和性能等方面取得了长足进步。尤其是在武装直升机和运输直升机的研制和生产方面，已经跻身世界先进国家的行列，形成了跨越式发展的局面。

我国直升机产业发展取得的成就，与我国在直升机领域对外交流与合作是密不可分的。随着我国对外开放广度和深度的加大以及直升机技术发展水平的日渐成熟，我国直升机分队在军用和民用领域遂行中外联演联训、联合巡逻、联合救灾和医疗救护、对外援助等大型涉外任务的机会日益增多。如今，在直升机领域的对外交流与合作已经成为国家外交的重要组成部分。无论是哪种形式的国际合作方式，都需要大批高素质外语人才和具有较强外语能力的直升机教学科研人员、飞行人员和机务维修人员。本辞典正是在这种背景下应运而生的。

本辞典编纂人员在直升机领域都具有较为丰富的对外合作与交流经验。他们或多年从事对外直升机装备、飞行和机务教学与研究，或多次参加中外联演联训和直升机分队国际维和任务。在执行涉外任务的过程中，储备了大量相关领域有价值的外语资料。通过搜集和整理国内外直升机技术资料，对从事直升机各学科门类研究的专家进行咨询、座谈和共同研究，形成了涵盖直升机各领域研究的较为完备的术语集合，为执行或即将执行涉外任务的直升机专业人员，院校和科研院所从事直升机专业的对外教学与研究人员，直升机专业英语学习爱好者的工作、学习和研究提供参考。

本辞典共收录英汉词条2万余条，均按照字母表的顺序排列，内容主要涉及直升机驾驶、直升机机械维护、直升机航电维护、直升机军械维护等专业。其中直升机驾驶专业术语主要涉及直升机飞行理论、直升机飞行气象、直升机空中领航等学科；直升机械维护专业术语主要包括直升机构造、发动机构造、直升机机械维护和操纵规程等学科；直升机航电维护专业术语主要涵盖直升机通信与导航设备、直升机仪表与电气设备、直升机航电装备及其管理和操作规程等学科；直升机军械维护专业术语主要针对机载武器和发射系统及其维护和操作规程等学科。

由于时间仓促，加之编者水平有限，本辞典中可能存在一些疏漏和错误，希望广大同行不吝赐教，以便进一步改进我们的工作，使本辞典将来能够以更加完善的面目呈现给广大读者。

编者

2022年4月

体例说明

一、编排方式

为查阅方便,所有词条均按英文字母顺序排列。拼法相同而词源、词义不同的词，分别立为词条。

二、释义

同一英文词条的多种不同汉语释义用分号隔开。例如：

riveting machine 铆接机；铆合机

retractable float 伸缩浮筒；收放式浮筒

三、符号

除一般符号外，本辞典中的常用符号有：

1. 圆括号“( )”

（1）圆括号内的词是对应词的解释、补充或强调说明。例如：rehearsal day（演习、作战）预演日，“演习、作战”指演习或作战的预演日；revolution per minute (RPM) 转速（转 / 分），“转 / 每分”指转速的单位为转 / 每分；range 靶场；靶区；射击场;（尤指）导弹及无人驾驶飞行器的试验场或武器试验场;（导弹等）射程；航程；距离；测距；距离试射；区域；幅度；范围；无线电指向标;【无】波段。其余以此类推。

（2）圆括号内的词可以替换前一个单词。例如：

reinforce 增援；支援 ；加强；增加强度，加固；炮（枪）身的粗厚尾部

reserve 预备队；预备役；后备部队；保留；保存；储备；预定；预备补给品；储（藏）量；伞兵的备用伞

（3）圆括号内还可放置译文的军事缩写形式或完整形式。例如：

revolutions per minute (RPM) 转速（转 / 分）

RPM (revolutions per minute) 转速（转 / 分）

2. 分号“ ；”

分号为间隔号，用于将同一英文词目的不同译文隔开。例如：

rotary knob 回转旋钮；旋转开关

rotating pin shaft 转动销轴；旋转销轴

3. 连字符号“-”

连字符号用于英语合成词中。例如：

run-down check 停车检查，停转检查

satellite-and-laser-guided bombs 卫星及激光导航导弹

4. 中括号【 】

（1）单数形式的名词词目释作某义时需用或常用复数形式时，在释义前予以标注。例如：

scramble 紧急起飞（航空母舰上或地面上飞机的紧急起飞）；驾驶员的紧急起飞动作；【复】截击；拦截目标；（电子信号）加扰；加密；用改变的频率通讯

roll control 滚转控制；横侧操纵，滚转操纵；【复】横侧操纵机构

（2）中括号内为各军兵种和专业技术分类等缩略语。本辞典内使用的缩略语及其含义如下：

【空】——空军
【陆】——陆军
【炮】——炮兵
【海】——海军
【象】——气象
【天】——天文
【航】——航空
【计】——计算机
【雷】——雷达
【通】——通信
【无】——无线电
【直】——直升机
【固】——固定翼
【口】——口语
【箭】——火箭、导弹
【令】——命令、口令
【美】——美军、美国
【英】——英军、英国
【俄】——俄军、俄罗斯
【地理】——地理
【舰船】——军舰
【防化】——防化

目　录

A

AAI（arrival aircraft interval）到达飞机间隔

AAI（average arrival interval）平均到达间隔

AAM（air-to-air missile）空对空导弹；空空导弹

AAR（airport arrival rate）机场到达率

AAS（aerodrome advisory service）机场咨询服务

AAT（assigned arrival time） 指定到达时间

AB（air base）空军基地

abandon 离机；弃机；抛弃；放弃

abandon the aircraft 离机（指机上人员跳伞）

abandon the takeoff 中断起飞

abandoned airport 废弃机场

abandoned takeoff 中断的起飞

abatable deposit 耐磨层

abatable layer 耐磨层

abatable tip 可磨耗叶冠

abbreviated flight plan 简略飞行计划

abbreviated precision approach path indicator（APAPI）简易精确进场航道指示器

abbreviated visual approach slope indicator system（AVASIS）简易目视进近下滑道指示系统

ABC（atomic, biological, chemical）warfare 原子生物 [细菌] 化学战

abeam 正横着的（地）；正对着机身中部的（地）

aberration 光行差；像差；色差

ABM（anti-ballistic missile）反弹道导弹

abnormal change 反常变化

abnormal clearance 间隙不正常

abnormal friction 异常摩擦

abnormal noise 异常噪声

abnormal oil consumption 滑油消耗异常

abnormal response time 反应时间异常

abnormal rubbing 不正常摩擦

abnormal variation of performance 性能不正常变化

abnormal vibration 异常摆动；振动异常

A-bomb 原子弹

abort 中断；中止；放弃

abort drill 中断操练

abort takeoff 中止起飞

abort takeoff drill 中断起飞操练

aborted engine start 发动机开车失败

aborted landing 中止着陆

aborted start 中止的起动

aborted takeoff 中止的起飞；中断的起飞

abortion 中止；中断

abortive mission 中止的任务

above ground level（AGL）地面高度；绝对高度

above mean sea level（AMSL）距平均海平面高度

above obstruction level（AOL）越障高度

above sea level（ASL）海拔高度；相对高度

abrasion 磨损

abrasion-resistant cap 耐磨帽

abrasive tip 耐磨蚀叶冠

abreast 朝同方向并列（的）；并肩（的）

abrupt manoeuvre 机动过猛

abs-initio trainer 初级教练机

absolute aerodynamic ceiling 绝对气动升限

absolute air superiority 绝对空中优势

absolute altimeter 绝对高度表

absolute altitude 地面高度；绝对高度

absolute angle of attack 绝对迎角

absolute ceiling 绝对升限

absolute command of air 绝对制空权

absolute control of air 绝对制空权

absolute delay 绝对延迟

absolute fix 绝对定位；可靠定位

absolute humidity 绝对湿度

absolute instability 绝对不稳定性

absolute parallax 绝对视差

absolute pressure 绝对压力
absolute pressure controller 绝对压力控制器
absolute ratings 极限参数
absolute stability 绝对稳定性
absolute stereoscopic parallax 绝对立体视差
absolute temperature 绝对温度
absorbent cotton 脱脂棉
AC（alternating current）交流电
AC circuit 交流电路
AC component 交流分量
AC generator 交流发电机
AC power center 交流电源中心
ACARS（aircraft communication addressing and reporting system）飞机通信寻址与报告系统
ACARS front end processing system 飞机通信寻址与报告系统前端处理系统
ACARS ground system standard 飞机通信寻址与报告系统 地面系统标准
ACAS（airborne collision avoidance system）机载防撞系统；空中防撞系统
ACCB（air conditioning control box）空调控制盒
accelerated flight 过载飞行；加速飞行
accelerated stall 过载失速
accelerate-go distance 加速前进距离
accelerate-stop distance 加速煞停距离
accelerate-stop distance available 可用中断起飞距离
accelerating arm 加速臂
accelerating arm base 加速臂座
accelerating pump 加速泵
acceleration 加速
acceleration check 加速检查；加速确认
acceleration control unit（ACU）加速控制装置；加速控制器
acceleration controller 加速控制器
acceleration controller lever 加速控制杆
acceleration error 加速度误差
acceleration switch 惯性开关
acceleration-deceleration unit 加减速单元
accelerative attitude 加速姿态
accelerator 油门；油门杆；加

速器

accelerator pump 加速泵

accelerometer 加速度表；加速度计

acceptable alternative product 可接受的代用品

acceptance test 接机考验；验收实验

acceptance test procedure 验收测试（试飞）程序

acceptance unit 接受单位

accepting controller 接受单位；管制员

accepting unit 接受单位；管制员

access 通路；入口

access control and signaling equipment 接续控制与信令设备

access link 传送线

access panel 检查口盖

access taxiway 进入滑行道

accessibility 可接近性；可达性

accession 增加新员；入伍

accessories 附件；附加装置

accessories and harness 附件和电缆

accessories drive 附件传动

accessories max intensity 附件最大用电量

accessory drive 附件传动装置；附件传动

accessory drive box 附件传动盒

accessory drive casing 附件传动机匣

accessory drive gear 附件传动机构

accessory drive train 附件驱动链

accessory drive unit 附件传动组件

accessory gear train 附件齿轮系

accessory gearbox 附件传动机匣；附件齿轮箱

accessory relay 附件继电器

accessory replacement 更换附件

accident 事故；遭难；失事

accident data recorder（tape）飞行事故记录器（记录带）

accident rate 事故率

accident report 事故报告

accident-free 无事故的

accommodation 居住舱室；膳宿供应；适应性调节（眼睛等）；招待设备【美】

accommodation coefficient 调节系数

accredited medical conclusion 经认可的医学结论

accumulator 蓄电池；蓄压器；储油器

accumulator jar 蓄电瓶

accuracy landing 准确着陆

ace 一流飞行员；王牌飞行员

ace pilot 特级飞行员；王牌飞行员（击落五架以上）；一流飞行员

acetone 丙酮

A-check A 检

acid battery 酸性电池

acid cell 酸性电池

acknowledge 收到应答

ACL（aircraft list）飞机列表

ACN-PCN（aircraft classification number-pavement classification number）飞机分级数 - 道面分级数

acorn 整流罩；整流板

acoustic 声学的；听觉的

acoustic splitter 消声隔板

acoustic velocity 音速

ACP（armed control panel）武装（火控）控制板

acquisition 目标定位；捕获；目标指示；天线定向；截获，获取（信号、情报等）

acquisition and tracking radar 搜索与跟踪雷达

acrobatic flight 特技飞行

across wind force spectrum 侧风力分布谱

across wind load 侧风负载

action line 作用线

action of spring 弹簧作用

action shift 动作转换

activation 激活；触发

active air defense 积极防空措施；主动防空

active clearance control 主动间隙控制

active connection 有效连接；主动连接

active control technology 主动控制技术

active control（s）主动控制

active countermeasures 有源对抗

active detection systems 主动探测系统

active flight plan 有效飞行计划

active frequency 有源频率

active front 活跃锋

active homing guidance 主动寻的制导

active jamming 积极干扰；有源干扰

active missile 主动寻的导弹

active power 有功功率
active runway 现用跑道；使用中的跑道
active system 有源系统；主动系统
active two-terminal network 有源二端网络
active-active system 双活系统；主动 - 主动系统
activity attributes 活动属性
activity code 活动编码
actual calculated landing time 实算着陆时间
actual circuit 实际电路
actual combat 实战
actual core speed 实际核心转速
actual fan speed 实际风扇转速
actual flight level 实际飞行高度
actual navigation performance 实际导航性能
actual runway length 实际跑道长度
actual time of arrival 实际到达时间
actual time of departure（ATD）实际离场时间
actual time over 实际经过时间
actuating arm 驱动杆
actuating cylinder 作动筒
actuating horn 操纵摇臂
actuating shaft 驱动轴
actuation 作动
actuator 作动筒；作动器；舵机
actuator disk 制动圆盘
actuator motor 舵机
ACU（acceleration control unit）加速控制装置；加速控制器
AD（air distance）无风距离
adaptation of components 部件的匹配
adaptation to requirement 需求调节
adapter 转接器；适配器
adaptive communications 自适应通信；自适应通信系统
adaptive differential pulse code modulation 自适应差分脉码调制
ADC（air data computer）大气数据计算机
Adcock antenna 艾特可克定向天线
ADCUS（advise customs）通报定制
added mass 附加质量
additional check valve 附加单向活门
additional function 附加功能
additional mass 附加质量
additional services 附加咨询服务

additive operational circuit 加法运算电路

adequate visual reference 充分视觉参考

ADF（adjustable delay factor）可调延迟因子

ADF（automatic direction finder）自动方位搜寻器；自动测向仪

ADG（air driven generator）空气驱动发电机

adiabatic 绝热的

adiabatic change 绝热变化

adiabatic compressible flow 绝缘可压缩流

adiabatic cooling 绝热冷却

adiabatic heating 绝热升温

adiabatic lapse rate 绝热直减率

adjacent zone 临近地域

adjustable 可调节的

adjustable cam switch unit 可调凸轮电门组件

adjustable horizontal stabilizer 可调水平安定面

adjustable propeller 可调螺旋桨

adjustable resistor 可调电阻

adjustable stabilizer 可调整稳定面

adjustable tail plane 可调水平安定面

adjustable vertical stabilizer 可调式垂直安定面

adjustable-pitch propeller 变距推进器

adjustable delay factor（ADF）可调延迟因子

adjuster 调整机构

adjusting fire 试射

adjusting mechanism 调整机构

adjusting washer 调整垫圈

adjustment coefficient 调节系数

adjustment fire 试射

adjustment screw 微调旋钮

adjustment washer 调整垫圈

ADR（airport departure rate）机场离站率

ADS（air data system）大气数据系统

ADS（automatic dependent surveillance）自动相关监视

ADS-B out（automatic dependent surveillance-broadcast out）监视信息发送

ADT（assigned departure time）指定离场时间

advance 推进

advance device 提前器
advance of phase 时段提前
advance ratio 前进比；进动比
advance unit 提前器
advanced air base 前进空军基地
advanced attack helicopter 先进攻击型直升机
advanced automated system 先进自动化系统
advanced automation training system 先进自动化培训系统
advanced boundary information message 高级边界信息报文
advanced communication navigation surveillance system 先进的通信导航监视系统
advanced surface movement guidance and control system 先进地面活动引导和控制系统
advanced trainer 高级教练机
advancing blade 前行桨叶（直升机）
advancing side 前行桨叶
advection 平流
advection fog 移流雾
adverse condition 不利因素
adverse weather 不利天气；恶劣天气
adverse weather aerial delivery system（AWADS）恶劣天气空投系统
adverse weather condition 不利气象
adverse weather flight 不利气象飞行
adverse weather window 恶劣天气观察窗
adverse yaw 不利偏航；反向偏航
advise customs（ADCUS）通报定制
advise intention 告知意图
advisor 顾问
advisory 情况通知；通报
advisory airspace 咨询空域
advisory area 咨询区
advisory circular 咨询通告
advisory light 辅助着陆灯
advisory message 注意信息
advisory notice 注意事项
advisory route 咨询航线
advisory service 咨询服务；咨询勤务
AEO（all engines operating）发动机工作正常
AEO（all engines operating）rating

AEO 状态（发动机工作正常）
aerated plastics 泡沫塑料
aerial 天线【英】
aerial barrage 空中气球拦阻网
aerial beacon 航空信标
aerial bomb 航弹；航空炸弹；空投炸弹
aerial bombardment 空袭
aerial camera 航空照相机
aerial cannon 航空机关炮；航炮
aerial crane 空中吊车
aerial delivery system 空投系统
aerial film 航空胶片
aerial imagery 空中成像
aerial photogrammetry 航空摄影测量（测绘）制图
aerial photography 航拍
aerial picket 空中巡逻
aerial refueling 空中加油
aerial refueling delay 空中加油延迟
aerial supply 空中补给
aerial support 空中支援
aerial survey 空中测绘
aero done 滑翔器
aero elastic stability 气弹耦合
aero elastic tailoring 气动弹性剪裁
aero elasticity 气动弹性力学
aero engine 航空发动机
aero foil 机翼；翼型
aero foil resistance 翼型阻力
aerobatic flight 特技飞行
aerobatics 特技飞行
aerobridge 登机桥
aerodrome 机场；航空站
aerodrome advisory service（AAS）机场咨询服务
aerodrome beacon 机场信标
aerodrome climatological summary 机场气候概要
aerodrome climatological table 机场气候简表
aerodrome control 机场管制
aerodrome control radar 机场监视雷达
aerodrome control radio 机场联络电台
aerodrome control service 机场管制服务
aerodrome control tower 机场管制塔
aerodrome control unit 机场控制单位（室）
aerodrome elevation 机场标高；场高

aerodrome flight information service（AFIS）机场飞行情报服务
aerodrome forecast（AF）机场（天气）预报
aerodrome identification 机场识别
aerodrome identification sign 机场识别标志
aerodrome obstacle chart 机场障碍物图
aerodrome operating minimum 机场最低飞航限度
aerodrome operational area 机场操作区
aerodrome practice 本场飞行训练
aerodrome rating 机场等级
aerodrome reference code 机场参考代码
aerodrome reference point 机场基准点
aerodrome station 机场电台
aerodrome surveillance（approach）radar（ASR）机场监视（进近）雷达
aerodrome taxi circuit 机场滑行路线
aerodrome traffic 机场交通
aerodrome traffic circuit 机场起落航线
aerodrome traffic frequency（ATF）机场交通频率
aerodrome traffic zone（ATZ）机场交通地带
aerodrome visibility operational level 机场能见度运行等级
aerodrome warnings 机场（气象）警告
aerodynamic axis of airfoil 气动力轴
aerodynamic balance 气动补偿；气动平衡
aerodynamic braking【空】气动力减速；【空】气动力制动
aerodynamic center 气动力中心
aerodynamic center（of wing section）气动力中心（翼剖面）
aerodynamic coefficients 空气动力系数
aerodynamic configuration 气动外形
aerodynamic damping 气弹耦合；气动弹性耦合；气动阻尼
aerodynamic drag 气动阻力
aerodynamic efficiency 气动效率
aerodynamic force 气动力

aerodynamic heating 气动加热
aerodynamic layout 气动布局
aerodynamic loading 气动载荷
aerodynamic trail 气动（凝结）尾迹
aerodynamic twist 气动扭转
aerodynamics 空气动力学
aerodyne 重航空器
aeroembolism 空气栓塞；空气栓塞症
aerofoil rotational resistance 翼型旋转阻力
aeromarine 水上飞机
aeromedical evacuation system 空中医疗后送系统
aerometeograph 航空气象记录仪
aerometeorograph 航空气象记录仪
aeronaut 气球驾驶员；轻航空器驾驶员
aeronautical administrative communication 航空行政通信
aeronautical advisory station 航空咨询电台
aeronautical approach chart 航空进场图
aeronautical beacon 航行信标
aeronautical binary phase shift keying 航空双相移键控
aeronautical chart 航空图
aeronautical communication architecture 航空通信结构
aeronautical communication service processor 航空通信业务处理器
aeronautical communication subgroup 航空通信分组
aeronautical fixed service 航空固定通信业务
aeronautical fixed station 航空固定电台
aeronautical fixed telecommunication network（AFTN）航空专用通信网
aeronautical ground light 航空地面灯光
aeronautical information 航空情报
aeronautical information circular（AIC）航空信息通报
aeronautical information manual（AIM）航空信息手册
aeronautical information publication（AIP）航空信息出版物
aeronautical information regulation and control 航行资料规划和管制（定期制航行通告）

aeronautical information regulation and control（AIRAC）航空信息规定与控制

aeronautical information service（AIS）航行情报服务

aeronautical light beacon 航行灯标

aeronautical material specification 航空材料标准

aeronautical message transfer service 航空报文移交业务

aeronautical meteorological station 航空气象站

aeronautical mile 航空英里

aeronautical mobile service 航空移动通信服务

aeronautical passenger communications 航空旅客通信

aeronautical product 航空产品

aeronautical public correspondence 航空公用通信

aeronautical radio aids 航空无线电导航设备

aeronautical radio incorporated（ARINC）航空无线电设备公司

aeronautical route chart 航路图

aeronautical satellite data link 航空卫星数据链

aeronautical station 航空（地面）电台

aeronautical surfaces 机场控制区设施（航行面）

aeronautical telecommunication service 航空电信服务；航空通信服务

aeronautical telecommunication station 航空电信站；航空电台

aeronautics 航空学

aero-otitis 航空性耳炎

aero-otitis media 航空性中耳炎

aeropause 大气航空边界；适航限层

aeroplane 固定翼飞机

aeropulse 脉冲式喷气发动机

aeroshow 航空展览

aerosinusitis 航空性鼻窦炎

aerospace 航空航天空间；航空航天的

aerostat 轻航空器；浮空器

aerostatics 空气静力学

aerothermodynamic duct（athodyd）冲压式喷气发动机

aerotow 空中拖航

AEW（air early warning aircraft）空中预警机

AEW&C（airborne early warning and control）空中预警与控制
AF（aerodrome forecast）机场（天气）预报
AFB（air force base）空军基地（美）
AFCS（automatic flight control system）自动飞行控制系统
AFCS（four-axis automatic flight control system）四轴自动飞行控制系统
affirm 是的；同意（陆空通话用语）
affirmative 是的；同意；是这样（陆空通话用语）
AFIS（aerodrome flight information service）机场飞行情报服务
AFSS（automated flight service station）自动飞行服务站
aft 在（或向）船尾；在（或向）机尾
aft core cowl 发动机后核心整流罩
aft fairing 后整流罩
aft fuselage 机身后部
aft left 后左
aft part 后部
aft thrust reverser control 后反推装置操纵
aft toilet drain 后厕所泄放口
after action review 事后回顾
after bottom center / after bottom dead center 下止点后；下死点后
after top center / after top dead center 上死点后；上止点后
afterbody 尾端；后机身
afterburner 加力燃烧室
afterburning 二次燃烧；加力燃烧；复燃
afterfiring 尾喷管喷火；（活塞式发动机）喷火
aft-fan 后部风扇；尾部风扇
AFTN (aeronautical fixed telecommunication network）航空专用通信网
AFWA（air force weather agency）空军气象局
aggregate demand indicator 综合指令指示器
AGL（above ground level）地面高度；绝对高度
AGM（air-to-ground missile）空地导弹
agonic line 零磁偏线
agravic illusion 失重错觉

agreed point 约定点
A-head 原子弹头
AHRS（attitude and heading reference system）姿态与航向参照系统
AI radar（airborne-intercept radar）机载截击雷达
AIC（aeronautical information circular）航空信息通报
AIC（ammunition identification code）弹药识别代号
AICBM（anti-intercontinental ballistic missile）反洲际弹道导弹
aid to navigation 导航辅助设备
aid worker 救援人员
ailavator 升降副翼
aileron 副翼
aileron angle 副翼偏角
aileron booster 副翼助力器
aileron buzz 副翼颤振
aileron drag 副翼阻力
aileron reversal 副翼反效
aileron station 副翼坐标点
AIM（aeronautical information manual）航空信息手册
AIM（air intercept missile）空中截击导弹
aiming error 瞄准误差
aiming point 瞄准点；标定点
aim-off 提前量
AIP（aeronautical information publication）航空信息出版物
air 空气；大气
air accident 飞行事故；空中失事
air accident investigation 飞行事故调查
air admission 空气进入
air alert 空中警戒
air almanac 航空天文历书
air ambulance 救护飞机
air arm 航空兵（指兵种）
air assault 空袭
air assault brigade 空袭旅；空中突击旅
air attack 空袭
air attack alarm 空袭警号；防空警报
air barrage 空中阻击
air base（AB）空军基地
air bleed 引气孔；引气
air bleed valve 放气活门
air blower 鼓风机
air brakes 空气减速装置
air bridge 廊桥【英】

air cargo 空运货物
air carrier 空运机构；航空承运人
air carrier airport 航空运输机场
air circulation 空气流动
air compressibility 空气压缩性
air condition system 空调系统
air conditioning control box (ACCB) 空调控制盒
air conditioning switch 空调开关
air conditioning system 空调系统
air conditioning system controller 空调系统控制器
air conditioning unit 空调组件
air contact（战术）空军引导组
air control 空中威慑；空中管制；引导控制
air control light 防撞灯
air controller 航空管制员
air cooled generator 空冷发电机；空冷发生器
air corridor 空中走廊
air cover 空中掩护
air current 气流
air current separation 气流分离
air cycle machine 空气循环机
air data 大气数据
air data calculator 大气数据计算器
air data computer（ADC）大气数据计算机
air data distributor 大气数分配器
air data inertial reference system 大气数据惯性基准系统
air data inertial reference unit 大气数据惯性基准单元
air data module 大气数据模块
air data module control switch 大气数据模式控制开关
air data reference panel 大气数据基准面板
air data sensor 大气数据传感器
air data sensor heater controller 大气数据传感器加温控制器
air data system（ADS）大气数据系统
air data system control switch 大气数据系统控制开关
air data unit 大气数据装置
air defense 防空
air defense artillery 防空炮
air defense emergency 防空紧急状态
air defense identification zone 防空识别区

air defense long-range radar 远程防空雷达
air defense suppression 防空压制
air defense system 防空系统
air defense warning «red» 红色防空警报；一级空袭警报；空袭紧急警报
air defense warning «white» 白色防空警报；三级空袭警报；解除空袭警报
air defense warning «yellow» 黄色防空警报；二级空袭警报；可能遭受敌机和 / 或敌导弹的攻击
air defense warning condition 防空预警等级
air delivery equipment 空投设备
air density 空气密度
air distance（AD）无风距离
air distribution duct 空气分配管
air driven generator（ADG）空气驱动发电机
air early warning aircraft（AEW）空中预警机
air evacuation 空中后送
air exhaust equipment 排气装置
air exhaust opening 排气口
air ferry 空运
air filter 气滤
air fire 空中火力
air fire plan 空中火力部署计划
air fleet 机群
air flow 气流，空气流量
air force 空军
air force base（AFB）空军基地【美】
air force weather agency（AFWA）空军气象局
air freighter 货机
air ground communication system 空地通信系统
air ground VHF subnetwork 空地甚高频子网
air impingement starter 气动冲击起动机
air inlet 进气道；进气口
air inlet casing 进气机匣
air inlet cone 进气帽罩；进气锥
air intake 进气道；进气口
air intake blanking device 进气道堵盖
air intake casing 进气机匣
air intake equipment 进气装置
air intake guard 进气罩
air intercept missile（AIM）空中

截击导弹

air interception 空中侦听

air interdiction 空中拦截

air liaison officer 空军联络官

air loading 装机，空运装载

air lock 气塞

air logistic support operation 空中后勤支援行动

air maneuver arm 空中机动兵种

air maneuver operation 空中机动作战

air marker 航行标志

air mass 气团

air material storehouse 航材库

air met（airman's meteorological information）飞行员气象信息；低空气象情报

air miss 空中接近；空中相撞事故征候

air mobile brigade 空中机动旅

air mobility 空中机动

air navigation 空中导航

air navigation aid 助导航设备

air navigation chart 空中航行图

air navigation facility 航空导航设备

air navigation plan 空中航行规划

air navigation radio aids 航空无线电导航设备

air navigation system 空中航行系统

air observation 空中观察

air observation post（air OP）空中观察站；空中观察所

air observer 空中观察员

air offensive 空中攻击

air officer 空军官员

air officer commanding（AOC）空军指挥官

air parcel 空运

air path 空气流路

air path cleaning 空气通道清洁

air path fouling 空气通道污积

air photographic reconnaissance 航空照相侦查；空中照相勘测

air plot 航空图表

air pocket 气潭；气穴

air portability 航空适运性

air position 无风位置；空中位置

air position indicator（API）空中位置显示器

air power 空军力量

air pressure 气压

air pressure adding 补充气压

air pressure gauge 气压表

air pressure tapping zone 引气压力区

air pressure transmitter 气压传感器

air priority 空中优先次序

air raid 空袭

air reconnaissance 空中侦察

air refueling 空中加油

air refueling control point（ARCP）空中加油控制点

air refueling initial point（ARIP）空中加油起始点

air report 空中报告

air report special（ARS）特别空中报告

air rescue 空中救护

air rescue center（ARC）空中救援中心

air resistance 空气阻力

air resonance 空中共振

air right 空权

air route 航线

air route certificate 航线证书

air route surveillance radar（ARSR）航线监视雷达；航路监视雷达

air route traffic control center（ARTCC）空中交通管制中心；航路管制中心

air scoop 进气口

air separation 气流分离

air speed 空速

air start 空中起动；空中发射；喷气发动机停车（熄火）后重新发动

air station 航空站

air step 机载客梯

air strike 空袭

air suction stroke 吸气冲程

air superiority 空中优势

air supply 空中补给

air supply and cabin pressure controller 空气供给和机舱压力控制器

air supply control and test unit 供气控制和测试组件

air support 空中支援

air supremacy 制空权

air surface zone 海上反潜区

air surveillance 对空监视

air surveillance radar（ASR）对空监视雷达

air survey 空中侦测

air system 空气系统

air tactical control（ATC）空中战

术控制

air tapping 空气的引出；引气；空气引出量

air tapping point 引气孔

air taxi 接地飞行；贴地飞行

air taxiing 接地飞行；贴地飞行

air taxiway 空中滑行道

air temperature 大气温度

air terminal 航空集散站

air traffic 空中交通

air traffic advisory service 空中交通咨询服务

air traffic clearance 空中交通许可

air traffic control（ATC）空中交通管制

air traffic control assigned airspace（ATCAA）空中交通管制指定空域

air traffic control beacon interrogator 空中交通管制信标询问器

air traffic control beacon interrogator（ATCBI）空中交通管制信标询问器

air traffic control center 空中交通管制中心

air traffic control clearance 航管许可

air traffic control communications system 空中交通管制通信系统

air traffic control instruction 航管指示

air traffic control radar beacon system（ATCRBS）空中交通管制雷达信标系统

air traffic control services 空中交通管制服务；航管服务

air traffic control specialist 空中交通管制专业人员

air traffic control system command center 空中交通管制系统指挥中心

air traffic control system command center（ATCSCC）delay factor 空中交通管制系统指令中心延迟因子

air traffic control unit 空中交通管制单位

air traffic controller 空中交通管制员

air traffic flow management 空中交通流量管理

air traffic management 空中交通管理

air traffic radar 空中交通管制雷达

air traffic service（ATS）空中交通服务

air traffic service（ATS）route 空中交通服务航路

air traffic service communications 空中交通服务通信

air traffic service reporting office 空中交通服务报告室

air traffic services airspaces 空中交通服务空域

air traffic services unit 空中交通服务单位

air transit route 航空转运路线

air transmitter 气体传导器

air transport 空运；航空运输

air transportation 航空运输

air tunnel 收风筒

air turbine starter 空气涡轮起动机

air turbine starter control valve 空气涡轮起动机控制活门

air umbrella 空中掩护；空中保护伞

air under pressure 压缩空气

air vector 空中向量

air vent 通气孔；换气口；通气口

air vent chamber 通大气腔

air way 航路

air waybill 航空运货单；空运单

AIRAC（aeronautical information regulation and control）航空信息规定与控制

airaudio 航空无线电设备

airborne 空中的；机载的

airborne alert 空中待命；空中警戒；空警戒备

airborne antenna system 机载天线系统

airborne assault 空降突击

airborne assault operation 空降突袭作战

airborne beacon processing system 机载信标信息处理系统

airborne brigade 空降旅

airborne cannon 航空机关炮；航炮

airborne cannon device 航炮装置

airborne cannon weapon system 航炮武器系统

airborne channel change 机载电台频道更换

airborne collision avoidance system（ACAS）机载防撞系统；空中防撞系统

airborne command system 空中指挥系统

airborne communication equipment 机载通信设备

airborne data bus 机载数据总线
airborne data link processor 机载数据链处理器
airborne data loader 机载数据装载机
airborne delay 空中延迟
airborne division 空降师
airborne division's aviation brigade 空降师空中旅
airborne early warning and control（AEW&C）空中预警与控制
airborne equipment 机载设备
airborne equipment log 机载设备履历本
airborne fire control radar 机载火控雷达
airborne flight information system 机载飞行情报系统
airborne force 空降部队
airborne frequency correction 机载电台频率校正
airborne information integrated processing system 机载信息综合处理系统
airborne interception 空中拦截
airborne lift 空中运力；空运
airborne link terminal 机载链路终端
airborne machine gun 航空机关炮；航炮
airborne operations 空降作战
airborne power unit（APU）机载动力装置；辅助动力装置
airborne radar 机载雷达
airborne radar calibration 机载雷达校正
Airborne Radarscope Used In Navar（Air Force）（NAVASCOPE）导航雷达使用的机载雷达显示器
airborne regiment 空降团
airborne self-protection jammer（ASPJ）机载自卫干扰机
airborne telephone 机载电话机
airborne transmitter 空用发射机
airborne troops 空降部队
airborne vibration 空中振动
airborne vibration monitor 机载振动监控器
airborne weather radar 机载气象雷达
airborne-intercept radar（AI radar）机载截击雷达
airbrake 阻流板；减速板
air-breathing 空气助燃的（发动

机）；吸气的

airburst 空中爆炸；空爆

air-conditioning system 空调系统

air-control system 飞行操纵系统

air-cooled engine 气冷式发动机

air-cooled oil cooler 气冷式滑油散热器

air-cooled turbine blades 气冷涡轮叶片

aircraft 航空器

aircraft abandonment 弃机（指机上人员跳伞）

aircraft approach category 飞机进近分类

aircraft arresting barrier 飞机（着陆）拦阻网

aircraft arresting cable 飞机（着陆）拦阻索

aircraft arresting gear 飞机（着陆）拦阻装置

aircraft arresting hook 飞机（着陆）拦阻钩

aircraft arresting system 飞机（着陆）拦阻系统

aircraft arresting wire 飞机（着陆）拦阻绳

aircraft basic operating weight 飞机基本重量

aircraft block speed 飞机轮挡速度

aircraft bolt 飞机螺栓

aircraft cable 航空缆绳

aircraft call sign 飞机呼号

aircraft capacity 飞机容量

aircraft captain 机长

aircraft carrier 航空母舰

aircraft classification 飞机分类；航空器分类

aircraft classification number（CAN）飞机分级数；飞机分类编号

aircraft classification number-pavement classification number（ACN-PCN）飞机分级数 - 道面分级数

aircraft clock 航空时钟

aircraft commander 机长

aircraft communications addressing and reporting system（ACARS）飞机通信寻址与报告系统

aircraft component 飞机部件

aircraft condition monitoring system 飞机状态监控系统

aircraft conflict 飞机冲突

aircraft conflict alert 飞机冲突警报

aircraft dimensions 飞机尺寸
aircraft dispatcher 航行调度员
aircraft disturbance 机体扰动
aircraft earth station 航空地面站
aircraft engine 飞机发动机
aircraft equipment 飞机设备
aircraft equipment list 飞机装备清单
aircraft external light 飞机外部灯
aircraft flight manual 飞机飞行手册
aircraft flight operation regulation 航空器飞行操作规程
aircraft fuel system 飞机燃油系统
aircraft gun 舰炮
aircraft heading 飞机航向
aircraft horizontal axis 飞机水平轴
aircraft horizontal flight path 飞机水平飞行轨迹
aircraft icing 飞机积冰
aircraft identification 航空器识别
aircraft illustrated parts catalog 飞机图解部件目录
aircraft information management system 飞机信息管理系统
aircraft inspection 航空器检查
aircraft installation delay 飞机安装延误
aircraft instrument 航空器仪表
aircraft integrated data system 飞机综合数据系统
aircraft intercom system 机内通话系统
aircraft list（ACL）飞机列表
aircraft load rating（ALR）飞机额定载荷
aircraft log 航空器记录；航空日志
aircraft log book 航空器记录簿
aircraft maintenance 航空器维护
aircraft maintenance manual 航空器维护手册；飞机维护手册
aircraft maintenance task oriented support system 飞机维护任务保障系统
aircraft major repair 航空器大修理
aircraft maneuver 航空器动作；航空器机动
aircraft manufacturer 飞机制造商
aircraft marking 航空器标识
aircraft mechanic 航空机械员；飞机机械员
aircraft minor alteration 航空器小改装
aircraft minor repair 航空器小修

aircraft mix 飞机种类
aircraft mode C intruder alert 飞机C模式警报
aircraft modification 飞机改装
aircraft movement surface condition report 飞机地面运动区状态报告
aircraft nut 航空螺帽
aircraft observation 航空观测
aircraft operating manual 飞机操作手册
aircraft oscillation 机体振动
aircraft owners and pilots association（AOPA）航空器所有者和飞行员协会
aircraft performance 航空器性能
aircraft performance limitation 航空器性能限制
aircraft plumbing 飞机管路
aircraft proximity（AIRPROX）飞机接近度
aircraft radio engineer 航空无线电工程师
aircraft rating 飞机等级
aircraft recovery manual 飞机修复手册
aircraft registration 航空器登记
aircraft repair 飞机修理
aircraft rigging 飞机调整
aircraft role equipment 机载任务设备
aircraft scrambling 飞机紧急起飞
aircraft separation assurance 确保飞机间隔
aircraft servo 飞机伺服
aircraft situation display 飞机状态显示器
aircraft speed mach number 飞机速度马赫数
aircraft stand 停机位
aircraft store 飞机挂载物
aircraft structure 航空器框架
aircraft surge launch and recovery（ASLAR）飞机松缆弹射起飞与回收
aircraft symbol 飞行器标志
aircraft system 航空器系统
aircraft tiedown 飞机系留
aircraft type 机型
aircraft utilization rate 飞机使用率
aircraft vectoring 飞机导引
aircraft washers 航空垫圈
air-defense unit 防空部队
air-driven pump 空气驱动泵
airdrop 空投

airdrop platform 空投平台
air-express 空运
airfield 军用机场
airfield capacity 机场容量
airflow 气流
airflow separation 气流分离
airfoil 机翼
airfoil classification 翼型分类
airfoil profile 翼型
airfoil section 翼剖面
airfoil thickness 翼型厚度
airframe 机身
airframe coupling 机体耦合
airframe movement coupling 机体运动耦合
air-fuel mixture ratio 油气混合比
air-ground communication 陆空通信
air-ground control radio station 空地指挥电台
air-ground data link system 空地数据链系统
airhead 空军前进基地;（在对方境内强行建立的）空降场
air-intercept zone 空中截击区；截击空域
airland battle 空地一体战；空降作战
air-landed operation 空降行动
air-launched weapon 机载武器
air-launched weapon system 机载武器系统
airlift 空中运力；空运；空运量
airlift capability 空运能力
airlift requirement 空运要求
airline dispatcher 航空调度员
airline modifiable information 航线可变更信息
airload 气动载荷；空运装载
airman 航空兵（指飞行员）
airman rating record 飞行员等级纪录
airmanship 导航术
airmass source region 气团源地
airmen’s meteorological information（AIRMET）低空气象情报
AIRMET（airmen’s meteorological information）低空气象情报
airmobile forces 空中机动部队
airmobile operation 空运机动作战行动
air-oil cooler 气冷式滑油散热器
air-oil separator 空气滑油分离器
airplane 飞机
airplane condition monitoring

function 飞机状态监视功能
airplane reference field length 飞机参考机场长度
air-plot wind velocity 航空标绘风速
airport 机场，飞机场；航空站
airport acceptance rate 机场容许进场率
airport access 机场联外运输
airport advisory area 机场咨询区
airport and airways surveillance radar 机场和航路监视雷达
airport apron 停机坪
airport arrival rate（AAR）机场到达率
airport beacon 机场灯标
airport capacity 机场容量
airport city 机场城市；空港城
airport classification 机场分类
airport configuration 机场布局
airport departure rate（ADR）机场离站率
airport diagram 机场略图
airport elevation 机场标高；场高
airport facility directory 机场设施目录
airport geographical position 机场地理位置
airport ground access road 机场联外道路
airport ground access system 机场联外运输系统
airport ground movement control system 机场地面交通管制系统
airport identification beacon 机场识别信标
airport information desk 机场信息查询台
airport landside service procedure 机场地勤程序
airport layout planning and design 机场布局规划与设计
airport lighting 机场照明设备
airport management 机场管理
airport marking aids 机场辅助标记
airport movement area safety system（AMASS）机场运动区安全系统
airport obstruction chart（OC）机场障碍图
airport obstruction charts（OC）机场障碍图
airport of departure 起飞机场
airport of destination 目标机场
airport planning manual 机场计划

手册

airport pressure 场压高度

airport radar service area（ARSA）机场雷达服务区

airport reference point（ARP）机场基准点；机场参考点

airport remote parking 机场远端停车

airport reservation office 机场预约办公室

airport rotating beacon 机场旋转灯标

airport runway 机场跑道

airport sign 机场标志

airport sketch 机场略图

airport surface detection equipment 机场场面探测设备

airport surface traffic automation 机场场面交通自动化

airport surveillance radar 机场监视雷达；机场搜索雷达

airport system planning 机场系统规划

airport taxi chart 机场滑行图

airport traffic area（ATA）机场交通区

airport traffic control service 机场交通管制服务

airport traffic control tower（ATCT）机场交通管制塔台

airport traffic pattern 机场起落航线

airport vehicle and equipment 机场车辆与设备

airport-to-airport distance 机场间距

AIRPROX（aircraft proximity）飞机接近度

air-raid siren 空袭警号；防空警报

air-release valve 放气阀

airscrew 螺旋桨

airship 飞艇

airside 空侧

airside facilities 机场控制区设施（航行面）

airspace 空域

airspace capacity 空域容量

airspace Class A A 类空域

airspace Class B B 类空域

airspace Class C C 类空域

airspace Class D D 类空域

airspace Class E E 类空域

airspace Class F F 类空域

airspace Class G G 类空域

airspace classification 空域划分

airspace conflict 空域冲突

airspace control authority 空域管制官

airspace control order 空域管制命令

airspace control system 空域管制系统

airspace denial 空域遏制

airspace flight 空域飞行

airspace hierarchy 空域等级

airspace management 空域管理

airspace structure 空域结构

airspeed 空速

airspeed bug 空速指示错误；空速显示故障

airspeed envelope 空速包络线

air-speed hold 空速保持

airspeed indicator（ASI）空速指示器

airspeed reference bug 空速参照错误

airspeed scale 空速值

airspeed selected value 空速选择值

airspeed tape 空速速度带；空速刻度带

air-spray atomizer 空气雾化器

airstream 气流

airtight sealing 密封框

airtime 留空时间；飞行时间

air-to-air collision 空中相撞事故

air-to-air guided missile 空空导弹

air-to-air missile（AAM）空对空导弹；空空导弹

air-to-ground 空对地

air-to-ground attack 空对地袭击

air-to-ground missile（AGM）空地导弹

air-to-surface guided missile 空面导弹

air-to-surface missile（ASM）空对面导弹

air-to-underwater missile（AUM）空对水下导弹

airway 航路

airway beacon 航路标灯

airway course light 航向灯

airway facilities 航路设施

airway planning standard 航路设计标准

airworthiness 适航性

airworthiness certificate 适航证

airworthiness directives 适航指令

airworthiness status 适航情况

airworthy 适航的

AIS（aeronautical information

service）航空情报服务
alarm in high sound 高音报警
alarm in low sound 低音报警
alclad 包铝
alcohol ice prevention 酒精防冰
ALD（available landing distance）可用着陆滑跑距离
aldis lamp 轻便信号灯
alert 警报
alert area 警戒区
alert height 警戒高度
alert notice（ALNOT）警戒通知
alert phase 告警阶段
alert service bulletin 警示服务通告
alerting service 告警服务
alignment 校准；对准
alignment and tracking procedures 对准与跟踪过程
alignment of element systems 单元系统对准
alignment time 对准时间
alkaline battery 碱性电池
alkaline cell 碱性电池
all engines operating（AEO）发动机工作正常
all stations 全站
all-engine（power units）operating takeoff 全发起飞
all-engine screen speed（V3） 全发掩蔽物速度
all-flying tail 全动平尾
all-gear-up landing 机腹着陆
allied forces 联军
allied troops 联军
all-moving slab 全动平屋
all-moving tail 全动尾翼；全动平尾
allocation 配置（部队）
allocation of attention 注意力分配
all-out attack 全面进攻
allowable cargo load 容许货物负荷
allowable crosswind component 容许最大侧向风速
allowable engine operating time 发动机许可使用时间
allowable gross takeoff weight 许可起飞总重
alloyed steel 合金钢
all-up-weight 空中总重量
all-wave antenna 全波天线
all-weather 全天候
all-weather air defense fighter 全天候防空战斗机；全天候战斗机

all-weather fighter 全天候防空战斗机；全天候战斗机

all-weather takeoff 全天候起飞

ALNOT（alert notice）警戒通知

along-track distance 沿航迹距离

along-track error 沿航线误差

Alouette“云雀”直升机（法国）

alpha 希腊语第一个字母（A，α）；迎角

alpha control range 迎角模式；迎角控制范围

alpha floor protection 迎角保护装量

alpha hinge 摆振铰

alpha mode 迎角模式；迎角控制范围

alpha protection 迎角保护装量

alphanumeric display 字母数字显示

alphanumeric keypad 文字与数字键区

alphanumeric symbol 文字与数字符号

ALR（aircraft load rating）飞机额定载荷

ALT（altimeter）高度表

ALT（altitude）高度；海拔高度

alternate 备降场

alternate aerodrome 备降机场

alternate air 备用气源

alternate airport 备用机场

alternate communications facilities 备用通信设施

alternate control load 交变操纵载荷

alternate navigation on CDU 控制显示组件的备用导航

alternating current（AC）交流电

alternating current motor pump 交流电机泵

alternating force 交变力

alternating internal forces 交变内力

alternating moment of flexure 交变弯矩

alternating stress 交变应力

alternative interrogator 可选择询问器

alternative landing field 备降场

alternator 交流发电机

altimeter（ALT）高度表

altimeter correction 高度表修正量

altimeter setting 高度表拨定

altimeter setting value 高度表拨正值

altitude（ALT）高度；海拔高度
altitude alert system 高度报警系统
altitude capture 高度截获（主显 / 飞行指引仪上）
altitude chamber 高度模拟室；高度舱；高空舱；高空室
altitude datum 高度基准
altitude difference 高度差
altitude dysbarism 高空减压症
altitude engine 高空发动机
altitude flight 高空飞行
altitude hold 高度保持（主显 / 飞行指引仪上）
altitude hold mode 高度保持模式
altitude hole 高度空白区
altitude intercept 高度截距
altitude interrogation mode 高度询问方式
altitude reporting switch 高度报告电门
altitude reservation（ALTRV）高空预留区
altitude reservation service 高空预留服务
altitude restriction 高度限制
altitude select 高度选择
altitude sickness 高空病
altitude slot 高度间隙
altitude tint 高度色层
altitude wind tunnel 高空模拟风洞
altocumulus 高积云
altostratus（As）高层云
ALTRV（altitude reservation）高空预留区
alumel 镍铝合金
alumel thermocouple 镍铝合金热电偶
aluminum 铝
aluminum alloy 铝合金
AM（amplitude modulation）调幅
AMASS（airport movement area safety system）机场运动区安全系统
amber 琥珀色
amber overspeed light 超转黄色灯
ambient 周围的；外界的
ambient air 环境空气；周围空气；大气
ambulance 救护车
ambush 伏击；埋伏；设伏地点
AME（authorized medical examiner）指定的检查医生
amended clearance 变更许可
America standard code for information

interchange 美国信息交换标准码

ammeter 电流表；安培计

ammunition 弹药；炮弹；子弹；航炮炮弹

ammunition container 弹箱

ammunition delivery board manipulator 送弹板操纵器

ammunition feeding and ejection system 供排弹系统

ammunition feeding cover 进弹机盖

ammunition feeding sheath 进弹套

ammunition identification code（AIC）弹药识别代号

ammunition loading pneumatic tube 装弹气压筒

ampere-meter 电流计；安培表

amphibian aircraft 水陆两栖飞机

amphibious assault vehicle 两栖突击车

amphibious combat vehicle 两栖战车

amphibious light tank 两栖作战轻型坦克

amplifier 放大器

amplifying circuit 放大电路

amplitude 幅值

amplitude modulation（AM）调幅

amplitude modulation equivalent 等效调幅

amplitude modulator 调幅器

amplitude shift keying 振幅移位键控

AMSL（above mean sea level）距平均海平面高度

AN（Army-Navy）connector 军标接头（航空专用 37° 扩口接头）

an anti-skid device 防溜装置

anabatic wind 上升风

analog audio 模拟话音，模拟音频

analog communication 模拟通信

analog integrated circuit 模拟集成电路

analog signal 模拟信号

analog switching circuit 模拟开关电路

analog voice signal 模拟声音信号

analog-digital conversion 模数转换

analogue instrument 模拟式仪表

analysis and demonstration test 分析演示实验

ANAPROP（anomalous propagation）反常传播，变则传播

anchor point 锚定点

anchor spring 支撑销；支撑弹簧

anemograph 风速表
anemometer 风速表
anemoscope 测风器；风速仪
aneroid 真空膜盒
aneroid altimeter 空盒高度表
aneroid barometer 膜盒气压计
aneroid capsule 真空膜盒
angel “天使”（军事俚语，指一种起假目标作用的扰乱反射体，常悬于降落伞或气球上使之缓缓下降）;（千英尺计的）飞行高度
angle constraint value 角约束值
angle displacement transmitter 角位移传感器（绝对式编码器）
angle of advance 前置角；超前角
angle of attack（AOA）攻角，迎角
angle of attack for infinite aspect ratio 无限展弦比翼型迎角
angle of attack indicator 迎角指示器
angle of bank 倾斜角
angle of climb 爬升角
angle of depression 倾角；眼高差，下偏角；垂度
angle of descent 下滑角
angle of dip 磁倾角
angle of downwash 下洗角
angle of elevation 仰角
angle of incidence 入射角
angle of incidence（AOI）桨距（桨叶安装角）；入射角
angle of lead 提前角
angle of pitch 桨叶迎角（飞机的）
angle of roll 滚转角
angle of stabilizer setting 安定面装配角
angle of view 视场角
angle of yaw 偏航角
angle part 角形件
angle position indicator 角位置指示器
angular acceleration 角加速度
angular contact ball bearing 向心推力球轴
angular deviation 角偏差
angular displacement 角位移
angular frequency 角频率
angular misalignment 角位移
angular momentum 角动量
angular speed 角速度
angular velocity 角速度
angular-momentum conservation 角动量守恒
anhedral 上反角

anisotropy 向异性
annihilation 歼灭；消灭
annular combustion chamber 环形燃烧室
annular combustor 环形燃烧室
annular duct 环形通道
annular indirect flow combustion chamber 环形折流燃烧室
annular reverse flow combustion chamber 环形回流燃烧室
annunciate 告示；通告
annunciator 信号器；报警器；指示器
annunciator light 信号盘灯
anodic treatment 阳极氧化处理
anomalous propagation（ANAPROP）反常传播，变则传播
anomaly 异常；【象】距平，【天】近点角
anoxia 缺氧
antagonism 对抗（状态）
antenna 天线【美】
antenna array 天线组
antenna control unit 天线控制组件
antenna coupler 天线耦合器
antenna feed 无线馈电点
antenna pattern 天线方向图
antenna tilt control 天线俯仰控制
anti-aircraft 防空用的；对空的
anti-aircraft artillery 防空高射炮
anti-aircraft fire 防空火力
anti-aircraft gun 高射炮
anti-balance tab 反补偿片
anti-ballistic missile（ABM）反弹道导弹
anti-chemical suit 防生化战衣
anticipated operation condition 预期运行状态
anticipation 预调
anticipator 提前器；预调器
ANTI-COLL light 防撞灯
anti-collision light 防撞灯
anti-collision light switch 防撞灯开关
anti-collision radar 防撞雷达
anticyclone 反气旋区
anticyclonic circulation 反气旋式环流
anticyclonic rotation 反气旋旋转
anti-drag wire（机翼机构的）反阻力张线
anti-freezing system 防冰系统
anti-G suit 抗过载飞行衣；抗荷

服（消除正加速度对人体影响的一种飞行服）；防失重服
anti-helicopter mine 反直升机地雷
anti-ice and leak detection controller 防冰与渗漏检测控制器
anti-icing 防冰
anti-icing device 防冰装置
anti-icing system 防冰系统
anti-intercontinental ballistic missile（AICBM）反洲际弹道导弹
anti-lock brake 防抱刹车
antinode 波腹
anti-radiation missile 反辐射导弹
anti-servo tab 反伺服补偿片
anti-ship missile 反舰导弹
anti-skid control valve 防滑控制活门
anti-skid device 防拖胎的装置，防滑的装置
anti-spin parachute 抗螺旋伞
anti-static wire 防静电线
anti-storm light 防风暴指示灯
anti-submarine 反潜的
anti-submarine helicopter procedure 直升机反潜程序
anti-submarine warfare（ASW）反潜战
anti-surface radar 反舰雷达
anti-surge function for manual control 手动防喘控制
antisymmetry 反对称
anti-tank（AT）反坦克的
anti-tank missile 反坦克导弹
anti-torque 反扭矩的
anti-torque device 反扭装置
anti-torque force 反扭矩力
anti-torque pedal 反扭转力踏板（脚蹬）
anti-torque rotor 反扭矩旋翼
anti-trade wind 反信风
anvil 铁砧云
anvil cloud 铁砧云
anvil crawler 铁砧爬行者
anvil dome 铁砧云冠
AOA（angle of attack）攻角，迎角
AOC（air officer commanding）空军指挥官
AOI（angle of incidence）桨距（桨叶安装角）；入射角
AOL（above obstruction level）越障高度
AOPA（aircraft owners and pilots association）航行器所有者和飞行员协会

AOR（area of responsibility）责任区

AP（assumed position）假定位置

Apache“阿帕奇”直升机

Apache Longbow“阿帕奇长弓”直升机

APAPI（abbreviated precision approach path indicator）简易精确进场航道指示器

APD（automated problem detection）自动冲突探测

APDIA（automated problem detection inhibited area）自动冲突探测禁止区

aperiodic compass 非周期罗盘；安定罗盘

apex 顶点

aphelion 远日点

API（air position indicator）空中位置显示器

apogee 远地点

apostrophe 撇号

apparent drift 视在漂移

apparent horizon 视地平线

apparent precession 视在进动

apparent solar day 视太阳日

apparent solar time 视太阳时

apparent sun 视太阳

apparent wander 视在漂移

Appleton layer 阿普顿电离层

applied force 外加力

applied force（effort）作用力

approach 进场着陆（飞行）；进近（飞行）

approach angle 进场角

approach area 进近区

approach chart 进近图

approach clearance 进场许可

approach control 进近管制

approach control facilities 进近管制设施

approach control office 进近管制室

approach control position 终端管制席位；进场管制席位

approach control procedure 进场管制程序

approach control radar 进场管制雷达

approach control service 进场管制服务

approach control unit 进场管制站

approach control zone 进场管制地带

approach coupler 进近耦合器

approach end of runway 跑道进近端

approach fix 进场点

approach funnel 进近净空区

approach gate 进场入口

approach in hooded cabin 暗舱进近着陆

approach indicator 进场指示器

approach intercept waypoint 进近切入点

approach light 进场灯

approach light beacon 进近灯标

approach light contact height 进场灯目视高度

approach light system 进场灯光系统

approach light system improvement program 进近灯光系统改进计划

approach light system with sequenced flashing light 顺序闪光的进近灯光系统

approach lighting system 进近照明系统

approach path 进场路线

approach plate 进场图

approach progress display 进近进程显示

approach sequence 进近顺序

approach slope ratio 进近下滑率

approach speed 进近速度

approach surface 进场面

approach time 进近时间

approaching 正在接近（陆空通话用语）

appropriate airworthiness requirement 适当的适航性要求

appropriate ATS（air traffic services）authority 空中交通服务有关管理部门

appropriate obstacle altitude 适当障碍高度

appropriate terrain clearance altitude 适当离地高度

approval request（APREQ）进近请求

approved 批准的

approved training 批准的训练

APREQ（approval request）进近请求

apron 停机坪

apron management service 停机坪管理服务

apron traffic 停机坪交通

apse line 拱线；拱点线

APU（airborne power unit）机载动力装置；辅助动力装置

APU controller 机载动力装置控制器

APU starter-contactor assembly 机载动力装置起动机接触器组件

aquaplaning 滑动

arc 孤线航迹

ARC（air rescue center）空中救援中心

arc navigation 圆弧导航

arched part 弓形件

ARCP（air refueling control point）空中加油控制点

arctic minimum 极区最低气温飞行气象条件

arcus 弧状云

area bombing 区域轰炸

area chart 区域图

area control center 区域管制中心

area control computer complex 区域管制计算机网

area control facilities 区域管制设施

area control service 区域管制服务

area forecast 区域预报

area forecast 区域天气预报

area navigation（RNAV）区域导航

area navigation（RNAV）approach configurations 区域导航进近配置

area navigation high route 区域导航高航路

area navigation low route 区域导航低航路

area navigation route 区域导航航路

area navigation system（ARNAV）区域导航系统

area of confusion 扰乱区

area of coverage（world area forecast system）天气预报系统覆盖范围

area of probability 概率区

area of responsibility（AOR）责任区

area of uncertainty 不确定区

area ratio 面积比

area rule 面积律

area target 面目标

area-increasing flap 增面襟翼

ARINC（aeronautical radio incorporated）航空无线电设备公司

ARIP（air refueling initial point）空中加油起始点

armada 舰队
armament 军事装备（复）；军备；武器，军械
armament system 武器系统
armature 电枢；衔铁
armed blockade 武装封锁线
armed control panel 武器（火控）控制板
armed control panel（ACP）武器（火控）控制板
armed forces 武装部队
armed helicopter 武装直升机
armed rating 锁定状态
armed reconnaissance 武装侦察
arming light setting 指示灯调定
arming vanes 引信旋翼
armor 装甲；装甲部队（总称）
armored brigade 装甲旅
armored column 装甲队伍
armored division 装甲师
armored forces 装甲部队（总称）；装甲兵
armored personnel carrier 装甲运兵车
armored vehicle 装甲车
armorer 军械士【美】
armor-piercing shell 穿甲弹
armourer 军械士【英】
arms control 武器管制
army aviation 陆军航空兵
army aviation air operation command 陆军航空兵空中指挥
army aviation brigade 陆军航空兵旅
army aviation center 陆军航空兵中心
army aviation combat formation 陆军航空兵战斗队形
army aviation combat preparation 陆军航空兵战斗准备
army aviation flight training brigade 陆军航空兵飞行训练旅
army aviation flight training regiment 陆军航空兵飞行训练团
army aviation heliborne command 陆军航空兵机降指挥
army aviation operations command 陆军航空兵作战指挥
army aviation regiment 陆军航空兵团
army aviation tactics 陆军航空兵战术
Army Blackhawk 陆军“黑鹰”直升机

army headquarters 集团军司令部

army light aircraft squadron 陆军轻型飞机中队

ARNAV（area navigation system）区域导航系统

around-pylons-eights 绕目标 8 字

ARP（airport reference point）机场参考点

array（天线）阵列；列阵；排列

arrester 限动器

arrester wire 阻拦索

arresting gear 制动装置；拦阻装置

arrestor brake 制动器

Arriel 阿赫耶（发动机）

Arrius 阿里尤斯（发动机）

arrival aircraft interval（AAI）到达飞机间隔

arrival center 到达中心

arrival delay 到达延迟

arrival message 到场电报

arrival or departure route 到达或离场航路

arrival report 到达报告

arrival route 到达航路

arrival sector 到达区

arrival sector advisory list 到达区咨询清单

arrival segment 到达段

arrival sequencing program 进场顺序计划

arrival stream filter（ASF）到达流筛选程序

arrival time 到达时间

arrow 箭标

arrow wing 箭翼

arrowhead formation 箭头编队

ARS（air report special）特别空中报告

ARSA（airport radar service area）机场雷达服务区

arsenal 军火库

ARSR（air route surveillance radar）航线监视雷达；航路监视雷达

ARTCC（air route traffic control center）空中交通管制中心；航路管制中心

artery 动脉

articulated point 铰接点

articulated rotor 铰接式旋翼

artificial failure 人为故障

artificial feel 人工感觉

artificial feel device 人工触感装置

artificial horizon 人工地平仪；人

工地平器件
artillery 大炮；炮兵
artillery fire 炮火
artillery projectile 火炮炮弹
artillery shell 火炮炮弹
As（altostratus）高层云
AS（automatic scan）自动扫描
ascending 上升
ascension 上升情况（指发动机）
A-scope A 型显示器；距离显示器
A-sector A 区
ASF（arrival stream filter）到达流筛选程序
ASI（airspeed indicator）空速指示器
ASL（above sea level）海拔高度；相对高度
ASLAR（aircraft surge launch and recovery）飞机松缆弹射起飞与回收
ASM（air-to-surface missile）空对面导弹
ASOS（automated surface observation system）自动地面观测系统
aspect 外观
aspect ratio 纵横比；展弦比
asphalt 沥青
asphalt concrete 沥青混凝土
ASPJ（airborne self-protection jammer）机载自卫干扰机
ASR（aerodrome surveillance（approach）radar）机场监视（进近）雷达
ASR（air surveillance radar）对空监视雷达
assault helicopter carrier 直升机突击航母
assault rifle 突击步枪
assembly 部件（能装箱运输的）；组装
assessment 评估
assigned altitude deviation 指定高度偏差
assigned arrival time（AAT）指定到达时间
assigned departure time（ADT）指定离场时间
assigned frequency 制定频率；规定频率
assimilation 消化
assisted takeoff 辅助起飞
association reference direction 关联参考方向
assumed autogiration 假设自转

assumed position（AP）假定位置
assumptions 假设条件
A-station 主站
asterisk 星标
astern 尾部区
astigmatism 散光
astrodome 天文观测室（安装在飞机机身顶部的透明圆顶形结构）
astronautics 航天学
astronavigation 天文导航
astronomical navigational tables 天文导航表
astro-tracker 天体定位仪；星体跟踪仪
ASW（anti-submarine warfare）反潜战
asymmetric airfoil 不对称机翼
asymmetric lift 不对称升力
asymmetric loading 不对称载荷
asymmetric thrust 不对称推力
asymmetry 不对称
asymptotic flare 渐进照明弹
asynchronous communication interface adapter 异步通信接口转接头
asynchronous link control 异步链路控制
asynchronous motor 异步电机
AT（anti-tank）反坦克的
at altitude 在高原
at own discretion 自己把握；自己决定
at your discretion 按你的判断
ATA（airport traffic area）机场交通区
ATC（air tactical control）空中战术控制
ATC（air traffic control）空中交通管制
ATC（air traffic control）airspace 空中交通管制分配的空域
ATC clearance 空中交通管制许可
ATC clears 空中交通管制许可
ATC frequency change service 空中交通管制改（换）频服务
ATC instructions 空中交通管制指令
ATC interfacility data communications 空中交通管制设施间数据通信
ATC radar tracker and server 空中交通管制雷达跟踪和服务器
ATC requests 空中交通管制要求
ATCAA（air traffic control assigned

airspace）空中交通管制指定空域
ATCBI（air traffic control beacon interrogator）空中交通管制信标询问器
ATCRBS（air traffic control radar beacon system）空中交通管制雷达信标系统
ATCSCC（air traffic control system command center）delay factor 空中交通管制系统指令中心延迟因子
ATCT（airport traffic control tower）机场交通管制塔台
ATD（actual time of departure）实际离场时间
ATF（aerodrome traffic frequency）机场交通频率
athwartship 垂直于纵轴
ATIS（automated terminal information service）自动终端信息服务
ATIS（automatic terminal information service）自动终端情报服务
atmosphere 大气层；大气压，大气单位
atmospheric air 大气
atmospheric circulation 大气环流
atmospheric gas 大气气体
atmospheric layer 大气层
atmospheric motion 大气运动
atmospheric temperature 大气温度
atmospheric visibility 大气能见度
atomic bomb 原子弹
atomic head 原子弹头
atomic warfare 原子战
atomiser 喷雾器
atomizer 喷雾喷嘴
atomizer-type burner 雾化燃烧室
atomospheric aberration 大气光行差
ATS（air traffic service）空中交通服务
ATS（air traffic service）route 空中交通服务航路
attached shock wave 附体激波
attachment 附件；附加装置
attachment point 连接点；夹持点；附着点
attack helicopter 攻击直升机
attack in coordination with ground force 协同地面进攻
attack jet 战机；战斗机；歼击机
attention signal 空袭预备警报信号
attenuation 衰减
attenuation coefficient 衰减系数
attenuator 衰减器

attitude 姿态

attitude acquire mode 姿态获取模式

attitude and heading computer 姿态和航向计算机

attitude and heading reference system 姿态和航向基准系统

attitude and heading reference system（AHRS）姿态与航向参照系统

attitude angle 姿态角

attitude control system 姿态控制系统

attitude direction indicator 姿态方向指示器

attitude director 姿态指引仪

attitude director indicator 姿态指示仪

attitude gyro 姿态陀螺仪；飞行仪表

attitude heading system 姿态航向系统

attitude hold mode 姿态保持模式

attitude hold system 姿态控制系统

attitude indicator 姿态指示器

attitude of flight 飞行姿态

attitude reconfiguration 姿态重新确认

attitude retention 姿态保持

attitude retention system 姿态保持系统

attitude stabilization system 姿态稳定系统

attrition rate 损耗率

attrition reserve aircraft 消耗储备飞机

ATTS（automated terminal tracking system）自动终端跟踪系统

ATZ（aerodrome traffic zone）机场交通地带

audio 音频的

audio alarm 音响报警

audio chime 音响报警

audio control panel 音频控制板

audio electronic control unit 音频电子控制单元

audio filter 音频滤波器

audio frequency 音频

audio integrating 音频集成

audio integrating system 音响综合系统

audio management unit 话音管理单元

audio monitor adapter 音响监控

接头
audio multiplexer 音频复用器
audio panel 音频仪表板
audio selector panel 音频选择面板
audio warning 声响告警
augmented turbofan 加力涡扇发动机
AUM（air-to-underwater missile）空对水下导弹
aural message 音频信息
aural synthesizer card 音频合成卡
aural warning 音响警告
aural warning generator（AWG）音响警报发生器
aural warning signal 音响告警信号
auricle bolt group 耳片螺栓群
aurora propagation 极光传播
Auster“奥斯特”（直升机）
authorization 授权
authorized agent 授权代理人
authorized frequency 核定频率
authorized medical examiner (AME) 指定的检查医生
authorized repair 授权修理
auto coordinator 转向协调仪
auto flight 自动驾驶仪飞行
auto hover mode 自动悬停模式
auto tiltometer 自动倾斜仪
auto/manual selector 自动 / 手动选择器
auto-bank unit 自动倾斜装置；自动倾斜仪
autobrake 自动刹车
auto-control 自动控制
auto-flight control 自动飞行控制
auto-flight system 自动飞行系统
autogiro 自旋转翼飞机；旋翼机
autogiro rotorcraft 自转式旋翼机
autogyro 自旋转翼飞机；旋翼机；自动陀螺仪
auto-inclination 自动倾斜
autokinetic illusion 自运动错觉
autoland 自动着陆
autoland approach 自动着陆进近
autoland status annunciation 自动着陆状态通告
autoland status annunciator 自动着陆状态信号牌
automated altitude reporting 自动高度报告
automated data processing 自动数据处理
automated digital data interchange 自动化数字数据交换

automated digital network 自动化数字网络

automated flight inspection system 自动飞行检查系统

automated flight service station（AFSS）自动飞行服务站

automated information transfer 自动信息传输

automated problem detection（APD）自动冲突探测

automated problem detection inhibited area（APDIA）自动冲突探测禁止区

automated radar summary chart 自动雷达综合图

automated radar terminal system 自动化雷达终端系统

automated surface observation system（ASOS）自动地面观测系统

automated surface observing systems（ASOS）自动地面观测系统

automated terminal information service（ATIS）自动终端信息服务

automated terminal tracking system（ATTS）自动终端跟踪系统

automated UNICOM 自动通用无线电通用服务

automated weather observing system（AWOS）自动化气象观测系统

automatic altitude reporting 自动高度报告

automatic altitude reporting system 自动高度报告系统

automatic approach and landing 自动进近和着陆

automatic balance 自动平衡

automatic centering device 自动对中机构

automatic control and landing system 自动控制和着陆系统

automatic data interchange system 自动数据交换系统

automatic dependent surveillance（ADS）自动相关监视

automatic dependent surveillance addressing 选址式自动相关监视

automatic dependent surveillance broadcast mode 广播式自动相关监视

automatic dependent surveillance function 自动相关监视功能

automatic dependent surveillance

unit（ADS unit）自动相关监视单元

automatic direction finder（ADF）自动方位搜寻器；自动测向仪

automatic fire 连发射击

automatic fire and overheat logic and test system 自动火警和过热逻辑测试系统

automatic flight control augmentation system 自动飞行控制增强系统

automatic flight control system（AFCS）自动飞行控制系统

automatic flight information service 自动飞行信息服务

automatic flight system 自动飞行系统

automatic frequency control 自动频率控制

automatic gain control 自动增益控制

automatic grenade launcher 自动榴弹发射器

automatic instrument landing approach system 自动仪表着陆进近系统

automatic landing system 自动着陆系统

automatic manifold air-pressure control 自动歧管气压控制器

automatic mechanical centering device 自动机械中立装置

automatic mixture control 自动混合比操纵

automatic mode 自动控制模式

automatic performance reserve 自动性能储备

automatic pilot 自动驾驶仪

automatic power reserve 自动功率储备

automatic relay 自动中继装置

automatic rifle 自动步枪

automatic riveting machine 自动铆接机

automatic scan（AS）自动扫描

automatic search jammer 自动搜索干扰机

automatic selective feathering 选择式自动顺桨

automatic slots 自动翼缝

automatic stabilizer trim system 自动安定面配平系统

automatic synchronization 自动同步

automatic temperature control 自动温度控制

automatic terminal information service（ATIS）自动终端情报服务

automatic test equipment 自动化测试设备

automatic throttle 自动油门控制

automatic throttle retarder 自动收油门装置

automatic toss 自动拉起投弹

automatic transfer 自动转换

automatic trim 自动配平

automatic VHF/DF（very high frequency/direction finder）自动甚高频测向仪

automatic voice network 自动化话音网络

autonomous engine starting 发动机自主启动

autonomy sensor 自主式机载传感器

autopilot 自动驾驶仪

autopilot computer 自动驾驶计算机

autopilot control system 自动驾驶操纵系统

autopilot control unit 自动驾驶控制单元

autopilot flight director computer 自动驾驶飞行指示计算机

autopilot flight director system 自动驾驶飞行指示系统

autopilot system 自动驾驶系统

autopilot yaw control channel 自动驾驶偏航（方向）控制通道

autorotation 自转飞行

autorotation k-factor 自动旋转系数

autorotation landing 自转着陆

autorotation to touchdown 自行旋转到着地

autorotative descent 自旋转下降

auto-throttle system 自动油门系统

autothrust 自动推力

autotilt 自动倾斜

autotrim 自动配平

auxiliary 辅助

auxiliary air intake 辅助进气门

auxiliary airport beacon 航站辅助灯标

auxiliary battery 辅助电瓶

auxiliary component 辅助元件

auxiliary flight control 辅助飞行控制

auxiliary fuel pump 辅助燃油泵
auxiliary fuel tank 辅助燃油箱
auxiliary generator control unit 辅助发电机控制组件
auxiliary hydraulic power control 辅助液压电源控制
auxiliary hydraulic pump 辅助液压泵
auxiliary power breaker 辅助电源断路器
auxiliary power contactor 辅助电源接触器
auxiliary power control 辅助电源控制
auxiliary power unit（APU）辅助动力装置
auxiliary pump 辅助泵
auxiliary rotor 尾桨
auxiliary tank 副油箱
auxiliary unit 辅助单位
auxiliary warhead 副战斗部
available landing distance（ALD）可用着陆滑跑距离
AVASIS（abbreviated visual approach slope indicator system）简易目视进近下滑道指示系统
AVCAT（aviation fuel）航空燃料
average arrival interval（AAI）平均到达间隔
average heading 平均航向
AVGAS（aviation gas）航空汽油
aviation 航空；航空兵
aviation accident 飞行事故；空中失事
aviation fuel（AVCAT）航空燃料
aviation gas（AVGAS）航空汽油
aviation gasoline 航空汽油
aviation medicine 航空医药
aviation occurrence 航空事故
aviation radio station/set 航空电台
aviation regiment 航空团
aviation routine weather report（METAR）航空例行天气报告
aviation safety 航空安全
aviation safety reporting system 航空安全报告系统
aviation standard 航空标准
aviation transportation association 航空运输协会
aviation turbine fuel（AVTUR）航空涡轮燃料
aviation VHF packet communications 航空甚高频分组通信
aviation weather 航空天气

aviation weather briefing service 航空天气简报服务

aviation weather forecast 航空天气预报

aviation weather information service（AWIS）航空天气信息服务

aviation weather processor 航空气象处理器

aviation weather service 航空气象服务

aviation-biphase shift keying 航空两相相移键控

aviation-quadriphase shift keying 航空四相相移键控

aviator 飞行员

avionic bay 航电舱

avionic equipment 航电设备

avionic standard communications bus 航电标准通信总线

avionics 航空电子设备；航空电子学

avionics equipment bay 机上电子设备舱

avionics rack 航电设备架

avionics system integration 航空电子仪表系统集成

avoidable accident 可避免的事故

avoiding turn 躲避转弯

AVTAG（fuel for gas turbine engine）航空汽油

AVTUR（aviation turbine fuel）航空涡轮燃料

AWADS（adverse weather aerial delivery system）恶劣天气空投系统

AWG（aural warning generator）音响警报发生器

AWIS（aviation weather information service）航空天气信息服务

AWOS（automated weather observing system）自动气象观测系统

axes 坐标轴

axes of an aircraft 飞机坐标系

axial compressor 轴向式压气机；轴流压气机

axial displacement 轴向位移

axial exhaust pipe 轴向排气管

axial flow 轴向流动

axial flow compressor 轴流压缩机

axial flow turbine 轴流式涡轮

axial hinge 轴向铰

axial load 轴向载荷

axial rotor 轴向转子

axial stator vanes 轴向静子叶片
axial wheel 轴向叶轮
axial-centrifugal compressor 轴向离心式压缩机
axial-flow compressor 轴流式压缩机
axial-flow state 轴流状态
axial-flow turbine 轴流式涡轮
axile bush 轴套
axis adjustment 校轴
axis flight director 轴飞行指挥仪
axis of freedom 自由度轴
axis of rotation 转动轴
axle shaft gear 驱动轴齿轮
azimuth 方位；方位角
azimuth altitude 地平经度
azimuth angle 方位角
azimuth blanking 方位消隐
azimuth card 方位盘
azimuth equidistant projection 等距方位投影
azimuth error 方位误差
azimuth guidance 方位制导
azimuth plane 方位平面
azimuth rate 方位变化率
azimuth resolution 方位分辨力
azimuth transmitter 方位台
azimuthal projection 方位投影

B

B&Z（boom and zoom）俯冲跃升攻击（战术）

B/A ratio 俯仰转动惯量与滚转转动惯量之比

babble 无线电通信干扰

back 叶背；功率曲线后区；飞机背部，机身背部；后座，后座舱；（风向的）逆时针旋转

back azimuth 后方位；背航道

back beam 反向波束

back bearing 反航向角；反向角

back course 反航线

back course approach mode 反航向进近模式

back course sector 反航向扇区；后航线扇区

back pressure 反压

back taxi/ back-taxi 反向滑行

backblow 反冲，后坐力

backfire 逆火；回火

backfiring 逆火；回火

backhaul 空载传输

backing 风向的逆时针变化

backing ring 衬环

backing up 倒车

backpack 背包；降落伞背包

backpack parachute 降落伞背包

backplane 底板

backscatter 反向散射

back-sheared anvils 反剪铁砧

backside of the power curve 功率曲线后区

backtrack 180° 掉头

back-up control 备份控制

backup emergency communications 备用紧急通信

back-up law 备用规律；备份规律

backup mode 备份模式

backward flight 后退飞行；后飞

backward scatter 反向散射；负翼阶

backwards movement 向后移动

backwash of the lead helicopter 长机尾流
bad weather 恶劣天气
bad weather aerial delivery system 恶劣天气空投系统
bad weather window 恶劣天气观察窗
baffle 折流板（或挡板）
bag fuel tank 软油箱
baggage 行李
baggage bay 行李架
baggage cabin 行李舱
baggage cargo loading manual 行李货物装载手册
baggage compartment 行李舱
baggage compartment load 行李舱负载
baggage door 行李舱门
baggage fire detection system 行李舱火灾探测系统
baggage-bay area 行李舱区域
bail out 跳伞【口】
bailer 跳伞者
bailout bottle 跳伞用氧气瓶
balance 平衡
balance area 补偿面积
balance check 平衡检测
balance equation 平衡方程
balance tab 平衡片
balance tabs 补偿片
balance weight 配重
balanced air force 均衡空中力量
balanced control surface 平衡的操纵面
balanced detector 平衡检波器
balanced equation 平衡方程
balanced mixer 平衡混频器
balanced modulator 平衡调制器
balanced runway concept 平衡跑道概念
balanced surface 平衡面
balancing of forces 平衡力
balancing procedures 均衡过程
balancing unit 配平机构
balancing weight 配重块
balked landing 放弃着陆；复飞；着陆失败
balked landing distance 复飞距离
ball bearing 滚珠轴承
ball screw 滚珠丝杠
ball valve 滚珠活门
ballast 压舱材料；压载物
ball-flexible rotor 球柔性旋翼
ballistic lane 弹道下空中走廊

ballistic missile 弹道导弹
ballistic trajectory 弹道
ballistic wind 弹道风
ballistics 弹道学
balloon barrage 气球拦阻网
balloonist 气球驾驶员
ball-screw 滚珠螺杆
band 频带
band of error 误差区域；误差带
band pass filter 带通滤波器
band width 频带宽度
bandage 绷带
band-pass 带通；通频带
band-pass filter 选带滤波器
bandwidth 带宽，频率范围；误差带；频段，波段；带宽频率；信息带宽
bang out 弹射，发射，抛出
bank 带坡度转变（飞行）；坡度（飞行）；倾斜飞行
bank angle 倾角
bank the aircraft 压坡度（飞行）
banking flight 倾斜（飞行）
banner cloud 旗状云
banner target 旗靶
banquette 射击踏垛
bar 巴（公制非法定气压计量单位）；人工地平线标志，地平指示杆
barber's pole 最大使用速度指针
barbette（军舰上的）露天炮塔；炮垛；炮塔座
bare base 无附属设备的基地
bare engine 光身发动机
barnstormer 空中自由飞行表演者
barogram 气压记录曲线（图）
barograph 自动气压计
barometer 气压表
barometric 大气压力的
barometric altitude 气压高度
barometric error 气压误差
barometric pressure 气压；大气压力
barometric pressure control（BPC）气压调节器
barometric pressure gradient 气压梯度
barometric scale 气压刻度
barometric setting value 气压设定值
barometric tendency 气压趋势
barostat 恒压器
barrack 军营；营房
barrage 气球拦阻网；压制干扰，

阻塞干扰；防空火网
barrage bombing 拦阻轰炸
barrage fire 拦阻射击
barrage jamming 压制干扰；阻塞干扰
barrel 炮管；枪管
barrel liner 炮管衬板
barrel roll 大圈横滚
barrel roll attack 大圈横滚攻击
barrette 排灯
barricade 阻拦网（航母飞行甲板上）
barrier 声障；拦阻网，跑道拦阻装置
barye 微巴（物理单位中用厘米 - 克 - 秒（CGS）表示的压力单位）
base 基地；底座
base course 基层；底层
base leg（起落航线的）第四边
base pin 管脚
base turn 三转弯；三四转弯（连续做三四转弯）
based aircraft 基地飞机
baseline/ base line 基线；原准机
basic air temperature（BAT）基础气温
basic empty weight 基本空重
basic encoding rule 基本编码规则
basic failure 基本故障
basic flight envelope 基本飞行包线
basic load 基本载荷
basic maneuvering envelope 基本机动飞行包线
basic operating weight（BOW）基本使用重量
basic principle 基本原理，基本原则
basic relational expression 基本关系式
basic runway length 跑道基本长度
basic trainer 初级教练机
basic transport runway or airport 基本运输跑道或基本运输机场
basic utility airport 基本应用机场
basic weight of aircraft 飞机基本重量
BAT（basic air temperature）基础气温
batch number 批号
battery bus 电池总线
battery charge condition 电池充电情况
battery charger 电瓶充电器
battery discharge condition 电池放

电情况

battery master 主电池

battery power 电池电量；电池余量

battle damage assessment（BDA）战斗损伤评估

battlefield 战场

battlefield electromagnetic environment 战场电磁环境

battleship 战舰

baulk/balk 着陆失败；阻止着陆

bayonet thermocouple 销接热电偶

bayonet type locking pin system 卡销紧锁系统

bazooka 反坦克火箭炮；火箭筒

B-code B 码

BDA（battle damage assessment）战斗损伤评估

BDC（bottom dead center）下死点

B-display B 型显示器

BDU（bomb disposal unit）未爆炸弹处理小组

beacon 烽火（台）；信标；信标发射机；信标台；导航台

beacon delay 信标应答延迟

beacon skipping 信标间断

beacon stealing 信标扰断

beacon tracking 信标跟踪

bead 垫圈，密封条；轮缘；焊缝；准星

beam 横梁

beam antenna 定向天线

beam attack 侧方攻击

beam capture 波束截获

beam deflection 波束偏转

beam resolution 波束分辨率

beam steering unit 天线方位控制组件

beam width 波束宽度

beam width error 波束宽度误差

beam-climber guidance 驾束制导

beam-riding guidance 驾束制导；波束制导

beam-riding missile 波束制导导弹

bear 后座驾驶员；领航员

bearing 方位角；轴承；方向（航空导航术语）

bearing angle 方位角

bearing capacity 承载强度

bearing chamber 轴承腔

bearing classification 方位角分级

bearing distance indicator 方位距离指示器

bearing failure 轴承故障

bearing of blades profile 叶片型面轴承
bearing range line 方位距离线
bearing sealing 轴承密封
bearing strength 承载强度
bearing support 轴承座支架
bearingless 无轴承
beat 跳振；差拍
beat frequency 拍频；差频
beat frequency oscillator 激频振荡器
beat up 突袭，低空突袭；惊扰
beaufort scale 蒲福风级
beep "滴"音；比普
beep button 比普按钮
beep switch 比普开关
beep trim 比普配平
beep trim switch 比普配平开关
before bottom center 下止点前
before top center 上止点前
before-flight inspection 飞行前检查
bell mouth 喇叭口
bellcrank 曲柄；摇臂
bell-mouth inlet duct 钟形口进气
bellows 波纹管
bellows-type 膜盒式
belly 腹部；机腹
belly bay 腹部，机身腹部；机腹着陆，用机腹接地
belly landing 机腹着陆；迫降
below cloud 在云下
below minimum 低于最低气象条件
belt ejection entrance 排链口
belt frame 连接隔框
bench check 内场检查
bench maintenance equipment 内场维修设备
bend（压缩机叶片）弯曲
bending deflection 弯曲变形
bending deformation 弯曲变形
bending load 弯曲载荷
bending moment 弯矩
bending rigidity 弯曲刚度
bending stiffness 弯曲刚度
bends 屈肢症；信号摆动
bent beam 弯曲波束
bent course 弯曲波束
BER（bit error rate）误码率，错码率
Bernoulli's equation 伯努利方程
Bernoulli's equation verification test 伯努利方程验证实验
Bernoulli's theorem 伯努利定理

best angle-of-climb speed（Vx）最佳爬升角速度

best climb speed（Vy）最佳爬升速度

best endurance speed 最佳续航速度

best glide speed 最佳下滑速度

best range speed（VIMR）最大航程速度

best rate-of-climb air speed（Vy）最佳爬升率空速

beta mode β 模式

between-flight maintenance preparation 再次出动机务准备

bevel gear 斜齿轮

beyond visual range（BVR）超视距

bias 系统误差；偏差；偏压（无线电）

biaxial helicopter 双轴直升机

biconvex aerofoil 双凸形翼型

bicycle undercarriage 自行车式起落架

big inclination 大坡度

big velocity 大速度

bimetal 双金属片

bimetal contact 双金属片触点

bimetal defective contact 双金属片接触不良

bimetallic strip 双金属条；双金属片

bimetallic strip 双金属片

bimetallic thermal lock 双金属片热锁

binary coded decimal 二进制编码的十进制

binary fractional notation 二进分式计数法

binding of the tube 导管包扎

bingo（驾驶员向指挥机构报告）剩余油量仅够安全返回

biochemical suit 防毒衣

biochemical weapons 生化武器

biohazard suit 防毒衣

biological warfare 生物战

biological weapons 生物武器

biphone 头戴双耳耳机

biplane 复翼飞机；双翼飞机

bipolar antenna 双极天线

bird hazard 鸟撞危险

bird ingestion 发动机吸进了鸟；鸟吸

bird strike 鸟撞

bi-stable relay 双稳态继电器

bi-stable type valve 双稳态活门

bistatic radar 收发分置雷达；双基地雷达

BIT（built-in test）自检测；机内自检

bit error rate（BER）误码率

BITE（built-in test equipment）自测设备

BKN（broken）破裂的；碎云，多云

BL（buttock line/butt line）纵剖线

black box 黑匣子；黑盒子

black fog 黑雾

Black Hawk“黑鹰”直升机

black squadron 维修中队

black-hole approach 黑洞进近

blackout 无线电通信中断；灯火管制

bladder-type tank 储气囊

blade 桨叶

blade aerial 刀形天线

blade angle 桨叶角

blade antenna 刀形天线；桅杆式天线

blade area 桨叶面积

blade attachment fittings 桨叶辅助配件

blade back 叶背

blade butt 桨叶根部

blade chord 桨叶弦线

blade connecting bolt 桨叶连接螺栓

blade creeping 叶片蠕变

blade cuff 翼尖罩

blade damper 桨叶减摆器

blade disc 整体式叶片盘

blade droop 桨叶下垂

blade element theory 叶素理论

blade face 叶面

blade flapping 桨叶挥舞

blade flutter 桨叶颤振

blade forks 叶把

blade grips 叶柄

blade hub 桨毂

blade incidence 桨叶迎角

blade loading 旋翼桨叶载荷

blade mean lift coefficient 平均升力系数

blade pitch angle 桨距（桨叶安装角）

blade pitch-changing mechanism 桨叶变距装置

blade radius 桨叶半径

blade root 桨根

blade root force 桨根力

blade root torque 桨根力距

blade section 桨叶截面

blade shank 桨叶柄，桨叶根部

blade span 翼展

blade stall 桨叶失速

blade station 桨叶部位标记

blade sweeping 叶片弯掠

blade tab 桨叶调整片

blade tip 桨尖

blade track 桨尖轨迹

blade tracking 桨尖轨迹检查；桨尖轨迹校正

blade twist 桨叶扭转

blade vortex interaction（BVI）桨涡干扰

blank blade 空白叶片

blank block 定位件

blank cap 盲堵

blank run 空转

blank screen 黑屏；空白的屏幕

blanket jamming 压制性干扰

blanketing 遮蔽，覆盖

blanking 遮盖物；消隐

blanking device 堵盖

blast 冲击波；强阵风，疾风

blast deflector 喷焰偏转器

blast fences 防吹墙

blast pad 防吹坪

blast pen 防冲击波掩体，防护掩体

blast-effect warhead 爆炸波效应弹头

bleed 放气；排气，泄放，排除；引气；减速

bleed air 引气；放气

bleed leak 引气渗漏

bleed management computer 引气管理计算机

bleed off 释放（空气）

bleed orifice 放气孔；泄放孔

bleed temperature sensor 引气温度传感器

bleed valve 放气活门

bleed valve actuator 放气活门作动器

bleed valve control unit 放气活门控制组件

bleed valve position microswitch 放气活门位置微动开关

bending air 引气；放出的空气

bleed-off taxiway 滑出道，出口滑行道

bending vibration 弯曲振动

B-license B 执照
blimp 软式小飞艇
blind bombing zone 盲目轰炸区
blind flying 盲目飞行（全凭仪表操纵的飞行）
blind nut 盲孔螺母
blind ranges 盲距
blind rivet 盲铆钉
blind speed 盲速
blind speeds 盲速
blind spot 盲区；（塔台）盲区；遮挡视线点
blind toss 盲投
blind transmission 盲发，盲信号传输
blind velocity 盲速
blind zone 盲区
blindfire 盲目射击
blind-flying panel 盲目飞行仪表板
blind-speed effect 盲速效应
blink（雷达）闪烁
blink zyglo 荧光探伤器；荧光探伤法
blinker 闪光警戒灯
blip（雷达的）反射脉冲
blister（机载雷达的）天线罩
blistering 起鼓
blitz 闪电战（尤指空袭）
blizzard 雪暴
block altitude 封锁高度
block diagram 框图
block speed（飞机）区间速度（根据飞行距离求得的真空速，并考虑到飞机起飞、爬高、下降、着陆等对速度的影响）；间距速率
block test 试车台测试
block time 轮挡时间
blockade 封锁
blockage 阻塞
blockage error-altimeter（高度表）堵塞出错
blockage error-VSI（升降速度表）堵塞出错
block-to-block time 轮挡间歇时间
blow back 挥舞后倾，挥舞锥体后倾；反吹
blow-back valve 襟翼过载阀；逆流阀
blowby 漏气；渗漏
blowdown tunnel 放气式风洞；暂冲式风洞
blowdown turbine 冲击式涡轮
blower 吹风器

blow-in doors（辅助）进气门

blowing 吹气

blown flap 吹气襟翼

blown parachute 开花式降落伞，脱落式降落伞

blown-boundary-layer control 吹气式边界层控制

blow-off valve 放气阀，排气阀

blowout 熄火

blowout plug 保险塞

blucket 后风扇叶片

blue alert 空袭警报

blue sector 蓝区

blue-line speed 蓝线速度

BMEP（brake mean effective pressure）制动平均有效压力

B-nut B 螺帽

board 登机【英】

board of inquiry 调查委员会

bob weight 配重

body axis 机体坐标系

body buttock line 机身纵剖线

body lift 机身升力

body reference plane 机身参考平面

body station 机身站位

body waterline 机身水线

bog 陷入软地

bogey 不明飞机

bogie landing gear 小车式起落架

boiloff 汽化；蒸发

bolt 螺栓

bomb 炸弹

bomb bay 弹舱

bomb burst “空中开花” 特技

bomb charging 装弹

bomb disposal unit（BDU）未爆炸弹处理小组

bomb impact assessment 轰炸评估

bomb impact plot 炸弹弹着点表示图

bomb loading 装弹

bomb rack 炸弹架

bomb release point 投弹点

bombardment 炮击；轰炸

bomber 轰炸机

bombing angle 投弹角

bombing errors 投弹误差

bombing height 投弹高度

bombing run 轰炸航路

bomblet 小型炸弹

bond line 胶层

bonding 电搭接；搭铁

bonding jumper 搭铁线；接地线

bonedome 保护头盔

Bonne map projection 彭纳地图投影

boom（空中加油用）伸缩套管；尾梁，尾撑；轰隆声，声爆；撑杆

boom and zoom（B&Z）俯冲跃升攻击（战术）

boom carpet 声爆接地区域

boom microphone 悬挂式麦克风

boom throw-forward 声爆接地距离

boost 推进

boost control 增压控制器

boost discharge pressure 增压泵出口压力

boost pressure 增压压力

boosted control 助力操纵装置

booster 增压器

booster pump 增压泵

booster pump switch 增压泵开关

booster transformer 升压变压器

boot 套管

bootstrap system 自举系统；自动持续系统

bora 布拉风

bore 气缸内径

bore cooling system 内腔冷却系统

bore cooling valve 内腔冷却活门

bore diameter 孔径

borescope 孔探仪；校靶镜

borescope inspection 内窥镜孔探检查

boresight 校靶镜

boresighting after test-fire 热校靶

boroscope 探孔仪

boroscopic inspection 孔探仪检查；内孔表面检查

boss 凸台；安装座

bottom bracket 底部托槽

bottom dead center（BDC）下死点

bottom rudder 下舵

bottom structure 底部结构

bottom transparent panel 底部透明面板

bottom view 仰视图

bounce 突然攻击；跳弹

boundary 边界

boundary condition 边界条件

boundary layer 边界层流；附面层

boundary layer control 边界层控制

boundary light 边界灯

boundary lubrication 边界润滑

boundary marker 边界标志

boundary-layer control 边界层控制
boundary-layer drag 边界层阻力
boundary-layer fence 边界层隔流片；翼刀
boundary-layer normal-pressure drag 边界层压差阻力
boundary-layer scoop 边界层进气口
boundary-layer separation 边界层分离
bow（压缩机叶片）叶尖弯曲
BOW（basic operating weight）基本使用重量
bow shock wave 弓形激波；头部激波
bow wave 弓形激波；头部激波；弓形波
bowser 加油车
box formation 箱型编队
box spanner 梅花扳手
box wrench 梅花扳手
BPC（barometric pressure control）气压调节器
braced type 桁架式
bracing 撑杆
bracing piece 支撑杆
bracket 支架
bracket of ammunition container 弹箱支架
brake 刹车（装置）；制动（装置）
brake and position sensor unit 刹车和位置传感器组件
brake apparatus 刹车装置
brake chute 减速伞
brake control unit 制动控制单元
brake control valve 刹车操纵阀；刹车开关
brake disc 刹车盘
brake fluid 刹车油
brake force coefficient 刹车作用力系数
brake horsepower 制动马力；刹车马力
brake hydraulic system 液压刹车系统
brake lines 刹车线
brake mean effective pressure（BMEP）制动平均有效压力
brake metering valve 刹车计量活门
brake parachute 减速伞，刹车伞
brake pedal 刹车脚踏板
brake pressure indicator 刹车压力指示器

brake status 刹车状况；制动系统状况

brake system control unit 刹车系统控制组件

brake temperature compensator 刹车温度补偿器

brake temperature monitoring system 刹车温度监控系统

brake temperature monitoring unit 刹车温度监控组件

brake transmitter 刹车传感器

brake unit 刹车组件

brake wear 刹车磨损

brake-specific fuel consumption（BSFC）制动比油耗

brake-turbine-type heat exchanger 制动涡轮热交换器

braking 刹车；制动

braking action 制动作用

braking action advisory 制动作用通报；刹车作用通报

braking action report 刹车作用报告

braking distance 刹车距离

braking propeller 制动螺旋桨；逆桨螺旋桨

branch circuit 支路

Brayton cycle 布雷敦循环

breach 突破；突破口

break 突然改变航向，急剧压坡度；脱离编队；失速状态；信号中断；突然退出空战；穿云；云洞

break away（气流）分离

break down minor accident 三等事故；一般事故

break lock 突破锁定；打破锁定

break-and-make switch 断通转换开关

breakaway height 下滑拉起高度

breakaway thrust 起动推力

breakdown 故障；衰弱；三等事故

breakdown switch 故障开关

breaker 断路器；断电器；自动保险电门

break-even load factor 盈亏临界装载因数

breaking of the line 线路断线

breaking of the pipe 管路损坏

break-off height/ altitude 下滑拉起高度

breakout 避开

break-over voltage 导通电压

breaks in overcast 阴间多云
breakthrough 突破；突围
break-up 分解；分裂
breather 通风器；通气装置；防毒面具
breather hole 分离孔
breather-pressurization valve 通气孔密封阀
breathing 空气分离
bridge 梁
bridge of resistor 桥接电阻
bridging 搭接（导线）
briefing 任务讲解
bright band 亮带
bright period 雨停天晴期间
bright radar indicator tower equipment 塔台高亮度雷达显示设备
brightness minus key 亮度减弱键
brightness plus key 亮度增强键
brightness rocker switch 亮度摇臂开关
brightness value 亮度值
brilliant 智能；灵巧（用于描述武器的术语）
broadband antenna 宽频带天线
broadcast 广播（一种与空中导航相关的信息传播形式）
broadcast circuit 广播电路
broadcast control 概略引导（一种空中拦截技术）
broadcast-controlled air interception 概略引导空中拦截
broken 多云的；裂云的
broken（BKN）块云
broken cloud 裂开云
brownout 灯火管制
brush 毛刷
brush support 电刷支架
brushless motor 无电刷发动机；无刷直流电动机
B-scope B 型显示器
BSFC（brake-specific fuel consumption）制动比油耗
B-station B 站
bubble horizon 气泡地平；视水平面
bubble point 起泡点
buckle 扣住，皮带扣
buckling 翘曲
buddy refueling 伙伴式空中加油
buffer 缓冲
buffer area 缓冲区
buffer strut 缓冲支柱
buffet 抖振；扰流抖振；颤抖

buffet boundary 抖振界限
buffeting 振动
bug 电键
build standard 构造标准；预备标准
built-in test（BIT）自检测；机内自检
built-in test equipment（BITE）自测设备
built-in test equipment manual 自检手册
built-on rotor blade 固接式桨叶
bulge 鼓泡
bulkhead 隔板；隔温板；隔框
bulkhead rib 框肋
bullet 子弹
bungee 橡筋绳；弹性束
bunker 燃料库；沙坑；地堡
bunker-buster bomb 地堡炸弹
bunt 半外筋斗；跃升半外筋斗
bunter plate 阻弹器安装板
burble 湍流漩涡；转捩
burble angle 转捩迎角
burble point 转捩点；转捩迎角；转捩马赫数
burbling 气流分离
burn off 熔化
burner 燃烧室
burning（压缩机叶片）烧伤
burnout velocity 燃尽速度
burn-through 烧穿
burn-through range 雷达作用距离
burr（压缩机叶片）毛刺
burst 点射
burst of gunfire 连发射击
burst RPM 爆炸转速
bus 总线（数据或电气的）；汇流线
bus bar 汇流条；母线
bus bar failure 电源总线失效
bus control unit 汇流条控制组件
bus power control unit 汇流条电源控制组件
bus tie 汇流条馈电线
bus tie breaker 汇流条连接断路器
busbar 汇流条；母线（电气）
bush ring 轴套
bush step 衬套台阶（发射管）
buster 最大连续功率飞行
busy-hour operation 忙时运营总量
butterfly tail 蝶形尾翼；V 形尾翼
buttock line/butt line（BL）纵剖线
button 按钮

button light 按键灯

butt-testing 靶垛试验

Buys Ballot's law 白贝罗定律

buzz 嗡鸣；近距飞行；颤振，振动

buzzer 蜂音器；警报器

BVI（blade vortex interaction）桨涡干扰

BVR（beyond visual range）超视距

bypass duct 外涵道

bypass engine 内外涵发动机

bypass jacket 旁通环道

bypass line 旁路线；旁通管道

bypass ratio 涵道比

bypass relief valve 旁通释压阀

bypass turbojet 内外涵涡轮喷气发动机

bypass valve 旁通阀

bypass valve 旁通活门

C2（command & control）指挥与控制

CA（certificate of airworthiness）适航证

cabane 顶架；翼间支架

cabin 机舱

cabin air conditioning 座舱空气调节

cabin altitude 座舱高度

cabin attendant 舱内乘务人员

cabin blower 机舱增压器

cabin conditioner failure 座舱空调故障

cabin configuration test module 客舱构型测试组件

cabin crew 客舱组员

cabin depressurization 客舱失压

cabin differential pressure 机舱压差

cabin fire extinguisher 座舱灭火器

cabin interphone 座舱内部话机

cabin interphone system 客舱内话系统

cabin lighting switch 客舱照明开关

cabin lighting system 客舱照明系统

cabin load 座舱负载

cabin log book 客舱记录本

cabin pressure acquisition module 座舱压力采集组件

cabin pressure altitude 舱压高度

cabin pressure control panel 座舱压力控制面板

cabin pressure control system 客舱压力控制系统

cabin pressure controller 客舱压力控制器

cabin pressure regulator 机舱（座舱）压力调节器

cabin pressurization 座舱增压

cabin pressurization safety valve 机舱（座舱）压力安全阀

cabin roof 舱顶

cabin service system 客舱服务系统

cabin supercharger 客舱增压器

cabin temperature controller 客舱温度控制器

cabin temperature sensor 座舱温度传感器

cabin ventilating and heating 机舱通风与供热

cabin ventilation system 机舱通风系统

cabin window 舷窗

cable drive mechanism 钢索式传动机构

cable tension indicator 钢绳张力指示器

cadet 军校学员

CAE（control area extension）管制区延伸

cage 锁定陀螺

caging 锁定陀螺

caisson disease 沉箱病

calculated altitude 计算高度

calculated landing time 计算着陆时间

calculated time of arrival 计算到达时间

calendar inspection 定期检修

calendar limit 日历寿命极限

caliber 内径；弹径；口径

calibrate 校正；校准

calibrated air speed（CAS）标定空速；校准空速；校正空速

calibrated airspeed 校正空速

calibrated altitude 校准高度

calibrated flow 校准流量

calibrated orifice 标定放油孔；限流孔

calibrated restrictor 降压节流嘴

calibration 校准

calibration card 校准表；校准卡片

California bearing ratio（CBR）加州承载比；加利福尼亚承载指数（该术语用于机场建筑）

calipers 卡钳

call an alarm 报警

call for release 请求放行

call mission 听召唤出动

call sign 呼号

call sign or identification 呼号或识别

call up 呼叫，呼唤（设施与飞机之间的最初联系的声音）

calling out 大声呼唤

calm 无风（或称静风）
calm wind 无风（静风）
calm wind run way 静风跑道
cam 凸轮
cam disc 凸轮盘
cam engine 凸轮发动机
cam lobes 凸轮凸角
camber 拱形；弧形；弯度
camber flaps 变弯度襟翼
camber line 中弧线
cambered wing 有弯度机翼，非对称剖面机翼
camera angle of coverage 照相机视角，照相机视界
camera axis 摄影光轴；光轴
camera axis direction 摄影光轴方向
camera cone（照相机）镜筒
camera cycling rate 照相机曝光频率
camera gun 照相枪；摄像枪
camera nadir 相底点
camera tilt 照相机倾斜角
CAMI（Civil Aerospace Medical Institute）民用航空医学协会
camouflage 伪装
camshaft 凸轮轴
CAN（aircraft classification number）飞机分级数
can-annular combustion chamber 环管燃烧室
can-annular combustor 管环燃烧室
canard 前翼
canard configuration 鸭式构型；前翼构型
canary 敌我识别信号
cancel 取消（空中交通管制术语，意思是“取消已经发放的许可”）
cancel distress 取消危机呼叫
cancel flight plan（CNFLP）取消飞行计划
canceling IFR 取消仪表飞行规则
candle 烛光；伞衣未全开的降落伞
canister 霰弹；霰弹筒
cannibalization 串件；拼修
cannibalize 拆用部件或零件；拼修
cannon 机关炮；加农炮；火炮
cannon container 炮箱
cannonball 炮弹
cannular 环管式

cannular combustion chamber 环管燃烧室
canopy 前舱罩；座舱盖；座舱罩
cantilever 悬臂梁
cantilever ratio 悬臂比
cantilever wing 悬臂翼
can-type combustor 筒形燃烧室
cap（screwed to skin）帽盖（用螺栓固定于蒙皮上的）
cap cloud 山帽云
capacitance-type 电容式
capacitor 电容器
capacity（机场）容量
capillatus 鬃状云
capsule acting on the jet 作用在喷嘴上的膜盒
captain 机长
captain's panel 机长仪表板
captain's windshield wiper 正驾驶风挡刮水器
caption（图片的）说明文字
capture 截获；捕捉；捕获
carabin 卡宾枪
carbon brake 碳刹车
carbon brush 碳刷
carbon fiber 碳纤维
carbon film 碳膜
carbon monoxide（CO）一氧化碳
carbon seal 石墨封严环
carbonaceous fire 炭火
carbon-deposit 积炭
carbonisation 炭化
carbonisation of unburnt fuel 积炭
carbonization of the remaining fuel 余油积碳
carburetion 汽化
carburetor 汽化器
carburetor air-heater 汽化器空气加热器
carburetor heat 汽化器加温
carburetor icing 汽化器结冰
cardan joint 万向接头
cardinal altitudes 基数高度；基数飞行高度层
cardinal directions 基本航向；基本方位
cardinal flight levels 基数高度；基数飞行高度层
cardinal headings 基本航向；基本方位
cardinal point effect 方位基点效应
cardinal points 基本方位；基点方位
caret wing 盖烈特翼（内凹多角

形剖面翼）
cargo 货物
cargo compartment 货舱
cargo door sill 货舱门基座
cargo hook 吊货钩
cargo hook load indicator 吊货钩负载指示器
cargo load 货载
cargo load factor 载货率
cargo parachute 投物伞
cargo plane 运输机
cargo release switch 卸货开关
cargo sling 吊货带
cargo ton-mile 货物吨英里，吨英里
cargo transportation 货物运输
cargo-hold 货舱
carnet 海关文件；通关卡
Carnoy cycle 卡诺循环
carpet bomb 地毯式轰炸
carpet bombing 地毯式轰炸
carriage 炮架；起落架；挂架
carriage of external load 外部负载挂架
carrier 载波
carrier frequency 载波频率
carrier sense multiple access 载波侦听多路访问
carrier wave 载波
carrier-to-interference ratio 信号干扰比
carrier-to-noise ratio 信噪比
CARS（community aerodrome radio station）机场公用电台
cartography 制图学
cartridge 弹药；子弹；弹药筒；弹壳
cartridge starter 起动药柱
cartridge starting 火药起动，火药起动机
cartridge-pneumatic starter 弹药气动起动器
CAS（calibrated air speed）标定空速；校准空速；校正空速
CAS（close air support）近距空中支援
CAS（crew alerting system） 机组警戒系统
CAS reset switch 机组警戒系统重置开关
cascade reverser 叶栅式反推装置
case regulated pressure 壳体调节压力
casing 外壳；机匣；壳体

casing oil tank 机匣油箱
cassette recorder 盒式磁带录音机
casted disc 铸造轮盘
castellanus 堡状云
castle nut 槽形螺母
castle nuts 梅花螺帽；槽顶螺帽
casualty evacuation 伤员后送
CAT（clean-air turbulence）晴空湍流；脱离空中乱流区
CAT “A” clear area takeoff A 类净空起飞
CAT “B” takeoff B 类起飞
CAT A clear area approach A 类净空进近
CAT B approach B 类进近
catafighter 弹射起飞的战斗机
catalytic ignition 催化点火
catalyzer 催化剂
catapult 弹射；弹射器
catastrophe 一等事故（机毁人亡事故）
category Ⅰ operations Ⅰ类仪表着陆
category Ⅱ operations Ⅱ类仪表着陆
category Ⅲ operations Ⅲ类仪表着陆
category of aircraft classification 航空器分类
cathedral angle 下反角
cathode 阴极
cathode ray tube 阴极射线管
cat’s paw 猫掌风
caution 警告；注意事项
caution advisory panel 信号板
caution and warning system 警告和提示系统
caution message 警告信息
caution note 注意事项
cautionary 警戒；注意
cautionary met report 警示报告
cautionary range 告诫范围
cautionary weather warning 不利天气告警；不利气象报告
CAVOK（ceiling and visibility OK）天气良好
C-band C 波段
CBR（California bearing ratio）加州承载比；加利福尼亚承载指数（该术语用于机场建筑）
CCD（cursor control device）光标控制装置；光标控制器
C-check C 检
CCW（counter-clockwise）逆时针
CDI（course deviation indicator）航道罗盘；偏航指示器

CDL（configuration deviation list）外形缺损清单

CDT programs (controlled departure time program）管制离场时间程序

CE（certificate of experience）飞机驾龄证书

CE（circular error）圆误差

CEB（curves of equal bearings）等方位曲线；等方位线

ceiling 顶棚（板）；升限；云底高

ceiling and visibility OK（CAVOK）天气良好

ceiling balloon 测云气球，云幕气球，云幂气球，测云气球

ceiling light 云幂灯

ceiling projector 云幂灯，运照灯

ceiling unlimited 晴朗无云；无限云幂

ceilometer 云高计

celestial altitude 天体高度

celestial azimuth 天体方位；天体方位角

celestial navigation 天文航行

cell 电池

Celsius 摄氏度

center electronic center 中电子舱

center of gravity（CG）重心

center of gravity limits 重心范围

center of lift 升力中心

center of pressure（CP）压力中心

center of pressure（of airfoil）压力中心（翼剖面）

center of thrust 推力中心

center radar ARTS presentation 中心雷达 ARTS 显示

center radar ARTS processing 中心雷达 ARTS 处理

center section 中段

center vent tube 中央通气管

center weather advisory（CWA）中心天气报告

centering device 对中装置，定心装置

centering unit 中立机构

centerline 中心线

centerline thrust airplane 共轴反桨飞机

center-of-airfoil moment 翼型力矩中心

center-of-gravity envelope 重心包线

center-of-gravity limits 重心限制

center-of-gravity margin 重心位置范围；重心位置余度

center-of-gravity travel 重心移动；重心位置变化

center-of-pressure coefficient 压力中心系数

center-of-pressure travel 压力中心移动

center-pod configuration 中间吊舱式布局

center's area 中心空域

centimetric radar 厘米雷达

centimetric wave radar 厘米波雷达

centimetric wavelength 厘米波长

central air data computer 中央大气数据计算机

central console 中央控制台（操作台）；中央操纵台

central electrode 中央电极

central firewall 中心防火墙

central flow weather processor 中央流量气象处理机

central flow weather service unit 中央流量气象服务单元（组件）

central fuselage 机身中部

central lateral control package 中央横向控制组件

central maintenance 中央维护

central maintenance computer 中央维护计算机

central maintenance computer system 中央维护计算机系统

central maintenance system（CMS）中央维护系统

central meridian 中央子午线

central processing unit 中央处理器

central refueling system 单点加油系统

central screen 中央显示器

central structure 中部结构

central tracon automation system 中央终端雷达进近管制自动系统

central warning panel（CWP）中央告警板

central warning system（CWS）中间告警系统

central weather processor 中央气象处理器

centralized strapping unit 中央搭（连）接组件

centrifugal breather 油气分离器

centrifugal compressor 离心压缩机

centrifugal compressor casing 中介机匣

centrifugal compressor diffuser 离心压气机扩压器
centrifugal flow compressor 离心流压气机
centrifugal flyweight 离心飞重
centrifugal force 离心力
centrifugal fuel injection 离心甩油
centrifugal fuel injection wheel 离心甩油盘
centrifugal governor 离心调速器；离心调节器
centrifugal impeller 离心叶轮
centrifugal pendulum vibration absorber 离心摆式吸振器
centrifugal pull 离心力
centrifugal pump 离心泵
centrifugal wheel 离心叶轮
centrifuge 离心机；离心作用
centrifuge out 甩出
centripetal force 向心力
CEP（circular error probability）圆概率误差
ceram 陶瓷
ceramics 陶瓷
certificate of airworthiness（CA）适航认证
certificate of experience（CE）飞机驾龄证书
certificate of inspection 检验合格证
certificate release to service 放行授权
certificated aircraft 适航飞机
certificated for single-pilot operation 已认证的单人驾驶飞机
certification 证明；证书
certification pilot 认证飞行员
certification process 认证过程
certification service 航空器认证
certification test 认证试验
CFIT（controlled flight into terrain）地形控制飞行
CG（center of gravity）重心
ch（channel）信道；通道；通路；频道（无线电）
ch lat（change of latitude）纬度差；纬度变化；纬度变化量
ch long（change of longitude）经度差；经度变化；经度变化量
chaff 干扰物
chaff burst 突发箔条
chaff dispenser 箔条投放装置；干扰物投放器
chafing 磨损

chamfer 倒角；倒棱

chammy 麂皮

chance light 可移动的机场泛光灯

chandelle 急上升转弯；战斗转弯

change of latitude（ch lat）纬度差；纬度变化；纬度变化量

change of longitude（ch long）经度差；经度变化；经度变化量

change over point 转换点

changeover button 换向按钮

changeover level 转换高度

changeover point（COP）【航】管制转换点

changeover switch 转换开关

channel（ch）信道；通道；通路；频道（无线电）

channel capacity 信道容量

channel change 改频（改波）

channel light 航道灯标

channel rate 信道速率

channel spacing 频道间隔

channel wing 半环形机翼；槽形机翼

characteristic curves（of airfoil）特性曲线（翼剖面）

characteristic system 特征系统

characteristic velocity 特征速度

characteristics 特性（指飞机、发动机及其部件的性能）

charge 装药；装料；充电；电荷

charger 充电装备；加注器；装弹机构；弹夹

charging gear 装载设备；装载机构

charging point 加注器（飞机上用的补充氧、空气、水力流体等的接头）

charging stand 灌充台（指一种对飞机空调系统的维修设备）

chart 图表；航图；曲线图；计算图

chart index 航图索引；图表索引

chart of synthetic static state 综合静态图

chart of synthetic trend 综合动态图

chart reader 图形记录仪

chart sheet 图幅；图纸

charted VFR（Visual Flight Rules）flyways 图示目视飞行规则航线

charted visual approach（CVA）图示目视进近

charted visual flight procedures（CVFP）图示目视飞行程序

charted Visual Flight Rules（VFR）flyways 图示目视飞行规则航线
chase 追踪；追击
chase aircraft 飞机试验伴航飞机
chase helicopter 追击直升机
chase pilot 伴飞驾驶员
chassis 飞机起落架；车底盘；炮底架
chassis height 机架高
chatter 震颤；波形划痕；信号干扰【口】；谈话器
check 检查或测试（飞机）；检查飞行；考核飞行；阻击
check action fuze 惯性引信
check altitude 检查高度
check and ratify 核定
check back signal 后检信号；校验返回信号
check climb（检验飞机性能的）检查性爬高；检验上升
check command 检验指令
check flight 检验飞行；技术检查飞行
check for leaks 渗漏检查
check plus minus 正负校验
check point 检查点；检查目标；检查哨；检查站
check post 检查哨
check ride 检查飞行【口】
check ring 挡圈
check round 试弹
check sheet 检验单；校核表
check system 检查系统
check the gear down 检查起落架是否放下
check three greens 检查起落架放下锁妥
check valve 单向活门；止回阀；检验阀
check while taxiing 滑行时检查
checked-out missile（发射前）测试过的导弹；检查过的导弹
checked-out rocket 测试完毕的火箭
checker 检验器；试验装置；检验员；检查程序
checking circuit 检查电路
checking routine 检查程序
checking station（飞机装配和水平测量的）检查点；（螺旋桨桨叶上的）参照站位
checklist 检查清单；目录册；备忘录
checklist technical documentation

技术文件检查单
checkoff 检查完毕
checkout 检查；考核；使飞行人员熟悉飞机的训练；检查飞行；脱离通信联络网
checkout and firing subsystem 校正与发射子系统
checkout and maintenance 检查及维修
checkout device 检验装置
checkout equipment 检测设备
check-out firing 检查点火试验
checkout, control and monitor subsystem 检验、控制和监视分系统
checkout-launch control set（导弹）检验与发射控制装置
checkpoint 航线检查点
check-post 检查哨
checks before takeoff 起飞前检查
chemical agent 化学战剂
chemical biological defense 生化防御
chemical biological incident response force 生化事件反应部队
chemical biological quick reaction force 生化快速反应部队
chemical biological radiological-defense 化、生、辐射防护
chemical low-altitude missile, puny（CLAMP）小型化学燃料低空导弹
chemical warfare 化学战
chemical weapons 化学武器
chest pack 胸式伞包；胸伞
chest set 胸挂送话器；胸挂受话器
chief aviation maintenance technician 航空保养主任技师
chief aviation pilot 正驾驶员；机长
chief master sergeant 军士长
chief mechanical engineer 机械师
chief of air corps 航空兵司令
chief of air defense 防空司令
chief of air reserve 空军后备队司令
chief of air staff 空军参谋长
chief of army aviation 陆军航空兵司令
chief of aviation 航空兵主任
chief of operations 作战处处长
chief of ordnance 军械主任
chief technician 主任机械师

chief test pilot 主任试飞员；首席试飞员
chin 机头下部
chin blister 下颌式观察窗
chin intake 下颌式进气道
chin radome 机身前下方雷达天线罩
chin turret 机头下部炮塔
Chinook“奇奴克人”（CH-47 中型运输直升机）
chip detector 金属屑探测器
chirp radar 线性调频雷达；脉冲压缩雷达
chisel window（供空中摄影或其他电子设备用的）机头斜口窗
chock 轮挡；制动垫块；楔形垫
chock plug 堵头
chock time（飞机）停放时间
choke coil 扼流线圈
choke piston 扼流式活塞
choked flow 阻塞流
choked inlet 堵塞进气道
chopper 直升机【口】
chopper-borne 直升机载的
chord 弦线；弦长；翼弦
chord component 弦分力
chord length 弦长
chord line 弦线；翼弦
chord load 弦载荷
chord position 翼弦位置
chordwise 弦向弦向；沿翼弦方向
chord-wise gravity 弦向重心位置
chronometer 航行表；精密计时器
chute 降落伞；应急离机滑梯；抛射道；滑运道；滑下管道
cine camera 电影照相机
cine camera firing 照相射击
cine-gun 照相枪
circle a field 在飞机场上空做圆周运动；环绕
circle flight 圆周飞行
circle marker 圆形（着陆）标志
circle of equal altitudes 等高圈；地平纬圈
circle of latitude 黄纬圈；纬度圈
circle of longitude 黄经圈；经度圈
circle of origin 基准圈（以起始子午线、赤道等作为坐标基准的圆圈）
circle of position 等高圈；位置圈
circle to runway（runway number）绕场至某跑道（跑道编号）
circle-to-land maneuver 目视盘旋

进近机动

circling 圆圈飞行；盘旋

circling approach 绕场进场着陆

circling approach minimum 进入绕场着陆航线最低气象条件

circling guidance lights 着陆航向环形指示灯

circling performance 盘旋性能

circuit 电路；线路；线路图；（飞行）起落航线；绕圈飞行

circuit breaker 断路器；电路自动保险电门

circuit breaker panel 断路器板

circuit card assembly 电路板组件

circuit element 电路元件

circuit flying 圆圈飞行；航线起落飞行

circuit height 起落航线高度

circuit model 电路模型

circuit pattern 起落航线

circuitous air attack 迂回空袭

circuitous route 迂回航线；环形航线

circuits and bumps 连续起飞着陆（训练）【口】

circular 通报；圆形的；循环的

circular air defense 环形防空

circular airgram 航空通电

circular approach 绕场进场着陆

circular deviation 圆周罗差；等值罗差

circular error（CE）圆误差

circular error probability（CEP）圆概率误差

circular error probable 圆概率偏差；径向概率误差

circular flight 圆圈飞行；起落航线飞行

circular lamination 环形叠片

circular polarization 圆极化；圆偏振

circular probable error 圆概率偏差；径向概率误差

circular scan 圆周扫描；环形扫描

circular taxiway 圆形滑行道

circular velocity 圆周速度；圆轨道速度

circulate 循环；流通

circulation index 环流指数

circulation map 交通路线图；运行图；巡回交通图

circulation of the atmosphere 大气循环

circumferential taxiway 环形滑

行道
cirrus cloud 卷云
cirrocumulus 卷积云；絮云
cirrostratus 卷层云
cirrus 卷云
city terminal 城市航空点；城市航站候机室
Civil Aeronautics Administration【美】民用航空局
Civil Aerospace Medical Institute（CAMI）民用航空医学协会
civil air movement 民航运输
civil air navigation aids 民航导航设备
civil air regulations 民用航空规则
civil air route 民航航线
civil and military airport 军民合用机场
civil military cooperation 军民合作
civil military operations center 军民作战中心
CLAMP（chemical low-altitude missile, puny）小型化学燃料低空导弹
clamping ring 卡箍
clamping screw 固定螺钉
CLAMS（counter-measures launch and monitoring system）干扰发射与监视系统
clamshell 蚌壳式结构；蚌壳式座舱罩
clamshell thrust reverser 双折流板式反推力装置；蚌壳式反推力装置
classic flutter 经典颤振；标准颤振
classification of maps 地图分类
classification of property 军用品分类
classification of traffic 通信业务分类
classification yard 铁路调车场；补给品分类场
classified document 机密文件；秘密文件
classified outbound（飞机）飞出执行机密任务
classified sensor 鉴别传感器；分类传感器
clean 净形的（指无外挂物或起落架、襟翼在收起位置的）；流线型的
clean aircraft 净形飞机（指处于飞行中起落架收起以及无任何外

挂件的飞机）
clean approach 净形进场着陆
clean fin 净形垂尾（指垂直尾翼前面没有背鳍或者进气道之类的东西）
clean flap condition 襟翼收起的飞行状态
clean gross 无外挂物时的（飞行）总重
clean landing 不用增升装置着陆
clean launch（导弹）成功的发射;（导弹）正确的发射
clean unstick 柔和的离陆（起飞）
clean up the aircraft 收起落架与襟翼
clean-air turbulence（CAT）晴空湍流；脱离空中乱流区
cleaning solvent 清洗剂
clear 退弹；批准；放飞；发给安全许可证；飞越障碍；在指定空域取得制空权或空中优势；晴空
clear air turbulence tracking system 晴空湍流跟踪系统
clear approach 自由进近（降落区）
clear down 准许下降（高度）
clear flight level 允许飞行高度
clear for descent 批准下降
clear for penetration【空】允许穿云
clear for takeoff 准许起飞
clear ice 纯净冰；透明冰；明冰
clear of clouds 穿云飞行；出云
clear of obstacles 无障碍物
clear of traffic 交通控制；净空
clear sky 晴；碧空
clear to alter height 允许改变高度
clear to fly radial 允许径向飞行
clear to land 准许着陆
clear to taxi 准许滑行
clear zone（在跑道两边的）保险道；保险地带；无物区
clearance 批准；航行调度许可证；超越障碍；最低表尺
clearance area 净空区域
clearance bar 停止飞行滑行的指示线
clearance beyond 继续飞行许可
clearance for flight 飞行许可（证）
clearance for holding 等待（着陆）许可
clearance for takeoff 起飞许可
clearance limit 放行范围；许可界限；净空区界线

clearance of gun 退弹；排除停射故障
clearance plane（极低空飞行的）安全高度面；（雷）离地高度面
clearance plane elevation 安全平面（海拔）高度
clearance plane indicator 安全高度面指示器（低空飞行用）
clearance plane indicator light（极低空飞行）安全高度面指示灯
clearance valid 许可有效
clearance void 许可无效
cleared area 跑道头净空区；无障碍地区
cleared flight level 许可飞行高度层
cleared for takeoff 许可起飞
cleared level adherence monitoring 放行高度保持监视
cleared through 许可通过
cleared to enter traffic pattern 准许加大航线
cleared to land 许可着陆
cleared to leave tower frequency 准许脱离指挥塔频率
cleared via flight-planned route 许可通过计划飞行航路
clearing engine 清理发动机
clearing field of fire 清除射界的障碍物
clearing hospital 前沿医院；前方医院
clearing turns 搜索转弯；净空转弯
clear-of-clouds conditions 云外飞行条件
clearway 净空道；禁止中途停车的直通公路
cleat 夹板；系缆角；加固角片
climagram 气候图
climagraph 气候图表
climate 气候
climate control（飞行器内部的）气候控制；气候调节
climate of air arm combat 航空兵作战气候
climatic geomorphology 气候地貌学
climatic zones 气候带
climatological information 气象情报资料
climatology 气候学
climb 爬升
climb allowance 上升用油量

climb and acceleration time 爬高与加速所需时间

climb angle 上升角；爬高角

climb as a pair 双机上升

climb assisting unit 飞机上升助推器

climb attack 上升攻击

climb attitude 爬高姿态（爬高时的仰角）

climb away 上升脱离（目标）

climb check 检验上升（检查飞行性能的上升）

climb corridor（空中）爬升走廊

climb flight level 爬升至飞行高度层

climb full bore 全速上升；全速爬高

climb in turns 盘旋上升

climb indication（地平仪上）上升角指示

climb rate 爬升率

climb section（飞行轨迹的）上升段；地平仪球形刻度盘的上升指示部分

climb steeply 大角度爬升

climb to altitude 爬升至高度；上升到高空

climb-and-dive indicator【空】升降速度表

climbing attitude 爬升姿态

climbing cruise 爬升巡航

climbing performance 上升性能；爬高性能

climbing rotational resistance 爬升旋转阻力

climbing takeoff 直接爬升起飞（指离地后无一段平飞）

climbout（起飞后的）起始上升；转入稳定上升状态；起落航线飞行中的上升

climbout tactics（在空投货物后的）爬高战术

clinodromic 提前飞行的；有固定提前角的

clinometer 倾斜仪；倾斜计；测角器；磁倾计

clinoscope 倾滑指示器；倾斜仪

clipped wing 切削机翼；切去翼尖的机翼

clock code position 钟面代号位置；钟面方位代号（指目标相对于飞机或舰船的位置，目前在正前方为 12 点钟）

clockwise 顺时针

clockwise rotation 顺时针旋；右旋

clog 堵塞；制动器

clogging indicator 堵塞指示器；阻塞指示器

close air assistance 近距离空中支援；紧密空中支援

close air support（CAS）近距空中支援

close air support control 近距离空中支援控制；紧密空中支援控制

close air support exercise 近距离空中支援演习；紧密空中支援演习

close antenna 环状天线；闭路天线

close approach indicator 接敌指示器

close control bombing 近距离控制轰炸；精密控制轰炸

close control radar 近接管制雷达

close controlled air interception 精确引导的空中截击

close controlled interception 精确引导截击

close envelopment 近距离包围

close escort 近距护航；近距护航队

close flight plan 结束飞行计划；冻结飞行计划

close formation 密集编队

close parallel runways 近距平行跑道

close pattern 火炮射击的密集度；弹着密集度

closed aerodrome 机场关闭；已关闭的机场

closed circuit system 闭路系统

closed high 闭合高压

closed loop support 闭合回路支援

closed pattern（机场上空的）起落航线

closed runway marking 跑道关闭标志

closed taxiway marking 滑行道关闭标志

closed traffic 封闭航线

closing angle 闭合角；接近角

cloud 云；云状物；（雷达）干扰云；机群；导弹群

cloud amount 云量

cloud bank 云堤；云带

cloud banner 旗状云

cloud base 云底

cloud cap 云幞；云帽
cloud classification 云的分类
cloud cover 云量；云掩护
cloud current 云流
cloud deck 云顶；云层
cloud height 云底高度；云高
cloud layer 云层
cloud seeding 人工播云；人造云；云的催化
cloud/collusion warning（机械雷达的）云层防撞警告
cloud-height indicator 云高指示器
cloud-height recorder 云高自记仪
club propeller 试车螺旋桨（指专供活塞式发动机台架试车用的螺旋桨）
cluster 组；簇；火箭发动机簇（组）；发动机组；集束信号弹；集束炸弹；伞群
cluster bomb 集束炸弹
cluster missile 集束助推器式导弹；集束发动机式导弹
cluster rocket launcher 多管式火箭发射装置
cluster switch 组合开关
clutch 离合器；离合器控制杆；起重机抓钩
clutch control【直】离合器控制
clutch transmission system 离合器传递系统
clutter 杂乱回波；地物干扰；混乱
CMS（central maintenance system）中央维护系统
CNFLP（cancel flight plan）取消飞行计划
CO（carbon monoxide）一氧化碳
coalition action 联盟行动;（多国）联合行动
coalition air operation 联盟空中作战
coalition force 盟军;（多国）联合部队
Coanda flap 孔安达襟翼
coarse pitch（螺旋桨的）高；大距;（螺栓等的）粗螺距
coarse scale 粗读刻度盘
coarse sight 概略瞄准（具）
coarse system of bearing 方位粗测系统
coast down 减退；下降；向下惯性飞行；沿下降轨道惯性飞行
coast guidance 惯性飞行段制导
coast period 惯性飞行阶段；被动

飞行阶段

coast phase control system 惯性飞行阶段控制系统

coast-in point 入陆点；海岸突破点；入侵海岸地点（指入侵飞机越过海岸飞往内地目标的地点）

coating 涂层；镀层；保护层；涂料；底漆

coating for infrared and light reflection 红外与光反射涂层

coaxial 同轴的

coaxial cable 同轴电缆

coaxial carrier system 同轴载波系统

coaxial line 同轴线路

coaxial propeller 共轴螺旋桨

coaxial rotor 同轴双旋翼（指直升机的布局）；同轴转子

coaxial rotors helicopter 同轴双旋翼式直升机

Cobra “眼镜蛇”（美国机载雷达，法国机载低空打击和地形跟踪雷达）

Cobra attack helicopter 眼镜蛇攻击型直升机

Cobra gunship helicopter 眼镜蛇武装直升机

Cobra night fire control system “眼镜蛇”导弹夜间发射控制系统

Cobra Stinger 眼镜蛇毒刺号战车

cocked missile 处于待发状态的导弹

cockpit 驾驶员座舱

cockpit air temperature regulator 座舱气温调节器

cockpit area microphone 座舱区扩音器

cockpit blind spot 驾驶舱盲点

cockpit check 座舱实习（使飞行员熟悉座内各种设备的一种训练或测验）；座舱仪表检查

cockpit conditioner failure 驾驶舱空调故障

cockpit controls 驾驶舱内的操纵杆

cockpit cover 座舱蒙布；座舱盖罩

cockpit cowling（座舱开口处的）驾驶舱整流罩

cockpit crew（驾驶舱）机组人员

cockpit defogging 座舱（驾驶舱）玻璃除雾

cockpit defroster 座舱玻璃除霜器

cockpit display map 座舱内的平

面装置显示器

cockpit display unit 座舱显示装置

cockpit dome light 驾驶舱顶灯

cockpit door 座舱门；驾驶舱门

cockpit drill 座舱练习（起飞动作）

cockpit flight signal 机内飞行讯号（给机组人员的声音或手势信号）

cockpit inspection 座舱检查（起飞前驾驶员对座舱内各种设备的检查，以确保能正常工作）

cockpit interior 驾驶舱内部设备

cockpit layout 座舱布局

cockpit lever 座舱操纵杆

cockpit panel 座舱仪表板

cockpit procedures trainer 座舱驾驶操纵程序练习器

cockpit radar scope 座舱雷达显示器

cockpit resourse management 飞机座舱资源管理；驾驶舱资源管理

cockpit temperature sensor 驾驶舱温度传感器

cockpit utility lighting 驾驶舱多功能照明

cockpit ventilation system 驾驶舱通风系统

cockpit voice recorder（CVR）座舱话音记录器

cockpit wind shield broken 驾驶舱风挡破裂

code beacon 信号灯标；闪光灯标；电码信标

code converter 代码转换器；码型转换器

coder 编码器

coefficient 系数；率

coefficient of drag 阻力系数

coefficient of lift 升力系数

coefficient of seepage 渗透系数

coefficient of skewness 偏差系数；偏态系数

coffin launching facility 长箱形发射设备（导弹水平放置，发射时竖起）

coherent optical radar 相干光雷达

coherent oscillator 相干振荡器

coherent pulse radar 相干脉冲雷达

coherent radar 相干雷达；同调雷达

coherent transponder 相干应答机

coil 线圈；靠拢（装甲部队在休息时收缩距离）

coke 积炭；将飞机做成细腰式机身

col 山口；垭口；高山口
cold advection 冷平流
cold air unit 冷气装置
cold front 冷锋
cold junction 冷接点
cold reverse thrust 冷反推力
cold rocket unit 冷式火箭发动机（指使用单元推进剂分解反应产生热能的火箭发动机）
cold section 冷段
cold tank system 冷却油箱系统
cold wave 寒潮；冷（空气）波；寒流
coleopter 环翼机（一种垂直起降飞机，该机机翼为圆环形，机身在中间，停放时机头向上直立）
collateral 附带的；旁系的
collateral damage 附带损伤（损坏）
collective 总桨距操纵杆；总距杆
collective axis 总矩轴
collective control 总矩控制
collective input 总矩输入
collective lever friction control knob 总距杆摩擦控制旋钮
collective pitch 总距
collective pitch angle 总距角
collective pitch control 总桨矩操纵
collective pitch lever 总桨距操纵杆
collective reference marker 总矩参考标志
collective stick 总桨距操纵杆；（直升机的）油门变距杆
collective training 协同（合成）训练；集体训练
collector 收集器；情报收集机构
collimated display 平视显示器
collimating marks 平行标志；照相机框标
collimator set 准直仪设备
collimator sight 视准式瞄准具；平行光管式瞄准具
collision（车、船、飞机等的）碰撞；（与目标）相遇；（导弹）击中目标
collision accident 撞机事故
collision beacon 防撞灯
collision landing 碰撞着陆；硬着陆
collision-avoidance system 防撞系统
collision-course radar 相遇航向攻击雷达；前半球攻击雷达
collision-course tactics 相遇航向截击战术

color tints 色层

column 驾驶杆；纵队；分队

column tactics 纵队战术

Comanche【美】“科曼奇”（RAH-66 侦察 / 攻击直升机）

combat 战斗；作战；格斗

combat air patrol 空中战斗巡逻；掩护地面的空中巡逻队

combat aircrew 作战空勤人员

combat arms 战斗兵种（在陆军中负责实施地面战斗任务的兵种，如步兵、炮兵、防空炮兵、装甲兵等）

combat aviation 战斗飞行；作战飞机（特指战斗机）；战斗航空兵

combat briefing 战斗简报

combat chart 登陆火力支援图；作战海图

combat load 战斗装载

combat power 战斗力

combat service support detachment 作战勤务支援分遣队

combat support airlift 战斗支援空运

combi 客货两用运输机

combination failure 综合故障

combination inertia starter 混合惯性启动器

combination influence mine 混合触发雷；复式感应雷

combined acceleration and speed control 加速度与速度综合控制

Combined Air Force Operation Base 联合空军作战基地

combined arms 诸兵种合成的；联合兵种

combined arms and services training 诸军兵种合成训练

combined-arms combat 合同战斗

combined-arms reserve 合成预备队

combining circuit 组合电路

combining display【雷】混合显示器

combining plate 结合盘

combustion air 燃气

combustion chamber 燃烧室

combustion chamber deterioration 燃烧室损坏

combustion chamber drain valve 燃烧室漏油活门

combustion duration 燃烧持续时间；火箭发动机工作寿命；主动段飞行时间

combustion efficiency 燃烧效率
combustion instability 燃烧不稳定性
combustion starter 燃气启动装置
combustor 燃烧室；冲压式发动机燃烧装置
combustor drain valve 燃烧室排油阀
combustor efficiency 燃烧室效率
command 指挥；指挥部；司令部；军区；防御区;【计】指令；命令
command & control（C2）指挥与控制
command airspeed 规定空速
command bars（姿态指示仪上的）姿态主指针
command dot 指令点
command ejection system 指令弹射系统
command guidance 指令制导
command post 战地指挥所；指挥部
command select ejection system 指令选择弹射系统
command structure 指令结构
commandant 指挥官；司令（尤指要塞或防区司令）；军事学校校长
commander 指挥官
commander air rescue group 航空救护大队司令
commander air rescue squadron 航空救护中队中队长
commercial air transport operation 商业航空运输运营
commercial aircraft 商用飞机；营业飞机
commission of aeronautical affairs 航空委员会
commissioned officers 军官（持有委任状的）
common data link 通用数据链
common deposition 普通沉淀物
common digitizer 通用数字转换器
common display cockpit 共同仪表座舱
common emitter 共发射极
common ground support equipment 通用地面支援设备
common nacelle system 共用短舱系统
common navigation aid system 通用导航辅助设备

common screwdriver 一字解刀；一字改锥；一字螺丝刀

common tools 通用工具

common traffic advisory frequency（CTAF）公共交通咨询频率

common-flow afterburner 共流式加力燃烧室

communication 通信；交流；通信工具；通信设备；通信系统

communication anti-jamming 通信抗干扰（措施）

communication channel 通信波道；通信信道；通信电路

communication data link system 通信数据链系统

communication management unit 通信管理器

communication navigation surve-illance 通信导航监视

communication network 通信网络

communication security 通信安全；通信保密

community aerodrome radio station（CARS）机场公用电台

compartment pods（组合）战斗吊舱

compartment ventilation 短舱通风

compass 罗盘；指南针；圆规；两脚规；界限；范围

compass angle 磁针方位角

compass antenna 罗盘天线

compass azimuth 磁针方位角；罗盘方位角

compass base 罗盘校正台；罗差校正坪

compass calibration 校罗差

compass card 罗经刻度盘；罗盘卡

compass controller 航向系统操纵台；罗盘系统操纵台

compass correction card 罗盘误差表

compass course 罗盘航线角；罗盘航向

compass deviation 罗差

compass deviation compensation 罗差补偿

compass error 罗盘误差；罗经差

compass heading 罗盘指向；航向

compass locator 导航台；指南针定位器

compass locator outer marker 外定位信标台

compass north 罗盘北（罗盘指针

指北的一端所标示的未经修改的方向）；磁北

compass points 基点；方位基点

compass rose 罗盘仪刻度盘；罗经盘；指南罗盘

compass swinging 罗盘误差校正；测定罗差

compensated gyro 补偿陀螺

compensating magnets（罗差）补偿磁条

compensation 补偿；平衡；校正

complement 编制人员；（飞机及船只的）乘员定额；装备定额；装备标准；各种额外的和配属的部队单位；补充

complete engine 整台发动机

complete shutdown of the engine 发动机完全停车

complete vehicle 装配完整的飞行器

completely assembled for strike 装配就绪待出击

completeness of combustion 燃烧完全度

complex aircraft 复杂飞机

complex electromagnatic environment 复杂电磁环境

complex material 复合材料

component 构件；组件；分队；联合部队所属部队；后勤部队的专职部队

component failure 机件故障；元件失效

component life 组件寿命

component life-indefinite（飞机）部件寿命不定

component maintenance manual 部件维护手册

component overhaul manual 部件大修手册

component repair and overhaul manual 部件维修与大修手册

component replacement 更换零部件

composite 复合材料；混合编队；复合电路

composite aerial photograph 联配航空照片；复合空中照片

composite engine 组合式发动机

composite flight 复合飞行；混合飞行

composite flight plan 复合飞行计划；混合飞行计划

composite forces 混编部队

composite formation 混合编队

composite material 复合材料；合成材料

composite power plant 复合发动机；组合式动力装置

composite propellant 混成发射药；混合推进剂

composite propeller blade 复合螺旋桨桨叶

composite route system 复合航路系统

composite tail cone 复合尾锥

composite takeoff 复合式飞机起飞；母子飞机起飞

compound aircraft（既有机翼又有升力旋翼的）复合式飞机

compound engine 复合发动机；复合式活塞发动机

compound helicopter 复合式直升机（配有辅助推进系统的直升机）

compressed air 压缩空气

compressibility 压缩性；压缩系数

compressibility burble 压缩性扰流；压缩扰流

compressibility correction 压缩性修正

compressibility drag 压缩性阻力

compressibility effect 压缩性效应

compressibility error 压缩性误差

compressible flow 可压缩气流；可压缩流；压缩气流

compression 压缩

compression failure 压缩失稳

compression lift 压缩性升力

compression of the earth 地球椭率；地球扁率

compression pressure ratio 压缩压力比

compression ratio 增压比

compression rib 抗压翼肋；加强翼肋

compression ring（活塞式发动机活塞上的）压缩涨圈

compression stall 压缩机失速

compression stroke 压缩行程；压缩冲程

compression test 压力试验；压缩试验；耐压试验

compression wave 压缩波

compressive load 压缩载荷

compressor 压缩器；压气机；压缩机

compressor blade 压气机叶片

compressor bleed pressure 压气机引气压力
compressor bleed valve 压气机放气活门
compressor case 压气机壳体
compressor casing 压气机机匣
compressor corrosion 压气机腐蚀
compressor disc 压缩机盘
compressor discharge port 压气机放气孔
compressor discharge pressure 压力机排气压力
compressor discharge temperature sensor 压气机出口温度传感器
compressor efficiency 压气机效率
compressor erosion 压气机磨损
compressor inlet pressure 压气机进口压力
compressor inlet temperature 压气机进口温度
compressor outlet 压气机出口
compressor pressure ratio 压气机增压比
compressor rear frame 压气机后框架
compressor stall 压气机失速
compressor stall-margin curve 压缩机失速边界曲线
compressor surge 压缩机喘振
compressor vane 压气机叶片
compressor wheel 压气机叶轮
compulsory reporting point（飞机在空中的）规定报告点；强制报告点
computation 计算；计算过程
computed air release point 计算空投点；计算投弹点
computed airspeed 计算空速
computed altitude 校准高度；计算天体高度
computed collision course 计算相遇航线
computed correction 计算修正量
computed course 计算航向；预先计算好的航向
computed-centerline approach 计算中心线进近
computer assisted design 计算机辅助设置
computer based instruction 计算机基本指令
computer display channel 计算机显示通道
computer human interface 人机

接口

concave slope 凹形斜面

concavity 凹面；凹处；凹度

concentration 集结；浓度；密度

concession 允许；特许；租让

condensation level 凝结面；凝结高度

condensation nuclei 凝结核

condensation trail 凝结尾迹；雾化尾迹

condenser 电容；电容器；凝结器；冷凝器；聚光器

condition monitoring 状态监控

conditional stability 条件稳定性

conditioned air ground system connection 空调系统地面接头

conducting wire 导线

conductivity 电导率；传导性

cone（飞行器的）头锥；锥形（光）束；（机场）风袋

cone angle 锥角

cone of confusion 锥形干扰区

cone of silence 静锥区

configuration 外形；形状；轮廓；结构；布局方案；配置形式

configuration deviation list 技术状况偏差单，构型偏差单（指飞机的附属文件，列出飞机使用时机上允许缺少的或允许带有某些缺陷的项目）

configuration deviation list（CDL）外形缺损清单

configuration setting 配置设定；组态设定值

configuration strapping unit 配置连接组件

confined area 限制区域

confirm 确认（空中交通管制术语）

conflict 冲突；战争

conflict alert 冲突警报信号；冲突战斗准备

conflict resolution 冲突解决

conflict search 冲突搜索

conformal map projection 正形地图投影

conformal projection 正形投影；保形投影

conformal-array radar 保形阵列天线雷达（指天线罩保持飞机蒙皮外形不变的一种阵列天线雷达）

congested area（空中交通的）拥挤地区

conical projection 圆锥投影

conical scanning 圆锥扫描

conical surface 锥面

coning angle 旋转翼的倒锥角

connected network 连通网络

connecting bar 连杆

connecting casing 连接机匣

connecting piece 连接件

connecting rod 连杆

connection 连接

connection confirm 连接确认

connection ear 连接叉耳

connection flange 连接法兰

connection plug 接线柱

connector 连接器；插塞；连接线

conscript 征募

conscription 征兵

consecutive fire 连续射击

consecutive operations 连续作战

console 操纵台；控制台；仪表板；悬臂架；落地式支架；落地式（无线电）接收机

consolidated cabin display 综合机舱显示器

conspicuity lights 防撞灯

constant altitude 等高度；固定高度；恒定高度

constant differential pressure valve 等压差活门

constant heading 固定航向

constant level flight 匀速平飞

constant pressure cycle 等压循环

constant pressure source 恒压源

constant pressure valve 定压活门

constant speed drive 恒速驱动；恒速传动

constant speed propeller 恒速螺旋桨（指通过自动变距实现恒速，即自动变距螺旋桨）

constant wind 稳定风

constant-bearing navigation 定向航行；恒向航行;（导弹）平行接近法引导

constant-displacement pump 定排量泵

constant-flow oxygen system 连续供氧系统；恒流供氧系统

constant-pressure chart 等压图

constant-speed drive unit（CSDU）恒速驱动装置

constant-speed unit（CSU）等速装置；恒速装置

consumption 损耗

consumption part 消耗件

contact 接触；连接;（飞机升空后再）接地；一次起落；能目视

地面（飞行）;（在雷达荧光屏上）发现敌人目标；回波

contact approach 目视进场着陆

contact battle 遭遇战

contact flying 目视飞行；地标飞行

contact lost（雷达目标、磁探目标等）失探；目标失踪

contact point（路线）会合点；接触点;（飞机）联络点

contactor 电流接触器

contain 包含；牵制；围困

containment capsule 密封舱

contamination 污染；污染物；混合

contingency command post 应急指挥所

contingency power 应急功率

contingency readiness exercise 应急战备演习

contingency relief force 紧急救援部队

continue off the flight deck 冲出飞行甲板后继续起飞

continue port 继续左转

continue wave radar 连续波雷达

continue wave radar（CW radar）连续波雷达；等幅波雷达

continued takeoff 相继起飞

continuity equation verification 连续方程验证

continuous BIT 连续自检

continuous communication watch 连续通信监视

continuous disturbance 连续干扰

continuous duty 连续工作状态；连续任务

continuous escort 全航线护送；全程护航

continuous flight 连续飞行；不着陆飞行

continuous flow oxygen system 连续流出式供氧装置

continuous power 连续功率

continuous rain 连续性降水

continuous reinforcements 陆续增援部队

continuous wave interference 连续波干扰

continuous wave interferometer 等幅波干扰仪

continuous wing 连续翼；整体翼；无接缝翼

continuous-flow fuel injection system

连续流燃油喷射系统

continuous-flow oxygen system 连续供氧系统

continuous-wave radar 连续波雷达

contour 等高线；轮廓；外形；概要；大略

contour flight 等高线飞行（指按地形的等高线低空飞行）

contour flying 沿地形飞行；地形回避飞行；掠地飞行

contour interval 等高线间距

contour line 等高线；轮廓线

contrails 凝（结尾）迹；拉烟；飞行云轨迹

contrarotating propeller 反向旋转螺旋桨

contrarotation 反转；反向旋转

control accelerometer 控制系统加速度表，控制系统加速度传感器

control and display unit 控制与显示设备

control and warning 控制与预告

control area 管制区域；管制区；空中交通控制区

control area extension 管制区延伸

control area extension（CAE）管制区延伸

control arm 操纵杆；操纵臂

control cabin window anti-icing system 驾驶舱风挡防冰系统

control cock 控制开关

control column 控制杆；操纵杆

control console 中央控制台

control display unit 控制显示组件；控制显示装置

control efficacy 操纵功效

control flight regulation 管制飞行规则

control force 驾驶杆力；作用在操纵面上的空气动力；操纵力；控制力

control force reversal 控制力反向；操纵力反向

control force system 操纵感力系统；载荷感觉装置

control gear 操纵装置；控制装置

control gear up and locked 起落架收上并锁住

control handle 操纵手柄

control instruments 驾驶仪表；控制仪表

control knob 控制按钮

control lag 操纵延迟；操纵滞后；调节延迟

control lever 操纵杆；控制杆
control lever retarded 操纵杆反应迟缓
control linear system 操纵线系
control lock 操纵锁
control loop 控制回路
control mode 控制方式
control moment 操纵力矩
control of acceleration curve 加速控制曲线
control of deceleration curve 减速控制曲线
control panel 控制面板；操纵台
control reversal 操纵反效；副翼操纵反效
control rod 控制杆；飞机操纵传动杆；飞机操纵杆；控制棒（为原子核反应器的重要组成部分之一）
control rotor（直升机的）水平旋翼；尾桨；舵桨
control sector 管制分区；扇形管制分区
control sensitivity 操纵灵敏度；舵面反应灵敏度
control software 控制软件
control stability augmentation system（CSAS）控制增稳系统
control stand 控制台；操纵台
control stick 飞机操纵杆；控制杆
control strip 控制航带；控制片
control surface 操纵面；舵面
control system 控制系统；操纵系统；调节系统
control system electronics unit 控制（操纵）系统电子组件
control system failure 控制系统失效
control team（战术）空军引导组
control tower 控制塔台；指挥（调度）塔台
control tracking station 控制跟踪站
control unit 控制装置；操纵机构；控制器；控制单元
control valve 制动活门；控制阀
control wheel steering 驾驶盘操纵
control zone（交通）管制空域；管制地带；禁区；机要区；交通管制区
control-and-reporting center（CRC）管制和报告中心
controls-fixed flight 握杆飞行
controls-free flight 松杆飞行

controllability 操纵性能；控制特性；可控性；可操纵性

controllable pitch propeller 可调螺距螺旋桨；变距螺桨

controlled airdrome 管制机场

controlled airspace 管制空域

controlled airspace（instrument restricted）管制空域（仪表限制）

controlled airspace（instrument/visual）管制空域（仪表 / 目视）

controlled departure time 管制离场时间

controlled descent 控制下降；操纵下降

controlled flight 受控制飞行；可控飞行

controlled flight into terrain（CFIT）可控飞行撞地

controlled interception 控制拦截

controlled mosaic 控制点镶嵌图；经过更正的镶嵌空中图片

controlled or uncontrolled route 管制或非管制航路

controlled separated flow 控制分离流；受控分离流

controlled slots 可控缝翼

controlled time of arrival 准时到达时刻

controller 控制器；操纵台；舵；调节器；调节仪表；控制员；操纵员；调度员；地域防空指挥官；制动装置

controller air 空中控制员

controller-pilot data-link communications（CPDLC）管制员 - 飞行员数据链通信

controlling plane 操纵面；控制面；舵面；控制飞机；制导飞机

control-wheel steering 驾驶盘操纵

convection 对流

convective condensation level 对流凝结高度

convective current 对流气流

convective instability 对流不稳定性

conventional airfoil wing 常规翼型机翼

conventional arms 常规武器；常规军备

conventional artillery 常规炮兵

conventional weapons 常规武器

converged fire 集中火力射击（炮兵）

convergence（气流）辐合；辐合度；集中火力射击；集火射向

convergence angle 收敛角；会聚角

convergence factor 收敛因子（航图上任何两条子午线之间的夹角对其实际经度变化的比率）；集火射系数

convergency 集中火力射击；集火；集火射向

convergent-divergent duct 收敛扩张管道

convergent-divergent inlet 收敛扩张进气道

conversation of angular momentum 角动量守恒

conversion 改装；改型；（直升机或垂直起落飞机）过渡飞行；转换；变换

conversion angle 转换角；外形变换角

conversion flight mode（垂直起落飞机的）过渡飞行方式

conversion instruction 转换指令

conversion table 单位换算表；换算表

converter 变流器；变频器；变流机

convertible helicopter 倾转旋翼直升机；变换式直升机

convertible hold 通用货舱

convertiplane 推力拉力换向式飞机；平直两用飞机

convex surface 叶背

conveyor 传送机；输送设备；搬运者；传达者

convoy route 护航航线；汽车运输队的行进路线

convoy routing 护航航线选定；护航定向

cooler 冷却器；冷却剂

cooling fan 冷却风扇

cooling fins 散热片；冷却片

cooling gills 散热（空气）导流板；散热鱼鳞片

cooling system 冷却系统

cooperative aircraft 合作飞机；有应答的飞机（指空中交通管制中，带有二次监视雷达应答机的飞机）

coordiantion fix 协调定位

coordinate system 坐标系

coordinated turn 协同转向

Coordinated Universal Time（UTC）协调通用时；协调世界时

coordinating altitude 协调高度

coordination 协调；协同

coordination of all three controls 三种控制动作（俯仰、偏航和滚转）的协调

COP（changeover point）【航】管制转换点

copilot 飞机副驾驶员；【口】自动驾驶仪

co-pilot 副驾驶员

copilot collective lever 副驾驶员总距杆

copilot glareshield 副驾驶员遮光板（防眩板）

copilot ventilation 副驾驶员通风系统

cordon 哨兵线；警戒线

core compartment 核心机

core compartment cooling valve 核心机冷却活门

core engine 核心发动机

corkscrew 螺旋；急盘旋下降；螺形飞行

corner reflector 角反射器

corner velocity 角点速度

corona 日冕；电晕

corrected altitude 修正高度

corrected azimuth 修正的方位角

corrected gyro 修正陀螺

corrected mean temperature 修正平均温度

correction 修正量；【通】更正；校正

correction angle 修正角

corrective action 校正作用；修正作用；（飞机）修正动作

corrective maintenance 故障检修；保养检修；校正性维修

corrective measure 排故措施；纠正措施

corrective torque 校正力矩；修正力矩

correctly banked turn【空】正常盘旋；协调转弯

correlation bombing system 相关轰炸系统

correlator 相关器；相关仪

corridor 走廊；通道；纵向狭隘地带；空中走廊；规定航路

corrosion 腐蚀

corrosion control 腐蚀控制

corrosion prevention manual 腐蚀防护手册

corrosion resistant steel 耐蚀钢；防锈钢

corrosive agent 腐蚀剂
corrugated sleeve 波纹套
costa 翼肋；结构肋（条）
cotter key 扁销键
cotter pin 保险销；开口销
cotter safety 开口销保险
counter 计数器；计算器；计数管；反；逆；相反的；对抗；还击
counter air 夺取空中优势作战；防空作战
counter clockwise 反时针方向；逆时针方向
counter weight mounting strip 配重固定条
counteract 抵消；阻碍；中和
counteraction 反作用；对抗作用；抵抗
counterattack 反击；反突击
counter-clockwise（CCW）逆时针
counter-clockwise rotation 左旋
counterintelligence 反情报；反间谍
countermeasures 干扰；对抗；对抗措施；对策
countermeasures airborne infrared 机载红外对抗措施
countermeasures aircraft 干扰飞机；电子对抗飞机
countermeasures chaff dispenser 电子对抗干扰物投放器
countermeasures detection device 电子对抗探测装置
countermeasures device 对抗装置；干扰装置
counter-measures launch and monitoring system（CLAMS）干扰发射与监视系统
countermeasures system 对抗系统
counterpart 对应部分；对手方
counterpoint 对照；相对物
counter-rotating propellers 反向转动螺旋桨
countersubversion operation 反颠覆行动
counter-surface-to-surface missile 反地对地导弹的导弹；反舰对舰导弹的导弹
counter-terrorism 打击恐怖主义；反恐怖主义
countervane 导流叶片；反向折流叶片
counterweapon 对抗武器；防御武器；截击导弹

counterweight 配重；平衡器

coupled approach 用仪表着陆系统自动进场着陆；无线电交连自动进场着陆

coupled engines 并车发动机（指两个发动机并车后带动一个螺旋桨）

coupled flight 耦合飞行（指自动驾驶仪与无线电台信号耦合操纵）

coupled flutter 耦合颤振

coupled mode 耦合方式

coupled oscillator 耦合振荡器

coupled power turbine 耦合动力涡轮；联动动力涡轮

coupled range finder 耦合测距仪

coupling analysis 耦合分析

coupling capability（与地面导航设备）交接进场着陆能力

coupling dynamics 耦合动力学

coupling shaft 联接轴

coupling sleeve 联轴套

coupling tube 联接管

course 航向；航向信标航向；航线；航迹

course angle 航线角；航向角

course bar 航道偏离针

course beacon 航线（指示）信标；航向信标

course bearing 航路角

course correction 航线修正

course deviation indicator 航向偏差指示器

course deviation indicator（CDI）航道罗盘；偏航指示器

course deviation scale 航道偏离刻度

course directing radar 航向指示雷达

course error 航向误差；航迹误差

course error offset 航线误差补偿

course indication 航向指示

course light 航线标灯；航路灯

course line 航线；航道线；方向线

course line computer 航线计算机

course line deviation 航线偏差

course pointer 航向指示器；航向指针

course recorder 航迹自动记录仪；（电罗盘）航向记录仪；航道记录图

course selector 航道选择器

course sensitivity 航向灵敏度

course setting error 航线设定误差
course-and-drift indicator 航向偏流指示器
course-and-speed computer 航向与速度计算机
course-control housing 航向操纵盒
cover 掩护；掩蔽；覆盖物；掩蔽物；蒙布；盖；罩；壳；外胎；护航飞机；空中侦察；空中照相范围；集中照射
cover search（空中照相侦察）拍摄地区选择
cover the flanks 控制侧翼
cover trace（飞行员用描图纸蒙在地图上描下的）侦察飞行航线简图
coverage 侦察程度；监视范围；作用距离；有效区（域）；视界；覆盖（层）；掩护；火力掩护地区
covered approach march 有掩护的战斗前进；掩护进军；掩护接敌行军
covered area 掩蔽地域
covertrace 侦察飞行航线简图
cowl 整流罩；（发动机的）包皮
cowl anti-ice valve 整流罩防冰活门
cowl flap 导流板；鱼鳞片
cowling 整流罩；（尤指）飞机引擎罩
cowling and fairing 整流罩
CP（center of pressure）压心
CPDLC（controller-pilot data-link communications）管制员 - 飞行员数据链通信
crab 航向法偏流修正；侧航修正；侧航；偏航；航空照相的横向偏差
crab angle 偏流修正角；偏航角；侧飞角
crack 裂纹；开裂
cracks（压缩机叶片）裂纹
craft 飞行器；船舶，艇；技能；技巧
crank（发动机等）冷转机柄；曲柄；重心不稳；易倾
crank case 机匣；曲轴箱
cranked trailing edge 曲折形后缘
cranked wing 曲折翼；倒海鸥式机翼
cranking 起动；摇转；冷转
cranking button 冷转按钮
cranking control button 冷转控制

按钮
crankpin 曲柄销
crankshaft 机轴；曲轴
crankshaft runout 曲轴跳动
crash 坠毁；飞机失事；系统性故障；事故；应急的；紧急的；突击性射击
crash assembly 防撞装置
crash landing 摔机着陆（尤指机腹着陆）；毁伤着陆；迫降
crash locator beacon 飞机坠地定位信标
crash pad 防撞垫，缓冲垫
crash position 迫降时机组人员位置；坠落时机组人员位置；失事地点
crash position indicator 失事地点指示器；应急无线电信标台
crash survival flight data recorder 抗毁飞行数据记录器
crash switch 坠毁开关；碰撞开关
crash tender 事故处理车；飞机失事救火车
crash-protected recorder 抗毁飞行记录器
crashworthiness 毁机安全性；适毁性
crashworthy fuel tank 防撞燃油箱
crash-worthy seat 防弹安全座椅
crater 弹坑；火山口；陨石坑
crazy flying 刀刃飞行；刀刃特技（飞机沿倾斜平面作等高的侧滑飞行）
CRC（control-and-reporting center）管制和报告中心
creep 潜伸；蠕动；匍匐；频率漂移
crescent wing 新月翼飞机；镰形翼飞机
crew 全体乘务员；班组；机组；空勤人员；地勤人员；机务人员
crew alerting system（CAS）机组警戒系统
crew briefing 机组简报
crew cabin 机组舱；乘员舱
crew capsule 乘员舱
crew check（飞机起飞前的）机组检查
crew check flight 机组检查飞行
crew chief 飞机地勤组组长；炮长；乘员组组长
crew door 乘员入口门；空勤组登机门
crew duty time 机组执勤时间；

机组人员值班时间

crew gunnery simulator 飞行中射击模拟器

crew interphone 机组人员机内通话器

crew maintenance 机组人员（进行的）维护

crew manual 机务手册

crew member 空勤组成员；炮手；操纵手

crew ratio 机组人数与飞机数之比

crew resource management（CRM）机组资源管理

critical altitude 临界高度；最大工作高度；最大射高

critical angle 临界角

critical angle of attack 临界攻角；临界迎角

critical case 危险情况（指飞机的动力装置和操纵系统的综合情况达到危险程度）

critical engine failure speed 发动机停车时的临界起飞决断速度（在此速度以上应使用其余发动机继续起飞，在此速度以下则应中断起飞）

critical engine inoperative 关键发动机不工作

critical frequency 临界频率；截止频率

critical height 临界高度

critical Mach number（Mcrit）临界马赫数

critical point 临界点；（飞行）航路临界点；转折点；要地；中间站

critical position 临界位置；极限位置；重要阵地

critical power unit 关键动力装置；关键动力组

critical pressure 临界压力

critical speed 临界速度；临界转速

critical speed of revolution 临界转速

critical speed of rotation 临界转速

critical stress area 临界应力区域

critical temperature 临界温度

critical zone 临界区（飞机为达到轰炸准确性必须保持直线飞行的地区）

CRM（crew resource management）机组资源管理

cropped surface 截尖翼面

cropped-fan engine 截尖风扇发动机
cross ahead interception 前交叉截击（指目标机将在截击机前方一定距离处通过）
cross bleed valve 交叉放气阀；交叉旋气活门
cross channel data link 交叉信道数据链接
cross feed 交叉输油（指飞机一侧的燃油向另一侧输送）；交叉供电；横进刀
cross firing 交叉火力；交叉射击
cross fitting 四通管
cross flow 交叉输油
cross monitoring 并行监控
cross pin 十字销
cross screwdriver 十字解刀；十字改锥；十字螺丝刀
cross section 横截面；剖面
cross track distance 偏航距离
cross trail angle（投弹）横偏移角
cross wind flying 侧风飞行
cross wind takeoff 侧风起飞
cross wind turn 一转弯
cross-beam 大梁
cross-beam rotor 交叉梁旋翼
cross-bleed 交叉放气；交叉旋气
cross-check 反复检查（飞行仪表）;（根据仪表）综合检查飞行情况;（飞行员的）注意力分配
cross-check technique（对飞行仪表）注意力分配法；综合判断法
cross-control stall 交叉操纵失速
cross-country 交叉地区；无公路地区；越野；横贯全国的
cross-country flight（中途着陆的）长途飞行
cross-country flying（中途着陆的）长途飞行
crosscoupling error 交叉耦合误差（陀螺自转轴与基准轴不重合引起的误差）
crossed control 交叉操纵
cross-feed system 交叉供油系统
crosshead 托架
crossing course 交叉航向；大航路捷径的目标航路
crossing point 交叉点；交点；航路捷径点；目标航路中点；过境点
crossing-runway approach 对交叉跑道进场着陆
crossover 交叉；跨越；上方交叉

飞过；航路捷径点；目标通过航路捷径

crossover piping 连接管道

crossover point 交叠点；防空武器射击转换点

crossover taxiway 连接滑行道；联络道

crossrange error 横侧偏差；偏离航线距离；弹道横向误差

crossrange maneuvering 横向机动

cross-section 截面；横断面

cross-sectional area 截面面积

cross-servicing 交叉勤务；互助勤务（指同驻一个基地的不同军兵种或不同国家的部队之间，在两次飞行间的例行维修和加油充气方面的相互支援）

cross-track bias 航线偏差；横向偏差

cross-track error 横侧偏差；偏离航线距离

cross-track position error 偏离航线距离测定误差

crosswind 侧风；横风

crosswind capability 侧风着陆能力；可以着陆的最大侧风速

crosswind figure of eight 侧风 8 字飞行

crosswind force 侧风力；横向风力

crosswind ground run 侧风滑跑

crosswind landing 侧风着陆

crosswind landing capability 侧风着陆能力；可以着陆的最大侧风速

crosswind landing gear 侧风起落架

crosswind leg（起落航线）第二边；侧风边

crosswind runway 侧风跑道

crosswind takeoff 侧风起飞

crosswind turn 一转弯

crown of holes 螺纹孔

cruciform 十字形

cruciform aircraft 十字翼航空器

cruciform airframe 十字形飞机机架；十字形翼弹体

cruise 巡航

cruise climb 巡航飞行状态上升；巡航攀升；巡航爬高

cruise control 巡航操纵；巡航控制

cruise control chart 巡航操纵图；巡航控制图

cruise descent 巡航下降
cruise flight 巡航飞行
cruise missile 巡航导弹
cruise performance 巡航性能
cruise relief autopilot 巡航自动驾驶仪
cruiser 巡航机；巡洋舰
cruising altitude 巡航高度
cruising attitude 巡航姿态（指飞行器在巡航段保持的姿态）
cruising boost 巡航增压
cruising level 巡航高度
cruising performance 巡航性能
cruising speed 巡航速度
cryptologic equipment 密码设备
crystal filter 晶体滤波器
crystal structure 晶体结构
crystal transducer 晶体变换器；晶体换能管；压电传感器
CSAS（control stability augmentation system）控制增稳系统
CSDU（constant-speed drive unit）恒速驱动装置
C-spar C 形梁
CSU（constant-speed unit）等速装置；恒速装置
CTAF（common traffic advisory frequency）公共交通咨询频率
cubic centimeter 立方厘米；毫升
cuff 增弯翼套
culture 人工地物；人造地形
culvert 涵洞；排水渠；地下沟；地下电缆道
culvert tail rotor 涵道式尾桨
cumuliform cloud 集状云；积云
cumulative time 累计时间
cumulonimbus 积雨云
cumulus 积云
cumulus form 积云层
cumulus fractus 碎积云
cumulus humilis 淡积云
current capacity 流量；电容量
current check 平时检查；例行校检
current flight plan 现行飞行计划；当前飞行计划
current 现行的；通用的；流行的；水流；气流；电流；趋势；潮流
current intensity 电流强度
current service message 当前勤务信息
current transformer 变流器；电流互感器

current transformer assembly 电流互感器组件

cursor 指针；指示器；（计算尺的）游标；（显示器）光标

cursor control device（CCD）光标控制装置；光标控制器

curvature correction 弹道弯曲修正角；弹道弯曲修正量；（地球）曲率修正

curve of pursuit 追踪曲线；跟踪曲线

curved approach 曲线进近

curved beam 弯梁

curved flight 曲线飞行

curves of equal bearings（CEB）等方位曲线；等方位线

curvic-coupling 圆弧端面齿；圆弧端齿

cushion augmentation devices 气垫增稳装置

cushion landing system 有（大能量）减震装置的着陆系统

cushioning effect 减震作用；地效作用

cut（发动机）关车；装定定时引信；（领航）位置线相交；位置线交角；近道；最短的路；切；割

cutaway 剖面；剖视图；切去；剪去；（飞行服的）露空部分

cut-back nozzle 斜切喷管

cut-in time 接通时间；（发动机）开车时刻；开始工作时间

cutlass 弯刀形（飞机）编队

cut-off 停车；关车；切断；阻碍；终止

cutoff altitude 停车高度；关闭发动机时的高度；停止供油高度

cutout valve 断路阀；切断活门

cutting pliers 平嘴钳

CVA（charted visual approach）图示目视进近

CVFP（charted visual flight procedures）图示目视飞行程序

CVR（cockpit voice recorder）座舱话音记录器

CW radar（continue wave radar）连续波雷达；等幅波雷达

CWA（center weather advisory）中心天气报告

CWP（central warning panel）中央告警板

CWS（central warning system）中间告警系统

cycle 循环；周期；周

cycle counter 循环计数器

cycle counting 循环计算；循环计数

cyclic 驾驶杆

cyclic control 闭环控制；循环控制

cyclic control column（直升机）驾驶杆

cyclic feathering 周期变距

cyclic pitch control 周期变距操纵

cyclic pitch control stick（直升机）驾驶杆

cyclic stick 周期变距杆

cyclic trim 周期配平

cyclogenesis 气旋生成；风暴的发展形成

cyclone 旋风；飓风；龙卷风；气旋

cyclone blade 漩涡叶片

cyclonic 气旋的；气旋式的

cylinder 气缸；动作筒；圆筒；圆柱（体）；圆柱形气瓶；毒气吹放钢瓶；（迫击炮）缓冲机

cylinder head 汽缸头

cylinder head temperature 汽缸头温度

cylinder liner 气缸套；气缸筒

cylindrical charge 圆柱形装药

cylindrical projections 柱面投影

D

D level maintenance 第四级维护

DA（delayed action）（引信）延迟作用，延期爆炸，定时爆炸

DA（division of armament）军械部（处）

DA（drift angle）（航行）偏航角；偏流角

DABS（discrete access beacon system）直接存取信标系统

DACT（dissimilar air combat training）异型机空战演练（有仿敌机型参与的空战演练）

daily inspection 日常检查

daily preflight check（first flight of the day）日常飞行前检查

daisy chain（依次开伞的）串联空投包；依次着陆机群；多机集体仪表进场

daisy clipping 掠地飞行

daisy cutter BLU-82 真空炸弹；着发杀伤炸弹；“雏菊切割器”炸弹

damage agent 战斗装药；杀伤体破坏剂

damage assessment 毁伤判定；损失估计；毁坏情况估计；（突击目标的）破坏效果评估；泄密损失估计；杀伤半径

damper 减震器；阻尼器；消焰器

damper flywheel 减震飞轮

damper oil 减震器油

damper regulator 气闸调节器

damper strut 缓冲支柱

damper tube 减幅管；阻尼管

damper valve 阻尼阀；缓冲阀

damper winding 阻尼绕组

damping 阻尼；减辐；衰减

damping coefficient 阻尼系数；挫折系数；减幅系数

damping factor 阻尼因数；挫折因数；减幅因数

damping moment 减振力矩
damping oscillation 减幅振荡
damping ratio of wind vane 风向标阻尼比
danger area 危险区；危险区域；危险空域（在该区内存在着危险，飞机（车辆等）禁止在该区活动）
danger beacon 危险灯标；危险信标
danger bearing 危险方位（角）（对于某一危险区能够安全行驶的最大及最小方位（角））
danger bearing angle 危险方位角（对于某一危险区能够安全行驶的最大及最小方位角）
DAR（defense acquisition radar）防空搜索雷达
daredevil【口】莽撞的飞行员
dare-to-die corps 敢死队
dark adaptation 对黑暗的适应性
dark trace tube 黑踪线管
dark vision 暗视（照明很差的条件下具有的优异视觉）
dark zone 暗区；预备区
darkening 变暗；变黑；灯光管制
dart 投射；发射；发射物；航空火箭;（机翼或尾翼的）；箭形标枪式地对地导弹镖;【通】对付地面雷达的电子对抗设备分系统
DART（dual-axis rate transducer）双轴速率变换器
DART（dynamic automatic radar tracking）动态雷达自动跟踪（系统）
dart configuration 后尾式布局
dash capability（短时间高速飞行的）冲刺能力
dash flight 冲刺飞行
dash lamp 仪表照明灯
dash light 仪表板灯
dashboard 仪表板
dashpot 阻尼器
dashpot throttle 阻尼器节流阀
data acquisition system 数据采集系统
data base unit 数据库组件
data bus 数据总线
data capsule（装于飞行器中能弹出后单独回收的）数据舱；测量仪器舱；数据资料采集舱
data communication 数据通信
data concentrator unit 数据集中组件；数据集中器

data control interface unit 数据控制接口单元

data encryption standard 数据加密标准

data flow 数据流；信息流

data link 数据链

data management center 数据管理中心

data processing function 数据处理功能

data recorder 数据记录器

data recording 数据记录

data signal display unit 数据信号显示装置

data terminal equipment 数据终端设备

data test facility 数据检测设备

data transfer unit 数据传递装置

data transmission device 数据传输装置

data-bus system 数据总线系统

datum 基准面；大地基准点；基准

datum airspeed 基准空速

datum line 基准线；零位线

datum point 基准点

dawn 黎明；拂晓

dawn raid 黎明突袭；拂晓进攻

day air defense fighter 昼间防空战斗机

day and night distress signal 日夜两用求救信号设备（一种手提式烟火与照明弹发射器）

day and night television system 昼夜电视系统

day applied tactics 昼间实用战术

day light visual flight 昼间目力飞行

daylight 白昼；白天

daylight beacon 昼标（白天导航用的无照明的建筑物）

D-bomb 深水炸弹

DCP（digitally controlled potentiometer）数控电位计

D-day 重大行动日；大规模进攻发起日；特定军事行动开始日

dead ahead 正前方；迎面；正面；直航向

dead air 闭塞空气；滞止空气区；静区

dead astern 正尾向；正向后；对正尾部；从正后方

dead center position 死点位置

dead engine 停车发动机；死发动

机（指空中停车以后不能再起动运转的发动机）

dead fluid zone 气流死区

dead load 静载；静重；本身的重量；恒荷载

dead load lever 起重杆

dead reckoning 推算定位法；推测领航法（依靠罗盘并根据速度、经过时间以及一已知位置之方向求得自己的位置）

dead side 下侧（飞机飞行编队远端的一侧）；跑道对侧（指正在使用的环形跑道对侧）

dead space 射击死角；视界死角；无信号区；死水区

dead spot 盲区；死区；静区；哑点；死点；死角

deadhead 送修（飞往维修基地或修理工厂）；搭机飞行（指飞行人员作为普通乘客搭乘飞机）；空载返航

deadheading 空机返航；空载返航；搭乘便机；（卡车等）空车返回

dead-reckoning position（DR position）航位推算位置；推算位置

dead-reckoning segment 航位推算段

dead-stick approach 停车进场

dead-stick landing 停车着陆（指多发动机全部停车）

deaerator 空气排除器；油气分离器

deaerator chamber 油气分离室

deaerator tray 油气分离器盘

debarkation 下飞机；（从飞机上）卸载

debarkation point 卸载点

debonding 脱粘

debooster 减速器；制动器；减压器；（电压或发动机增压的）限压器；（助力操纵的）回力机构

debrief 盘问（外来者等）；听取（飞行员等）汇报；询问（敌情、任务执行情况）

debriefing 情况汇报；飞行后讲评；天气咨询；保密训令（命令离职人员不得泄密）

debris 碎片；废墟；岩屑；碎屑；空间残骸

decade counter 十进制计数器

decalage（双翼或多翼飞机的）翼角差；机翼安装角差；差倾角

decanter 倾析器

Decca 台卡（导航）系统（一种相位式无线电测距差系统）

Decca chain“台卡”导航台链

decelerate 减速;（发动机）减小转速

decelerating transition 水平飞行向垂直飞行的过渡；反过渡；减速过渡状态

deceleration control unit 减速控制器

deceleration device 减速装置（使飞机在飞行中或降落时减速的装置，如减速伞、襟翼、航空母舰上的钩绊装置等）

deceleration error 减速误差

deceleration parachute 减速伞

deceleration to hover 减速到悬停状态

decelerative attitude 减速姿态

deceleron 减速副翼（副翼和减速板的组合）

decelostat 刹车压力自动调节装置;（机轮）防拖胎装置

deception 欺（诱）骗；欺骗措施；造假；迷惑；伪装；遮盖；诱（欺）骗性干扰

deception jamming 诱（欺）骗性干扰

decibel 分贝

decimetric wave radar 分米波雷达

decision addressing beacon system 决断寻址信标系统

decision altitude 决断（绝对）高度（指飞机沿下滑道进场着陆时一种规定的高度）；决断海拔

decision height 决断高度（飞行员必须对进场与着陆或复飞与执行另一次进场作出决断的高度，其依据取决于能否充分地看到跑道）

decision height（DH）决断高度（实际高度）

decision point 决定时空点;（起飞）决断点

decision speed 起飞决断速度（指在此速度以上，一台发动机停车，可继续起飞；小于此速度，应中断起飞）

deck 甲板;（大型飞机）层舱；驾驶舱；最低飞行高度;（云）盖;（多翼机的）一层机翼;（卡片）组、叠

deck alert 地面待命

deck park 甲板停放区（指在航母飞行甲板上停放飞机或车辆的区域）

deck run（航母上，非弹射起飞的）甲板起飞滑跑道距离

deck-level 极低空的；超低空的

deck-level attack 超低空攻击

deck-level flight 超低空飞行

declared alternate 申报备降机场

declared destination 申请的着陆机场；申报目的地

declared distance 公布距离；申报距离

declared thrust 公告推力（指飞机工厂印发的飞机推力值）

declared weight 申请起飞重量；申报重量

declination 磁偏角；偏角；赤纬

declutch 分离（离合器分开）；放空挡

decode 译码；解码；脱密

decoder 译码器；解码器；判读器；译码员

decompression 还原；减压；失压；释压

decompression chamber 减压舱；减压室

decompression sickness 减压病；高空病

decoy 假目标；诱饵；诱惑

decoy aircraft 假飞机；诱饵飞机

decoy airdrome 假飞机场

decoy bird 诱饵导弹；假导弹

decoy bomb 诱惑炸弹；制导炸弹

decoy camouflage 假目标伪装

decoy discrimination radar 诱饵（假目标）鉴别雷达

decoy drone 诱惑用无人驾驶飞机；无人驾驶飞机诱饵

decoy ejection mechanism 诱饵（假目标）弹射装置

decrab 做修正转弯以对准跑道；消除偏流修正角

dedicated runway（永久性的）仪表着陆跑道

deep air support 纵深空中支援；远距离空中支援（所攻击的目标距己方较远）

deep maintenance 深度维护

deep penetration 深远突破；纵深突破（如飞机飞入敌深远地区的行动）；深度穿透

deep stall 深度失速（指T型尾翼、发动机位置靠后的飞机失速

后，迅速发展到纵向操纵失效）
deepening 低压下降
deep-penetration attack 深入后方空袭（袭击）
deep-penetration mission 穿插任务
deep-penetration raid 深入敌后袭击
default value 默认值；设定值；缺省值
defect detection 缺陷检测
defect report 故障报告；缺陷报告
defective component 故障件；有缺陷的机件
defective contact 触点不良；接触不良
defective material summary 残损器材综合
defective sealing 封严失效；密封不良
defective ventilation 通风不良
defence works 防御工事
defense acquisition radar（DAR）防空搜索雷达
defense in coordination with ground force 协同地面部队防御
defense of great depth 大纵深防御
defense of key points 重点防御
defense of obstacle 障碍物掩护
defense outpost 防御前沿地区；防御前哨基地；防空哨
defense penetration 突防
defense priority 防御的轻重缓急；防务优先顺序
defense priority rating 防空的优先次序
defense readiness posture 戒备态势
defense strategic measures 防御性战略措施
defense visual flight rule（DVFR）防空目视飞行规则
defense visual flight rules（DVFR）防区目视飞行规则
defense visual flight rules flight plan（DVFR flight plan）防空目视飞行规则飞行计划
defense warning 防空预警
defensive bombing 防御性轰炸（目的在于削弱敌军的飞机场及其他类似的空军设施）
defensive circle 圆阵防御
defensive combat spread 防御性战斗横队（指双机疏散横队，间隔 1.6~3.2 千米，并略有距离和高度差）
defensive counter air 防御性防空

作战；防御性反空袭；防御性反航空兵作战

defensive echelons 防御的梯次区分（分为警戒地域、前方防御地域和预备队地域）

defensive electronics 防御电子设备

defensive spiral 防御性盘旋下降（指高过载加速的盘旋并俯冲，以摆脱敌攻击）

defensive split 防御性半滚倒转（指飞行分队分组向不同平面分离，迫使敌机放弃其中一组）

defined point after takeoff 起飞后限定点

defined point before landing 着陆前限定点

defined trajectory 规定的弹道

definition（雷达、电视）清晰度；分辨力

deflation sleeve 放气套筒；放掉气的袖靶

deflation valve 放气活门

deflected slipstream 偏转滑流；推力偏转

deflected slipstream aircraft 桨流偏转式飞机（一种短距起落飞机）

deflection 偏转；偏度；偏差；偏流修正量横偏；方向角；瞄准角

deflection angle 偏转角

deflection yoke 偏转系统；偏转装置；偏转线圈组

deflection-modulated display 偏转调制显示器

deflector 折转板；导流板

deflector rod 拨杆

deformation 变形

defuel 放（燃）油；抽油；（从油箱）排出燃料

degradation 降级；减低；老化；降化

degradation failure 降解故障；老化故障，退化失效；逐渐恶化的故障

degradation of ordnance equipment 军械装备降级

degraded mode 简化工作状态；简化工作方式（自动装置工作方式之一）

degraded performance 性能退化

degraded surface 降级道面（指机场积雪、积冰或蓄水使道面情况恶化）

degree of level 水平度
degree of readiness 战备程度
dehydration 脱水
deicer 除冰器；除冰设备；防冰器
deicer boot 去冰橡皮带（飞机翼面等前缘的橡皮带，可以充气使冰碎裂）；螺旋桨底座前缘的橡皮带（喷洒酒精于其上，以防结冰）
deicing 除冰
deicing fluid 防冰液
deicing system 除冰系统
delamination 脱层
delay 延迟；延迟火力支援（火力支援的报告用语）
delay rate 延误率
delay time 延迟时间；滞后时间
delayed action（DA）（引信）延迟作用，延期爆炸，定时爆炸
delayed clearance（起飞或进场的）延迟许可
delayed drop 延迟开伞
delayed fall 延迟开伞降落
delayed flap approach 延时襟翼进近
delays & cancellation（飞行）延误和取消
delivery altitude 投弹高度；投射高度
delivery approach（低空）空投进入（飞机按空投航线接近地面）
delta 三角形；三角翼；三角翼飞机；三角洲
delta aircraft 三角翼飞行器；三角翼飞机
delta hinge（直升机旋翼叶片的）挥舞铰；水平铰；水平关节
delta ratio 三角比率
delta wing 三角翼
demand assignment 按需分配；请求分配
demand mask 肺式供氧面罩；呼吸式面罩
demand oxygen system 肺式供氧系统
demarcation line 分界线；标界线
demister（尤指风挡玻璃的）除雾器；除雾装置
demisting switch 除雾开关
demodulation 解调；反调制
demodulator 解调器；反调制器；检波器
demolition aerial bomb 航空爆破

炸弹

demolition belt 爆破地带；爆炸地带（埋有地雷、爆炸物或其他障碍物的地带，用以保护自己及防止敌人活动）

demonstration flight 表演飞行；示范飞行；佯动飞行；示威飞行

demonstration force 佯攻部队

demonstration in force 强力佯攻；威力佯攻

DEMS（dynamic environment measurement system）（航天飞机上的）动力环境测量系统

denominator 分母

denotation illumination 照明弹

dense fog 浓雾（能见度 0 级，距离＜ 50 米）

dense mine fields 大密度雷场

dense propellant 重推进剂

dense smoke 浓烟

densimeter 密度表

densitometer 显像密度计；光密度计；海水密度计

density 浓度；密度；埋雷密度（指雷战中每米正面的平均雷数）

density altitude 密度高度（以对应于标准大气之密度的高度来表示的大气密度）

density error（空气）密度误差

density of fire 火力密度

density of forces 兵力密度

density ratio 密度比

dent（压缩机叶片）压痕；凹痕；压坑

departing aircraft 离场（起飞的）飞机；离开机场区的飞机

departure 离开；出发；起飞；离场；翼尖失速；翼下沉；失去安定性与操纵性；偏离（航向）航线偏移量；偏航距离

departure airfield 出发机场；出动机场；（空降部队）出发机场

departure alternate（飞行计划中指定的）备降场；起飞备用机场

departure area 出发地域；展开区

departure area diagram 起飞机场区域飞行图

departure area for assault 机降突击出发地域

departure clearance 离场许可；起飞许可

departure control 离场管制

departure end 起飞跑道末端；飞机起飞离地端；跑道起飞离地端

departure instruction 出发前指示；起飞前指示；起飞调度规则

departure lane 飞离航道

departure line 出发线

departure message 起飞报；飞离通报；飞离通知

departure pattern 离场起落航线

departure point 出发点；航线起点；脱离空域点；失速转变为螺旋的起点

departure procedure 离场程序；出境程序；离场程序（指离场飞机从爬高到航线高度的空中管制程序）

departure profile 离场航线剖面

departure route 离场航路；（机场区域的）离境航线

departure runway 离场使用跑道；起飞用跑道

departure strip（空中交通管制的）离场记录条

departure traffic 离场交通量

departure trajectory 起飞段弹道（从起飞到关闭发动机的一段）

dependent parallel approaches 相依并行进近

deplaning zone 下机地域（机降兵或装备与补给品下机的地域）

deploy 开伞；展开；疏开；部署；调度；配置；转场

deployed defense 仓促防御（已展开的部队在地面无防御工事或仅有仓促构成的防御工事的情况下进行的防御）

deployed for defense 防御展开

deployment 布防；展开；部署；开伞；（弹头）投放；（航天器）打开太阳电池板

depot 兵站；补给站；供应站；仓库；基地；维修与供应基地；兵员补给站

depot maintenance 仓库保养；后方维修（美空军装备维修的最高一级，翻修、全面重新装配、大修、改修等）；工厂修理

depot modification（飞机或其他装备在）航空维修基地改型

depot overhaul 后方大修；翻修；进厂修理

depot repair 后方大修；翻修；进厂修理

depressed pole 下倾极；下极；下天极

depressed sightline 下俯瞄准线

depressed trajectory 压低的弹道
depression 减小；降低；抑制；减压；抽空；（襟翼等的）偏度；下垂；下降；低气压；偏侧角
depressurize 减压；降压；释压；（座舱）解除密封
deprimered fuse 摘火引信；假引信
depth bomb 航空反潜炸弹；深水（航空）炸弹
depth perception 深度识别；深度感觉力
derated engine 降功率使用的发动机
derivative 方案；（飞机等的）改型
descending attitude 下降姿态
descent 下降；降落
descent angle 下降角
descent fuel 下降燃油（指从下降顶点到待机高度或进场高度所消耗的燃油）
descent indicator 下降速度表
descent instrument penetration 仪表穿云下降
descent lane（进入着陆时）下降航道
descent mode 下降工作状态
descent path 下降路线；下降段轨道；弹道降弧
descent speed 下降速度
descriptive name 图例名称；（地图）图注（文字）说明
desiccant 干燥剂
design cruising speed 设计巡航速度
design diving speed 设计俯冲速度
design gross weight 设计总重量（指在设计计算中使用的预期最大起飞重量）
design landing mass 设计着陆质量（重量）
design load 设计负载（负荷）
design maneuvering speed 设计机动速度
design maximum weight 最大设计重量
design takeoff mass 设计起飞质量（重量）
design taxiing mass 设计滑行质量（重量）
designated firing duration（火箭发动机的）设计燃烧时间；计算工作时间
designated flight instrument system

指定的飞行仪表系统
designated target 指定的目标
designation 指定；指明；标记；名称；番号；选派；任命
designator 选择器；指示器；命令符；指示符
desired course 期望（预定）航向；期望（预定）航线
desired fan speed 所需风扇转速
desired heading 期望（预定）航向；应飞航向
desired radio height 预期无线电高度
desired track 期望航迹；预定航线；所需航迹
desoldering 脱焊
dessicant 干燥剂
destination 目的地；目的港；航线终点；（飞行计划中的）着陆机场；（轰炸）目标
destination tower 着陆机场塔台
destroy（beam）【空】侧方击毁（空中截击语，指侧向攻击并摧毁目标）
destroy（cutoff）【空】切半径击毁（空中截击语，指雷达引导指令切半径攻击，截击并摧毁）
destroy（frontal）【空】正面摧毁
destroy（stern）【空】尾后摧毁
detach 派遣；派出；分遣；拆卸武器零件
detachable wing 可卸机翼
detached shock wave 离体激波；脱离之震波
detachment 支队；分遣队（派至他处执行任务的分队）；独立小分队；由各单位或本单位部分人员组成的临时编组
detachment point 分离点；拆卸点
detachment warfare 前哨战
detail of airborne aircraft 空中机群（分队）
detailed inspection 详细检查
detailed parts（结构或机件的）详细部件（一般在使用和贮存中不拆下）；整体部件；非拆散零件
detection device 检测装置
detection run 侦察距离
detection unit 探测部件；敏感元件；火箭导引头
detector（电讯）检波器；探测器；检测器；检毒剂
detent plunger 锁机顶塞
detergent 清洁剂

deteriorate 恶化；变坏

deterioration report 天气转坏的报告

determined limit 规定的极限

detonation 爆炸；（发动机）爆震；爆燃

detonation altitude 爆炸高度

detonation by influence 感应爆炸

detonator 起爆装置；雷管；起爆管

detotalizer 剩余量表（显示所剩燃油总量的燃油流量表；也指飞机上显示所剩弹药量的仪表）

detotalizing counter 剩余量计数器

detour 迂回；绕道；曲折；迂回路

develop in depth（向）纵深发展

developed area 展开面积；展开面

developed area ratio（螺旋桨）展开盘面比

development missile 研制的导弹；试验导弹

development test hop 试验机的试验飞行

development vehicle 研制中的运载器；发展型飞行器；飞行器试验样品

deviated pursuit course 前置跟踪航向；前置追击航向

deviation 偏差；偏差角；频率偏差；航线偏差；非常规飞行；飞行偏差；目标偏差

deviation amplifier 偏转放大器

deviation angle 偏向角；自差角；测向仪偏差角；落后角；脱离角

deviation card（磁罗盘的）误差表；罗差表

deviation limit 偏移极限

deviation of the wind 风偏向

deviation report（飞行途中）改变航线报告

deviation scale 偏离刻度

deviation table 罗差表；自差表

deviometer 偏航指示器

DEW（directed-energy weapon）定向能武器

dew point 结露点；露点；露点温度

dew-point temperature 露点温度；露点

dextrorotation 右旋

dextrorotation rotor helicopter 右旋旋翼直升机

DF（direction finder）测向仪；

无线电罗盘

DF（direction finder）approach procedure 测向仪进近程序

DF（direction finder）guidance 测向仪引导

DF（direction finder）loop 无线电罗盘环形天线

DGPS（differential GPS）差分全球定位系统

DH（decision height）决断高度（实际高度）

diagonal aeroengine 斜形（V 形）航空发动机

diagonal-flow compressor 斜流压缩机

diagram 图表；线图；表；图解；简图

diagram of the circuit 电路图

dial 表盘；刻度盘；拨号盘；数字盘

dial compass 罗盘仪

dial indicator 度盘式指示器

dial instrument 表盘式仪表

dial micrometer 厚度千分仪

dial plate 表盘；标度盘；分划盘；指针板

diameter 直径；光学镜头的放大倍数

diameter of bore 枪炮口径

diamond landing 菱形队形着陆

diamond landing gear 四点式起落架（指机身中心线上的自行车式起落架，加上左、右机翼下的辅助起落架）

diamond man 菱形编队末尾飞机飞行员；菱形编队的末尾飞机

diamond nine 九机菱形队形

diamond wing 菱形剖面翼

diamond-four takeoff 四机菱形编队起飞

diamonds 菱形激波

diamond-shaped formation 菱形队形

diaphragm 挡药板；隔板；薄膜；光圈；震动片；皮膜；膜

diaphragm pressure switch 膜片压力开关

diaphragm shell 炸药与弹头分装的弹壳

diaphragm spring clutch 膜片弹簧离合器

diaphragm type 膜片式

diesel ramjet 柴油冲压式喷气发动机

difference in elevation 高度差
difference in latitude 纬度差
difference in longitude 经度差
differential aileron 差动副翼
differential control 差动控制
differential correction（导航系统的）差分修正；微分改正
differential effect of range 距离表定修正量
differential effect of time of flight 飞行时间表定修正量
differential effects 弹道偏别影响
differential GPS 差分全球定位系统
differential pinion 差速小齿轮
differential pressure 压力差；差压
differential pressure indicator 压差指示器；压差表
differential pressure switch 压差开关
differential range 压差范围（保持密封舱内外有一定压力的范围）
differential screw jack 差动螺旋千斤顶
differential spoilers 差动扰流板
differential steering unit 差速转向机构
differential tail 差动平尾
differential tracking（若干目标）分别跟踪
differentiation 微分器
diffuser 扩散器；漫射体；扬声器纸盆；发动机进气道
diffuser area ratio 扩压器（出口/进口）面积比
diffuser efficiency 扩散器效率
diffuser intake 扩压器进口
diffuser vane 扩压器导向叶片
digit 数字；数位
digital 数字的
digital air data computer 空气数据数字计算机；数字式飞行数据计算机
digital air-data computer system 数字式航空数据计算机系统
digital attitude control system 数字式姿态控制系统
digital automatic flight control system 数字自动飞行控制系统
digital autopilot 数字式自动驾驶仪
digital avionics information system 数字式航空电子设备信息系统
digital bomb navigational system

数字式炸弹导航系统
digital cabin pressure control system 数字式座舱压力调节系统
digital circuit 数字电路
digital communication 数字通信
digital display 数字显示；数字显示器
digital engine control unit 数字式发动机控制装置
digital filter 数字滤波器
digital flight control system 数字飞行控制系统
digital flight data acquisition unit 数字式飞行数据采集组件
digital flight data recorder 数字飞行数据记录仪
digital flight data recording system 数字式飞行数据记录系统
digital information transfer system 数字信息转换系统
digital map 数字地图
digital map system 数字地图系统
digital mapping 数字制图；计算机地图绘制
digital read-out 数字读出
digital signal 数字信号
digital signal processor 数字式信号处理器
digital to analog converter 数字模拟变换器
digital volume adjustment knob 数字音量调节旋钮
digital-analog converter 数字 - 模拟转换器
digitalized battlefield 数字化战场
digitally controlled potentiometer（DCP）数控电位计
digital-to-analog 数字 / 模拟转换
dihedral 反角（指机翼或尾翼平面与横轴之间的锐角）；上反角；翼差角
dihedral effect 上反角效应（指变后掠翼飞机侧滑引起滚转）
dihedral of the rotor 旋翼轴倾斜度
dihedral wing 上反机翼；上反翼
diluent air 稀释空气
diluter-demand oxygen system 肺式自动供混合氧设备；混合氧断续供氧系统
dilution 冲淡；稀释
dilution air vent 稀释空气孔
dilution effect 稀释效应
dimension 尺寸；尺度；维；面积；容积；范围；方向

dimmer 变光器；减光器

dimpled tire 窝纹轮胎

diode 二极管

diode module 二极管组件

dip 磁倾角；罗盘磁倾角；机翼下倾下降；摇晃机翼；摆动机翼

dip angle 倾角；磁倾角；倾斜角；弹着角；落角；（空）急降俯角

diplexer 天线分离滤波器；双工器

diplomatic authorization（国际通航）外交认可；外交批准

diplomatic clearance 入境许可；飞行外交入境许可；外交签证；入境签证

dipole antenna 偶极天线；偶数发射天线

dipstick 量尺；测量杆；测深标尺；油尺；液面标尺

direct access radar channel 直接存取雷达信道

direct approach 直线进场（着陆）；直接路线；直接方法；直接手段（以使用武力为主的战略）

direct channel 直接信道；直通信道；直通线路

direct combat support 直接战斗保障；直接战斗支援

direct conducting resistance 正向导通电阻

direct current 直流（电）

direct descent 直接下降

direct fire 直接瞄准火力；直瞄火力；直接瞄准射击；目瞄射击；直接点火

direct flight 直线飞行；直航

direct fuel injection system 直接燃油喷射系统

direct lift control 直接升力控制；直接升力控制机构（系统）

direct radar guidance 雷达直接制导；视觉跟踪雷达制导；三点法雷达制导

direct radar scope recorder 雷达直接显示记录器

direct route 直接航线

direct side-force control 直接侧力控制

direct support 直接支援

direct tracking 直接跟踪

direct transit network 直接传输网络

direct troops and indirect troops 直

接部队与间接部队

direct trunking 直通中继

direct vision panel 目视仪表板

direct vision sight 直视式瞄准具

direct war loss 直接战争损失

direct wave 直接波；直射波；直达波

direct-acting engine 直接式发动机

direct-acting finder 直接作用（旋转式）选择器

direct-ascent powered-flight simulation 直接爬升动力飞行模拟

direct-challenge system 直接询问系统

direct-climbing target 直行离远上升目标（因爬升而远离射手的空中目标）

directed aircraft 僚机；被引导的飞机

directed beam 定向波束

directed energy weapon 定向能武器

directed inspection 深入检查（特指对飞机结构完整性的检查）

directed overshoot 着陆目测高；着陆时越过指定地点；（轰炸、射击时）越过目标；远弹；偏转过度

directed path 定向路径；指引的飞行轨迹；受控飞行轨迹

directed slipstream 定向滑流（指将螺旋桨或风扇滑流吹过整个机翼以实现短距离起降）

directed-energy weapon（DEW）定向能武器

direct-indicating compass 直读式罗盘

direction 航向；方向；引导；指示；说明（书）

direction balance 方向平衡

direction bearing 无线电定向

direction control 方向控制；方向操纵性；方向操纵面

direction finder 测向器；定向器；无线电罗盘

direction finder（DF）测向仪；无线电罗盘

direction finder（DF）approach procedure 测向仪进近程序

direction finder（DF）guidance 测向仪引导

direction finder（DF）loop 无线电罗盘环形天线

direction finding course 引导航向

（飞机飞向引导站应取的航向）

direction marker 指向标

direction of rotation（回转活塞）旋转方向

direction of secondary attack 助攻方向；次要作战方向

direction of travel 行进方向；航行方向

direction of turn 转弯方向

directional antenna 定向天线；方向性天线

directional beacon 定向信标

directional control（飞机等的）方向控制；方向操纵性；方向操纵面

directional control gear 航向控制装置

directional control pedals 方向控制踏板

directional control valve 定向控制活门；定向控制阀

directional gyro 陀螺方位仪；方向回转仪；航向陀螺仪；回旋式罗盘；（自动驾驶仪、轰炸瞄准具中的）方向稳定陀螺

directional gyroscope 航向陀螺仪；陀螺半罗盘

directional instability 航向不稳定性

directional light 定向灯光信号；航向信号灯

directional radio beacon 定向无线电信标

directional stabilizer 方向稳定器；航向稳定器；垂直安定面

directional steering 方向操纵；方位控制

direction-finding system 无线电测向系统

direction-vision panel 目视仪表板

dirty 起落架落下；空气动力减速板外露；襟翼向下

disarmament 解除武装；裁军

disarmed mine 不能爆炸的地雷；失效雷；止炸雷；解除引爆装置的雷；上保险的雷

disassemble 拆卸；分解（武器）

disastrous accident 致命事故；机毁人亡事故

disc 桨盘；圆板；盘状物

disc area 桨盘面积

disc attitude 桨盘姿态

disc brake 盘式制动器

disc clutch 盘式离合器

disc loading（直升机旋翼的）桨

盘载荷；旋转面负荷

disc pointer 仪表盘指针

discard-at-failure maintenance 无法维修时可抛弃的维修

discarding 报废

discarding sabot 可脱弹壳；炮弹软壳

discharge（枪炮）发射；放电；撤职；退伍；退役；卸载；退弹

discharge nozzle 排水喷嘴；泄放喷嘴

discharge operation 卸载作业

discharge port 排水口；输出口；排泄口

discharge temperature 排气温度

discharge valve 排水活门；输出活门；喷射活门

discharger 排气装置；排出装置；发射装置；发射架枪；掷弹筒；放电器；卸货者；卸载器

discharger cup 枪榴弹发射器（装在枪管上）；掷弹筒

discharging gear 卸货装置；减压装置；放出机构

discing（螺旋桨的）反推旋转（指飞机在地面时，螺旋桨以低矩运转，以形成阻力）

discone antenna 圆盘锥形天线

disconnect 断开；脱开；拆接

disconnect switch 隔离开关；断开开关；断绝开关

disconnectable item 可拆分物件

disconnecting device 解脱装置；断开装置

disconnection 断开

discontinued approach 中断进场（着陆）；复飞

discontinued installation 停止使用的设施；撤销的军用设施

discontinued takeoff 间断起飞

discrepancy 误差

discrepancy report squawk 误差报告信号

discrete 分离的；离散的

discrete access beacon system（DABS）直接存取信标系统

discrete address beacon system 离散地址信标系统

discrete circuit 分立电路

discrete command 断续指令

discrete frequency 离散频率；特殊频率

discrete noise 离散噪声

discrete signal 离散信号

discrimination 判别；辨别；鉴别；识别；分辨力；鉴别力

disembark 下飞机；卸载

disengage 脱离战斗；脱离接触；退出战斗

dish-out 飞机向压坡度方向大角度侧滑下降；在盘旋（转弯）中侧滑下降

disintegrating link belt 可分散的弹带

disintegrating-belt mechanism 散链机

disintegration 分解

disk friction wheel 圆盘摩擦传动轮

disk loading 转盘受力

disk type braking system 盘式刹车系统

disk type filter 片式滤清心子

dismounted flight training（飞行）徒步练习（指手持飞机模型在地面演习战术动作）

dismounting skills 拆卸技能

disorientation 迷航；失定向；迷失方向；定向障碍症

dispatch 调遣；调度；分配

dispatch deficiency 放飞允许偏差

dispatch delay 调度延迟；放飞延误

dispatch deviation 调度偏差；放飞偏差

dispatch deviation procedure guide 派遣偏差指南

dispatch reliability 调度可靠性；正点放飞率

dispatcher 调度员

dispenser 投放器；集束弹箱；储液器

dispersal 疏散；疏散

dispersion 施放；散布；分散；喷洒；弹着散布；射弹散布面；炸点散布面

dispersion error 射弹散布偏差；射弹散布误差

dispersion of helicopter formation 直升机编队分散

dispersion pattern 分散模式；射弹散布面；射弹散布样式

displace 移动；转移；置换；转移阵地；变换阵地

displaced central（trajectory）（弹道）中心移位

displacement 位移；阵地变换；航迹间隔（空中截击中，目标航

迹与截击机航迹之间的间隔）
display control 显示控制
display control unit 显示控制器
display control panel 显示控制面板
display dimming control panel 显示器亮度控制面板
display equipment 显示设备
display mode 显示方式
display select panel 显示选择面板
display unit 显示组件；显示器
disposition 部署；配置（部队）；处置；控制
disposition for ambush 伏击部署
disposition for anti-airborne defense 反空降部署
disposition for attack 攻击部署
disposition for counter-strike 反突击部署
disposition for defense 防御部署
disposition in depth 纵深配置
disposition in directions 按方向部署
disposition in directions and echelons 按方向成梯队部署
disregard 作废；报废
dissimilar air combat training （DACT）异型机空战演练（有仿敌机型参与的空战演练）
dissipation 耗散；消散
dissipation trail 消散尾迹，蒸发尾迹（尾迹不凝结）
dissymmetry 不对称；反对称
dissymmetry of lift 升力不对称性
dissymmetry of rotor thrust 旋翼推力不对称
distance available 可用距离
distance bar（飞机与其牵引车之间的）牵引杆
distance between vehicles 车间距离
distance flight 远程飞行；长途飞行
distance gauge 测距仪；测距规
distance markers（跑道两侧所标明的）长度数字标记；（雷达荧光屏上的）距离标记；距离标识器；距离标志
distance measuring equipment 测距设备；测距仪；测距装置
distance measuring equipment/normal 标准测距设备
distance measuring equipment/precision 精密测距设备

distance of landing roll 着陆滑跑距离

distance of take-off run 起飞滑跑距离

distance off course 偏航距离；偏离航道距离

distance run 已航行的距离；已飞行的距离

distance to go 待飞距离；未飞距离

distance-measuring equipment 测距设备

distance-of-turn anticipation 转弯提前量

distant early warning radar 远程预警雷达

distillation 蒸馏

distortion（燃气涡轮）流速或温度变化；失真（度）；扭曲；歪斜

distortion of image 图像失真

distress 遇险；危急；失事；遇难

distress call 遇险信号；求救信号；遇险呼叫

distress frequency 求救（无线电）频率；遇险求救频率；呼救频率

distress landing 强迫着陆；紧急着陆

distress phase 危险阶段（在空中救援时情况恶化，被救援之飞机与机上人员均处于危险状态之时期）

distress radio frequency 急救（遇难）无线电频率；求救无线电频率

distress signal 危急信号；遇难信号；遇险信号；呼救信号；遇难信号器材

distributed load 分布载重；分配负荷

distributed operations 分布式作战行动；分布式作业

distribution 分配；散布面；分布（指弹着点或火力）；分发（指命令或补给品）；分配人员

distribution box 配电箱；分线箱

distribution depot（补给品）分发仓库

distribution of airports 机场布局

distribution of firepower 火力分配

distribution of force（s）兵力配置；兵力部署

distributor 分配器；配电盘；分布器；配电器

disturbance（大气）扰流；扰乱；扰动（对飞机飞行来说，常指迎

角的意外变化）；飞行失控
disturbing moment 扰动力矩
ditch 沟渠；渠；壕；海上迫降
ditching 水上（紧急）迫降（失事飞机在水上操纵降落）
ditching drill（机上人员的）水上迫降应急程序；水上迫降救生练习；水上迫降应急出口
dive 俯冲（直升机）；下潜；潜水；突然从视野中消失
dive angle 俯冲角
dive brake 俯冲减速板；俯冲制动器；空气动力减速板
dive flap 俯冲减速襟翼；（改出）俯冲减速板
dive-bomber type approach 俯冲轰炸式进场（陡下滑线进场）
dive-bombing attack 俯冲轰炸
divergence 气弹发散；操纵反效；气动弹性不安定；扩散；偏离
divergent duct 扩张管
divergent nozzle 扩散喷管；扩散喷嘴
divergent passage 扩散型通道；扩压通道
diverse vector area 多向引导离场区域
diversion 转向；中途改变；迂回航道；改航；偏离规定航线；飞向备降机场；改变任务、目标或目的地；牵制
diversion air field 备降机场
diversion airfield 临时机场；备降机场
diversion order 到备降机场着陆的命令
diversion traffic 转向备降机场的空中交通
diversion valve 转向阀
diversionary landing 牵制登陆
diversionary missile 诱饵导弹；牵制性导弹
divert 转向；使转向；转换；偏航；改航；偏离规定航线；飞向指定的备降机场；改变目标、任务或目的地
diverter 偏向器；推力转向器；分流电阻；避雷针
divided landing gear 分轴式起落架；分置式起落架
dividing line（战斗）分界线
dividing plate 分度盘；分划板
division 师；航空分队；部门；分配；分派；分度；刻度；隔

板；航道分叉处

division aviation 师航空兵

division of airspace 空域划分

division of armament（DA）军械部（处）

division reserve position 师预备队阵地

doctrine（军事上的）主义；（基本作战）原则；作战条令；作战理论

document 记录（文件）；资料；文献

dog 机场导航指向标

dog fighter（空中）格斗飞机；夺取空中优势战斗机

Dog House（苏联）“狗窝”反导捕获跟踪雷达系统

dog tooth（机翼前缘的）锯齿

dogleg 折线（飞行航线）；航迹转向（用于修正航天器轨道倾角）；折线形变弹道（轨道）

dogleg attack 折线攻击

dolly（铆接用的）顶铁、抵座；轮式吊车；航空机载数据传输设备；战术数字信号传输信道

Dolphin Helicopter“海豚”直升机

domestic air route 国内航线

domestic airway 国内航路

dominating height 制高点

dominating point 制高点

door bundle（伞兵跳伞前先投下的）空投装载；伞投包（在飞行时人工投掷的包）

door load 舱门装载

door warning light 舱门（未锁住）警告灯

door-hinge rotor 折页铰接式旋翼；挥舞铰旋翼

dope 航空涂料；涂布油；添加剂；抗爆剂；吸收剂

doping（给飞机布质蒙皮）上涂料；（向活塞发动机气缸）注油；掺杂

Doppler beam sharpening 多普勒波束锐化

Doppler correction 多普勒修正

Doppler effect 多普勒效应（波源与观察点之间传输长度的时间变化率所引起的一种频移现象）

Doppler error 多普勒误差

Doppler hover 多普勒悬停（指依靠多普勒雷达信息保持零地速悬停）

Doppler navigation 多普勒导航
Doppler navigation radar 多普勒导航雷达
Doppler navigation system 多普勒导航系统
Doppler radar 多普勒雷达
Doppler ranging 多普勒测距系统；多普勒测距
Doppler shift 多普勒偏移；多普勒收发频率差
Doppler very high frequency omnidirectional range 多普勒甚高频全向信标
Doppler VHF omnirange 多普勒甚高频全向无线电信标
dorsal 机身背部的；机背
dorsal fin（飞机的）背鳍
dorsal spine（飞机机身的）背脊骨架
dorsal tank 背部油箱
dorsal turret 上炮塔
double acting engine 复动引擎（双作用引擎）
double delta 双三角形机翼
double drift 双偏流（求风速）法；曲折飞行（求风速）法
double pod 双联吊舱；双发动机吊舱
double point 双先头部队
double pole, double throw switch 双极双投开关
double pole, single throw switch 双极单投开关
double sideband amplitude 双边带调幅
double side-by-side landing gear 带有并列机轮的自行车式起落架
double slotted flap 双开缝襟翼
double taper 双梯形度翼
double transmission 双波发送
double valve 双向活门
double wedge 菱形翼型
double wedge wing 双楔形机翼；菱形机翼
double-acting set 双重作用装置；双向作用装置
double-entry centrifugal compressor 双入口离心压缩机
double-hinged rudder 双铰链方向舵
double-nozzled 双喷口的
double-redundancy 双冗余
double-root blade 双根叶片
double-slotted flap 双开槽缝式襟

翼（指襟翼上带小翼片，放襟翼后，翼片与襟翼间和机翼与襟翼间各有一个开缝）

doublet pendulum 双线摆式吸振器

double-tapered wing 双锥形机翼

double-wedge airfoil 双楔形翼型

down lock 起落架放下位置锁；下位锁；起落架释放固定锁

down the runway 沿跑道；在跑道入口外面；顺滑跑方向

down the wind 顺风

down to refueling 下降高度进行加油

down track fix 下航道定位点

down wind turn 二转弯

down-and-locked 起落架放下并锁定

downburst 突发下降气流；颓爆风；下击暴流；下冲气流

downdraft 下沉气流；下曳气流

downdraft carburetor 下向汽化器；下吸式汽化器

downed aircraft 击落的飞机；迫降的飞机（特指因战伤而迫降的飞机）

downed aircraft locator 飞机击落地点定位信标

downed airman 被击落飞机的机组成员；弃机着陆机组成员

download 下向载荷；（作战任务完成后或中断任务后）卸除载荷

down-looking radar 下视雷达

downpour 暴雨（＞100 毫米 / 日）

downstream clearances 下游飞行许可

downstream pressure 下游压力

down-time（因故障造成）停机时间；故障时间；非战斗状态时间；弹药装车时间；补给品拨发时间；军械修理时间

downward course 下降航向；下降航迹

downward flight 下降飞行

downward visibility 下方能见度；下方视界

downwash 气流下洗；下洗流；（升力螺旋桨或旋翼的）滑流

downwash airflow 下洗气流

downwash airstream 下洗气流

downwash angle 下冲角

downwash in wing wake 翼后下洗；机翼尾流中的下洗

downwash speed 诱导速度

downwash velocity 诱导速度

downwind 顺风；顺风飞行；（起落航线的）第三边；反着陆航向（飞行）

downwind landing 顺风降落

downwind leg（起落航线）第三边；顺风边

downwind takeoff 顺风起飞

downwind turn 二转弯

downwing side 机翼下偏侧；（转弯时）下垂机翼一侧

DR position（dead-reckoning position）航位推算位置；推算位置

drag 拉力；阻力；空气阻力；低空通过（低空飞越某一地区并仔细观察以准备着陆）

drag accelerometer 纵向负加速度表

drag at incidence 有迎角时的阻力

drag brace 承阻拉杆；后撑杆；承阻杆

drag brake 空气动力减速装置；减速板

drag characteristics 阻力特性

drag chute 制动降落伞；尾伞；阻力伞

drag coefficient 气动阻力系数；阻力系数

drag curve 阻力曲线

drag damper 阻力铰减摆器

drag force 阻力

drag hinge（直升机旋翼叶片的）摆振铰；阻力铰；垂直关节

drag in 带油门拖进场（指低空长五边进场着陆）

drag maneuver 诱饵机动（指双机组中有一架飞机引诱敌机，使其进入友机的射击范围）

drag parachute 减速伞；阻力伞；着陆制动伞

drag point 返回点

drag producing device 增阻装置；空气动力减速装置

drag strut 阻力支柱

drag wire 阻力线；（机翼或气球上的）阻力张线

drag-chute failure 减速伞故障；减速伞放不出

drag-free flight 无阻力飞行（指在真空中飞行）

drain 排水沟

drain and vent lines 放泄与通气管路

drain chamber 泄油腔
drain line 排放管；泄油管路
drain pipe 放油管；放泄管
drain plug 放油堵盖；放油堵头
drain point 放泄点
drain valve 放泄活门；放水活门；放油活门
drain valve retainer 排放活门护圈
drainage 排油；泄放
drainage pipe 放泄管；排水管道
draw bar 拉杆；连接杆；架尾棍；拖杆
draw-off line 外吸管
dressed 装挂齐全的（指飞机上所有设备和系统管路都完整）；配备齐全的（指发动机的外部附件、管路和控制装置都完整）
drift 偏航；航迹偏移；偏流角；偏差；偏移；偏移速度；（陀螺等的）漂移
drift angle 流压差；偏（流）角；（航行）偏航角；航差角；偏离角（指导弹偏离预定轨道的角度）
drift angle（DA）（航行）偏航角；偏流角
drift angle indicator 偏差角指示器
drift axis（航向陀螺的）框架垂直轴；漂移轴
drift bar 偏流针
drift coefficient 偏流系数
drift computer 偏航计算器；偏流计算器
drift correction（陀螺）漂移修正；偏流修正（量）
drift indicator 偏航角指示器
drift meter 偏流计
drift off the runway 偏离跑道（侧风影响飞机偏出轨道）
drift sight 偏流观测器；偏流计；偏航计
drift-down 在巡航中下降；逐步下降（指从航线剖面最高点逐步在航线上降低高度）
drifting fog 平流雾；吹雾
drifting snow 低吹雪
drill 训练；操练；教练；教官；编队飞行；训练守则（飞行特定阶段所采取的正确程序）
D-ring D 形吊索；D 形拉环；（降落伞上的）开伞拉环
drip pan 油样收集器；承油盘
drip valve 滴油阀
drip-stick gauge 滴流杆式油量表

drive axle 传动轴
drive coupling 主动联轴器
drive gear 驱动齿轮；主动齿轮
drive ratio 传动比
drive rod 驱动杆
drive shaft 传动轴；主动轴；驱动轴
drive system 传动系统；拖动装置
driven gear 从动齿轮
driven part 从动部件
driven shaft 从动轴
drizzle 毛毛雨（＜0.1 毫米 / 日）
drogue（空中加油管的）漏斗形接头；刹车伞；稳定伞；减速伞；锥形拖靶
drogue parachute 漏斗形减速伞；制动（降落）伞
drone 遥控无人驾驶器；靶机
droop 前缘襟翼；舱门下垂量；（直升机）螺旋桨叶片阻降器
droop balk 前缘襟翼连锁装置
droop compensator 转速下降补偿器
droop nose 下垂机头
droop restraint（旋翼桨叶）下垂限制器
droop stop（旋翼桨叶）下垂限动块
drooping ailerons 下偏副翼（指在放下襟翼时能随之下偏的副翼）；襟副翼
drooping leading edge 下偏前缘
drop altitude 空投高度（指离平均海平面的高度）；空投真高（指离地面的高度）
drop height 空投（相对）高度；投弹高度
drop in 着陆时失速附地
drop interval（依次空投之间的）空投时间间隔；（炸弹的）连投间隔
drop it 停止攻击；退出战斗
drop line 垂落线
drop message 空投文电（自飞机向地面或舰上部队投下的文电）；空投的通信袋；空降通报；空投信件
drop mission 空投任务；空降任务
drop tank 可空投油箱；副油箱
drop zone 空投场；空投区；伞降地域；伞降区
drop zone（DZ）空投场；伞降地域

drop-dead halt 完全停机
drop-launching 飞机发射；机上发射
droplet cloud 滴状云
dropsonde 空投式无线电探空仪
drum bearing 绞盘轴承
dry 空弹；干燥的；（机翼内）无燃油的；（悬挂接头）不能挂副油箱的；（发动机）不加力
dry air 干燥空气
dry clutch 干式离合器
dry contact 干结合（指空中加油机与受油机已接通加油管但还未放油）
dry crank 冷转
dry crank of the engine 发动机冷转
dry crank push button 发动机冷转按钮
dry engine 干发动机（指未加燃油、滑油等的发动机，或无加力燃烧室的发动机）
dry fog 干雾
dry plate clutch 干片式离合器；干板式离合器
dry power 干功率（指不开加力，通常等于军用马力或最大冷推力）
dry sump 干机匣（指底部只装有小收油池的发动机匣）；干收油池；干油池
dry takeoff（发动机）不喷水的起飞
dry weight 干重
dry-and wet-bulb hygrometer 干湿球湿度计
dry-bulb temperature 干球温度
dry-type forced-air-cooled transformer 干式压气冷变压器
D-scope D 型显示器
D-spar D 形梁
DU failure 显示组件故障
dual 双座飞行；随教员飞行；带飞；（飞机的）双套操纵装置；复式操纵装置；指导飞行小时数或次数
dual action bellcrank 双臂曲柄
dual action starter 双重作用起动机
dual cockpit 双操纵座舱
dual control 复式（双重）操纵；复式（双重）操纵装置；双重（双套）控制
dual drive 双套传动装置；双速传动装置
dual engine 双联发动机

dual firing system 复式发火系统（可防止点燃爆破装药时瞎火）

dual flight control 复式飞行操纵（装置）

dual flight time 复式飞行时间

dual generator failure 双联发电机故障

dual landing light system 复式着陆灯系统

dual linear actuators 双重线性作动筒

dual position sensor 两位传感器

dual rotor 双旋翼

dual rotor system 双旋翼系统

dual wrench 两用扳手

dual-action spring 双作用弹簧

dual-axis rate transducer（DART）双轴速率变换器

dual-mode antenna 收发天线

dual-mode seeker 双模式导引头；复式导引头（两种制导系统并用，如雷达与红外线并用）

dual-redundant computer 双余度计算机

dual-rotor helicopter 双翼直升机

dual-rotor system 双旋翼系统

dual-thrust motor 双推力发动机

dual-thrust nozzle 双推力喷管

duct 线槽

duct-burning turbofan 外函道燃烧加力涡轮风扇发动机；外函加力涡轮风扇发动机

ducted cooling 涵道式冷却；管道式冷却

ducted propulsor 涵道式推进器

ducted-fan engine 涵道风扇发动机；涡扇发动机

ducted-fan tail rotor 涵道风扇式尾桨

ductility 延性

due date 预定日期；支付日期

due in 待领数；待收

due out 待发

due-in 待领数已申请但尚未领到的补给品

due-in-from-maintenance 保养待领数量

dummy airdrome 假飞机场

dummy approach 模拟进场（着陆）

dummy dive 假俯冲

dummy drop 假空投

dummy hot run 冷开车；假热开车（指由起动机带动发动机运转，燃烧室只点火而不供油）

dummy missile 模型导弹；教练导弹

dummy missile firing 教练导弹发射

dummy pilot 假飞行员；飞行员模型

dummy plate 隔板

dummy pod 假吊舱；空心容器

dummy run 模拟射击；模拟俯冲轰炸（=dry run）；诱惑性攻击进入

dummy vehicle 假飞行器；飞行器模型；未装仪器设备的火箭

dump（弹药装备等）临时堆集所（场）；倾卸；堆放；应急放油；投弹；空投物资

dump valve 放油阀；倾泄活门；泄放活门

duplex burner 双油路喷嘴

duplex communication device 双向通信装置

duralumin 铝合金；杜拉铝；硬铝

duration flying 持久飞行

duration of ascent 爬高时间；上升时间

duration of down time 停机持续时间；非战斗状态持续时间

dust 灰尘

dust devil 尘卷；尘卷风；沙尘暴

dust haze 尘霾

dust storm 尘暴

dust strip 沙质起落道；土质简易机场

dutch roll 横滚；飘摆

duty ratio 作用比

D-valve D 形活门

DVFR（Defense Visual Flight Rule）防空目视飞行规则

DVFR（defense visual flight rules）防区目视飞行规则

DVFR flight plan（defense visual flight rules flight plan）防空目视飞行规则飞行计划

dwell 小停顿；保压；（航空自动武器闭锁机构工作时的）闭锁时间

dwell at 持续射击（火力支援语，指对目标持续进行无限期射击）

dwell on 对着某个目标不停射击

dwell time【雷】照射目标时间；等候时间（货物在转运存储区等候内运的时间）；停留时间；闭锁时间

dynamic air route planning 动态航线计划

dynamic air traffic control 动态空中交通管制

dynamic air ventilation 动态通气

dynamic air war game 空战动态推演；动态空战演习（演习双方将信息输入电子仪表中进行空战演习）

dynamic aquaplaning 动力滑水（指机轮在跑道浅积水面上作不接触跑道的高速滑动）

dynamic automatic radar tracking（DART）动态雷达自动跟踪（系统）

dynamic balance 飞行中力或力矩的平衡；动力平衡

dynamic damper 动力阻尼器；动力减震器

dynamic environment measurement system（DEMS）（航天飞机上的）动力环境测量系统

dynamic height 动力高度

dynamic impact energy 撞击动能

dynamic lift 动升力；气动升力

dynamic load 动力负荷；动载荷

dynamic mean sun 动平太阳

dynamic pressure 动压

dynamic pressure system 动压系统

dynamic response 动态响应

dynamic restriction 动态约束

dynamic rollover 动态滚转

dynamic stability 动力稳定（度）；动态稳定性；动安定性

dynamic stall 动态失速

dynamic stress 动应力；动态应力

dynamic thrust 动态推力；飞行推力

dynamic trim 动力配平（指飞行中用调整片使飞机配平）

dynamic trimming 动力调整；动力配平

dynamics 动力学；（直升机的）动力转动部件

dynamite 炸药；硝化甘油炸药；（用炸药）爆破

dynamometer 测力计；功率计

dynamotor 电动发电机

dysbarism 减压障碍；气压病；减压症

DZ（drop zone）空投场；伞降地域

E

E layer E 层；E 电离层

EA（electronic attack）电子攻击

EA（enemy aircraft）敌机

EADI（electronic attitude director/display indicator）电子式姿态指引仪

ear defender 护耳器

ear protector 护耳器

early ignition 早点火；提前点火

early in the flight 飞行起始阶段

early turn 提前转弯（未到航路转弯点提前转弯，以修正航迹，避免越出航路边界）

early warning 预警；早期警报

early warning aircraft 预警机

early warning radar 预警雷达

early-warning aircraft 预警机

early-warning radar 预警雷达

earphone 飞行帽耳机；耳机；听筒；受话器

earth axes 地面坐标系；地球坐标系

earth capture 地球捕获（进入地球引力场）

earth convergence 地球汇聚度；地球汇聚角

earth gravity assist 地球重力借力飞行

earth gyro 地球陀螺仪

earth induction compass 地磁感应罗盘

earth reference 地球坐标系；地球基准

earth referenced navigation 大地参考导航

earth system 搭铁系统；接地系统

earth wire 搭铁线；接地线；地线

earthing 接地

earthing strip 地线；接地片；接地条

earth-wire 接地线

EARTS（en route automated radar tracking system）航路自动化雷达跟踪系统

EAS（equivalent air speed）等效空速

EAT（expected approach time）预计进近时间；预期到达时间

ebullism 体液沸腾；起泡

ECAM（electronic centralized aircraft monitoring system）电子集中式飞机监视器

eccentric cam 偏心凸轮

eccentric disc 偏心盘；偏心垫圈

eccentric error 偏心误差

eccentric gear 偏心转动；偏心齿轮

eccentricity 偏心率；偏心距

ECCM（electronic counter countermeasures）电子反对抗措施；电子反干扰措施

echelon 梯队；梯次配置；指挥层次；指挥机构；（飞机或仅对）编成梯队；在梯队中占位；僚机飞到长机后 45° 的位置

echelon clouds 梯状云

echelon of maintenance（维修）保养等级；保养梯次

echo 回波；回声

echo box 回波谐振器；回波测试装置；空腔谐振器

echo intensity 回波强度

echo killer 回波抑制器；反射信号抑制器

echo power 回波功率

echo pulse 回波脉冲

echo signal 回波信号

ecliptic system of coordinates 黄道坐标系

ECM（electronic countermeasures）电子对抗

economical cruise mixture 经济巡航混合气

economical cruising speed 经济巡航速度

economical velocity 经济速度

economizer（供氧）节约器；节油阀；节热器；废气预热器

ECS（electronic combat squadron）电子作战中队

Ecureuil 松鼠（法国直升机）

EDCT（expected departure clearance time）预计离场批准时间

eddy 旋流；涡流；漩涡

eddy current torquer（修正陀螺的）涡流修正机；涡流转矩装置

edge alignment（螺旋桨的）边缘对位距离（桨叶的某一剖面中心线到前缘的距离）

E-display E 型显示器；距离仰角显示器

EEC（electronic engine control）电子发动机控制；发动机电子控制

EECU（engine electronic control unit）发动机电子控制单元

EET（estimated elapsed time）预计经过时间；预计已飞时间；估计的航程时间

EFAS（en route flight advisory service）（flight watch）中途飞行咨询服务；航路飞行咨询服务（飞行监视）

effect 作用；效应；影响；效果；实行；引起；命中

effective air path 有效无风航迹；有效航空航道

effective angle of attack 有效迎角；有效冲角

effective angle of incidence 有效倾角

effective casualty radius 有效杀伤半径

effective ceiling 有效升限

effective damage 有效损破损；有效破坏

effective helix angle 有效螺旋角

effective horsepower 有效马力

effective interception 有效截击

effective mean pressure 平均有效压力

effective miss-distance 有效误差距离（弹着点虽有一定误差，但仍足以造成原定的破坏程度）

effective pitch 有效螺距（螺旋桨转一圈，飞机前进的距离）

effective pitch ratio（螺旋桨的）有效相对桨距

effective profile drag 有效翼型阻力

effective propeller thrust 螺旋桨有效推力

effective radiated power 有效发射功率；有效辐射功率；有效辐射能量

effective range 有效射程；有效距离；有效航程

effective runway length 有效跑道

长度

effective span 有效翼展

effective tractive effort 有效牵引力

effective wind 有效风值

effective wing area 机翼有效面积

effective wing span 有效翼展

effectiveness 有效性；效率

efficiency 效率；功效

efficiency of catch 捕获效率

efficiency of loading 负载效率；（车辆）装载效率

EFIS（electronic flight instrument system）电子飞行仪表系统

egress 战斗机冲出目标区 / 战斗区的动作；在紧急情况下逃出机舱或座舱的行动

EGT（elapsed ground time）地面停留时间

EGT（estimated ground time）估计地面停留时间

EICAS（engine indication and crew alerting system）发动机指示和机组警告系统

eight “8”字形特技飞行

eight around pylons 绕双标杆的“8”字飞行（飞行高度低于标杆高度）

eight on pylons 绕双标杆的“8”字飞行（转圈时保持翼尖指向标杆）

ejection 弹射；喷射；发射；抛出；退弹壳

ejection angle（座椅）弹射角

ejection capsule 弹射舱；弹射回收记录仪器盒

ejection chute 抛壳道；抛壳套；退壳槽；弹射降落伞

ejection seat 弹射座椅

ejection slot（枪、炮的）排壳槽；抛壳口

ejection test vehicle 弹射试验飞行器

ejection training 弹射训练

ejector 喷射器；引射器；（子弹）退壳钩；（炮的）退弹壳器；顶出器；顶杆

ejector nozzle 喷射器的喷口

ejector release unit 弹射器投放装置

Ekman layer 埃克曼层

elaborate camouflage 周密的伪装

elapsed ground time（EGT）地面停留时间

elapsed time 经过时间；已飞时间
elastic axis 弹性轴
elastic clutch 弹性离合器；弹性联轴节轴承
elastic component 弹性元件
elastic damper 弹性减摆器
elastic deformation 弹性变形
elastic element 弹性元件
elastic force 弹性力
elastic hinge 弹性铰
elastic limit 弹性极限；弹性限度
elastic restraint 弹性约束
elastic restriction 弹性约束
elastic stability 弹性稳定性
elastic stop nut 弹性防松螺帽；弹性锁紧螺帽
elastic washer 弹性垫圈
elasticity 弹性；伸缩性
elastomeric hinge 弹性铰链
elastomeric lag damper 粘弹减摆器
elbow 弯头管；弯管；弯头
elbow bend（管子）弯头；弯管
elbow joint 弯头接头；肘接
electric altimeter 电气高度表；电测高度计
electric beacon（电）灯信标；电光信标
electric capacity altimeter 电容式高度表
electric charge 电荷
electric conductivity 导电性；电导率
electric control panel 电气控制板
electric current 电流
electric element 电器元件
electric energy 电能
electric engineering equipment 电工设备
electric equipment 电器设备
electric field 电场
electric firing lock 电点火装置
electric fittings 电气配件（零件）
electric fuse 电（子）引信；无线电引信
electric gyro 电动陀螺仪
electric hoist 电力起重机
electric lamp 电气照明
electric motor driven 电动机驱动的
electric motor with electromagnetic brake 具有电磁刹车的电动机
electric network frequency 电网频率
electric optical system（EOS）观

瞄装置

electric power monitor 电力监视器

electric pressure indication 电压指示

electric propeller 电控螺旋桨

electric pump 电动泵

electric stall warner 电动失速警告器

electric tachometer 电转速表

electric tachometer generator 电动测速机

electric technology 电工技术

electrical actuator 电动作动筒

electrical bay 电气舱

electrical check 电路检查

electrical contact 电触点；电触头；电气接触器

electrical control panel 电子控制面板

electrical control signal 电动调节信号；电动控制信号

electrical fire 电起火

electrical flight control system 电飞行控制系统

electrical flight information system 电子飞行信息系统

electrical flight instrument control display 电子飞行仪表显示器

electrical flight instrument control panel 电子飞行仪表控制板

electrical flight instrument system 电子飞行仪表系统

electrical flight path line 电航迹线

electrical gear 电气设备；电气装置

electrical generator 发电机

electrical load control unit 电载荷控制组件

electrical load management system 电载荷管理系统

electrical magnetic plug 电磁堵头

electrical power generating system 发电系统

electrical power service panel 电源勤务面板

electrical power system control panel 电源系统控制面板

electrical primary flight control system 初级飞行电控系统

electrical signal 电信号

electrical system 电气系统

electrical tactical map 电子战术地图

electrical teletachometer 电遥测转

速表

electrical trim unit 电动配平机构

electrical welding machine 电焊机

electrical wiring 安装电线；电线线路

electrically heated clothing 电加温飞行服

electrically signaled flying controls 电信号飞行操纵装置；电传飞行操纵系统

electrically suspended gyroscope 电悬浮陀螺仪

electrode 电极；电焊条

electrohydraulic servo valve 电动液压伺服阀；电动液压伺服活门

electromagnetic 电磁的；电磁场的

electromagnetic clutch 电磁离合器

electromagnetic compatibility 电磁兼容性；电磁适合性；电兼容性

electromagnetic countermeasures 电磁干扰；电磁对抗；反电磁措施

electromagnetic coupling 电磁耦合器

electromagnetic cover and deception 电磁掩护与欺骗

electromagnetic environment 电磁环境；电磁辐射场

electromagnetic filter 电磁滤波器

electromagnetic firing（火箭发动机）电磁点火

electromagnetic interference 电磁干扰；电磁干涉

electromagnetic interference（EMI）电磁干扰

electromagnetic pulse 电磁脉冲

electromagnetic sensor 电磁传感器；电磁敏感元件

electromagnetic spectrum 电磁频谱；电磁波谱

electromagnetism compatibility 电磁兼容性

electromechanical transducer 机电传感器；机电换能器；机电变换器

electromotive force 电动势

electromotive mechanism 电动机构

electron beam 电子束；电子注

electron beam process 电子束加工

electronic attack（EA）电子攻击

electronic attitude direction indicator 电子姿态方向指示器

electronic attitude director（display）

indicator 电子姿态指引仪；电子指引地平仪；电子垂直位置指示器
electronic attitude director indicator（直升机的）电子指引地平仪
electronic attitude director/display indicator（EADI）电子式姿态指引仪
electronic autopilot 电子式自动驾驶仪
electronic azimuth marker（在雷达示波器上）电子方位标志
electronic barrier 雷达警戒线；电子警戒线；电子壁垒；电子干扰措施
electronic beacon 电子信标
electronic blackout 电子设备工作中断
electronic box 电子箱
electronic centralized aircraft monitoring system（ECAM）电子集中式飞机监视器
electronic combat and reconnaissance 电子作战与侦察
electronic combat squadron（ECS）电子作战中队
electronic control assembly-pitch and yaw 电子控制组件 - 俯仰与偏航
electronic cooling system 电子冷却系统
electronic counter countermeasures（ECCM）电子反对抗措施；电子反干扰措施
electronic countermeasures（ECM）电子对抗
electronic course deviation indicator 电子偏航指示器
electronic crash locator（飞机）失事地点电子定位器
electronic deception 电子欺（诱）骗；电子迷惑
electronic display unit 电子显示装置
electronic engine control 发动机电子控制
electronic engine control（EEC）电子发动机控制；发动机电子控制
electronic equipment 电子设备
electronic flight display 电子飞行显示器
electronic flight instrument 电子飞行仪表

electronic flight instrument system (EFIS) 电子飞行仪表系统

electronic fuel control 电子燃油控制

electronic horizontal situation indicator 电子平面状态显示器；电子水平状态显示器

electronic instrument system 电子仪表系统

electronic intelligence (ELINT) 电子情报

electronic jamming 电子干扰

electronic jamming equipment 电子干扰设备

electronic line of sight 电子视线；电子瞄准线；电子直线（电磁波所走的直线，常指雷达发射天线波束对准目标的直线）

electronic order of battle 电子战斗命令；电子战斗序列

electronic panel 电子面板

electronic pilotage 雷达地标领航；电子领航

electronic pilotage indicator 电子领航指示器

electronic propulsion control system 电子推进控制系统

electronic protection 电子保护

electronic reconnaissance equipment 电子侦察设备

electronic repair division 电子修理分队

electronic route guidance system 电子航路导航系统

electronic scanning 电子扫描；雷达搜索

electronic standby instrument 电子备用仪表

electronic support measure (ESM) 电子保障措施；电子支援措施

electronic tachometer 电子转速表

electronic tactical deception 战术电子欺骗

electronic technology 电子技术

electronic warfare (EW) 电子战

electronic warfare equipment 电子对抗设备

electronic warfare support (EWS) 电子战支援

electronic warfare support measures 电子战支援措施

electronic warning 电子警戒

electronics 电子设备；电子学

electro-optical guidance 光电制导；

光电制导系统

electro-optical system（EOS）昼夜观瞄装置；昼夜观瞄系统

electrooptics 光电（场）学

electroplating 电镀

electrostatic discharge 静电放电

electrovalve 电磁活门；电动活门

electrovalve control relay 电磁阀控制继电器

element 要素；成分；单元；零件；元件；构件；电池；分队

element leader 带队长机；双机长机

elementary trainer 初级教练机

elephant ear 两侧进气道；耳形凸缘；悬垂升降舵；（火箭或导弹壁上开口处的）加固板

elevated heliport 高海拔直升机机场

elevated target acquisition sensor 空中目标搜索传感器

elevating clamp handle 高低机紧定器手柄

elevating force 上升力；升力

elevating gear 升降装置

elevating plane 升降翼

elevation 正视图；纵剖面图；标高；海拔；仰角；高低角；射角；上升；提高

elevation sight 高低瞄准具

elevation view（飞行器）垂直机动时仪表指示情况

elevator 升降舵；升降机

elevator angle 升降舵偏转角

elevator control 升降舵操纵

elevator control lever 升降舵操纵杠杆

elevator control mechanism 升降舵操纵机构

elevator illusion 升降舵错觉

elevator tab 升降舵调整片

elevator travel 升降舵偏度

elevator trim 升降舵配平

elevator trim indicator 升降舵调整片指示表

elevator trimmer 升降舵配平机构；升降舵调整片；升降舵修正片

elevon 升降副翼（副翼和升降舵的一种复合装置）

elimination 消除；删除

ELINT（electronic intelligence）电子情报

ELT（emergency locator transmitter）

求救定位信号发射器；紧急示位发射机；应急定位发射机

embark 搭机；搭载

embarkation 登机；装载；装运

emergency 紧急情况；意外状态；失事；事故；后备的；临时的；应急通信业务

emergency air 应急压缩空气

emergency alighting area 应急降落区

emergency altitude 应急高度；最低安全高度

emergency ceiling 应急升限；最大升限（多发引擎飞机的一个引擎有故障时，飞机能飞的最大安全高度）

emergency coordination center 紧急情况协调中心

emergency deployment readiness exercise 紧急部署战备演习

emergency descent 紧急下降；应急下降

emergency distance 紧急制动距离；中断起飞滑跑距离

emergency egress 应急离机；应急出舱

emergency escape device 紧急离机装置；紧急脱离装置

emergency escape equipment 应急离机设备

emergency evacuation 紧急撤运；紧急后运；应急撤离

emergency exit 紧急出口；安全门；紧急出口

emergency extension system 应急放下系统

emergency flotation gear 应急浮筒式起落架；（陆上飞机着水用的）应急漂浮装置

emergency kit 应急包

emergency landing 迫降；紧急降落；紧急着陆

emergency light 应急灯；紧急信号灯

emergency lighting supply 应急照明电源

emergency locator 应急定位器

emergency locator beacon 应急定位信标

emergency locator system 应急定位系统

emergency locator transmitter（ELT）求救定位信号发射器；紧急示位发射机；应急定位发射机

emergency locator/transmitter（用于航空救生系统）应急定位 / 发射机

emergency medical service 紧急医疗勤务

emergency mission support 紧急任务支援

emergency operating center 应急操作中心；紧急行动中心

emergency operations center 紧急作战中心

emergency parachute 应急降落伞

emergency phase 紧急阶段

emergency power unit 备用电力装备；应急电力装置

emergency procedure 紧急情况处置程序；特殊情况处置；应急措施

emergency safe altitude（ESA）紧急安全高度

emergency scramble 紧急起飞；紧急升空

emergency shutdown 紧急关车；应急停车

emergency static port 紧急静压口

emergency static pressure valve 应急静压阀

emergency transmitter 应急发射机

EMI（electromagnetic interference）电磁干扰

emitter 发射机；放射体；发射源

empennage 尾部；尾翼

emplacement 放列动作；放列；发射阵地；炮火掩体；发射阵地

empty dispenser 空子母弹箱

empty launcher 空发射装置（未装火箭的发射装置）

empty magazine 空弹仓；空弹匣

empty round 空弹壳

empty weight 空重；自重

empty weight to gross weight ratio（飞机）空载与总载重比

empty-weight center of gravity（飞机）净重重心

en route 在航路上；途中

en route air traffic control service 航路空中交通管制服务

en route automated radar tracking system（EARTS）航路自动化雷达跟踪系统

en route beacon 航线信标

en route climb 航路爬升

en route climb performance 航路爬升性能

en route descent 沿航路下降

en route fire suppression 途中火力压制

en route flight advisory service（EFAS）（flight watch）中途飞行咨询服务；航路飞行咨询服务（飞行监视）

en route holding 在航线上等待

en route minimum safe altitude warning 航路最低安全高度警告

en route penetration 中途突防；中途进入敌区

en route phase 航路飞行阶段

en route spacing program 航路间隔计划

en route time 途中时间

encode 加密

encoder 编码器；编码装置；编码员

encryption 加密；编成密码

end exercise 演习结束

end of descent 下降结束；下降终点

end-bending blade 端弯叶片；过弯叶片

endlap 终端（后部）重叠照片（两空中侦察摄影带之终端相重）

endurance（油量）续航；续航力；耐航性；持久性；持续时间；耐疲劳度；寿命

enemy aircraft（EA）敌机

energy of heat 热能

energy of light 光能

energy of motion 动能

energy of position 势能

energy spectrum analysis 能谱分析

enfilade 纵向射击；纵射炮火

engage 交战；进入战斗；攻击；攻击指定目标；接通；接合；（齿轮等）啮合

engagement 交战；接战；攻击；射击；截获（目标）；（飞机被地面拦阻装置）钩住；截住

engagement altitude 接通高度；投入战斗高度；开火高度

engagement area 交战地区；拦截区

engagement speed 接合速度（拦阻钩钩住拦阻索或飞机碰上拦阻网时的速度）

engaging gear 啮合的齿轮

engaging lever 换挡杆；接合杆；起动杆

engaging/engagement speed 啮合

速度；接合速度

engine 发动机；引擎

engine access hatch 发动机检查舱口

engine accessories 发动机附件

engine adjustment 发动机调整

engine air intake screen 发动机进气道滤网

engine air intake tube 发动机吸气管

engine air particle separator 发动机空气沙尘分离器；引擎空气颗粒分离器

engine alternator 发动机交流发电机

engine and gearbox unit 发动机和变速箱组

engine and warning display unit 发动机和警告显示组件

engine assembly 发动机组件；发动机总成

engine attachment 发动机连接；发动机附属件

engine bay 发动机舱

engine bearer 发动机座

engine bleed air 发动机放出（或引出）空气

engine bracket 发动机托架

engine braking 发动机制动

engine capsule 发动机密封舱

engine casing 发动机机罩；机舱棚

engine chamber 发动机燃烧室

engine change 更换引擎

engine compartment 发动机短舱；发动机舱；发动机室

engine condition monitoring 发动机状态监视

engine configuration 发动机配置

engine control 发动机操纵；发动机调节；发动机操纵机构（装置）

engine control lever 发动机操纵杆

engine control stand 发动机操纵台

engine control system 引擎控制系统

engine cowling 发动机罩

engine cradle 发动机架；发动机地面托架；（测扭矩用）发动机摇架

engine cranking 发动机冷转

engine cranking motor 发动机起动马达

engine crankshaft 发动机曲轴
engine cruising speed 发动机巡航转速
engine cut out takeoff 单发停车起飞
engine cutoff 发动机停车；发动机断开
engine cycle 发动机循环
engine efficiency 发动机效率
engine electronic control 发动机电子控制；电子式发动机控制
engine electronic control unit（EECU）发动机电子控制单元
engine endurance 发动机（工作）寿命
engine enemy 发动机不利因素
engine exhaust 发动机排气
engine failure 发动机失效
engine failure during approach 进近时段发动机故障
engine failure during takeoff 起飞时段发动机故障
engine failure in flight 飞行时段发动机故障
engine failure speed 发动机故障速度
engine feed line 发动机燃料供给管路
engine fire detection system 发动机火警探测系统
engine fire extinguishing system 发动机灭火系统
engine flight hours 发动机空中工作小时数
engine fluid 发动机防冻液
engine front mounting 发动机前支座
engine front support 发动机前支架
engine fuel inlet union 发动机燃油进油接头
engine fuel system 发动机燃油系统
engine functioning 发动机工作
engine gauge unit 发动机仪表盘装置
engine gearbox 发动机变速箱
engine governor auto switch 发动机控制器自动开关
engine governor manual switch 发动机控制器手动开关
engine ground handling 发动机地面操作
engine ground strap 发动机接地片

engine health monitoring 发动机状态监控；发动机完好性监控

engine ice 发动机结冰

engine in flight restart 飞行中发动机重新开车

engine indication and crew alerting system（EICAS）发动机指示和机组警告系统

engine inlet pressure 发动机进气压力

engine installation 发动机安装

engine instrument 发动机仪表

engine interface and vibration monitoring unit 发动机接口和振动监控器

engine interface unit 发动机接口组件

engine lifting 发动机吊装

engine logbook 发动机飞行日志簿

engine maintenance unit 发动机维护装置

engine manual 发动机手册

engine menu 发动机菜单

engine mount 发动机安装；发动机机架

engine mounting 发动机安装；发动机机架

engine nacelle 发动机短舱

engine normal start 发动机正常启动

engine off 发动机停车

engine oil system 发动机滑油系统

engine operating efficiency 发动机工作效率

engine out takeoff 单发停车起飞

engine parameter display 发动机参数显示器

engine performance indicator 发动机性能指示器

engine pickup 发动机上的传感器

engine pod 发动机吊舱

engine pressure ratio 发动机压力比

engine pressure ratio limit 发动机压力比极限

engine preventive maintenance 发动机预防性保养

engine priming 发动机起动注油

engine rating 发动机状态；功率状态

engine restart 发动机重新启动

engine rig test 发动机（整台）台架试验；发动机装配试验

engine ring cowl 发动机环状整流罩

engine room 发动机舱
engine rotating assembly 发动机转动部件
engine run-up（起飞前）发动机试车
engine run-up time 发动机试车时间
engine scuffing 发动机磨损；发动机拉缸
engine serial number 发动机系列号
engine setting 发动机设置
engine shield 发动机护罩
engine shorting-out 发动机停火
engine shroud 发动机挡板
engine shutdown 发动机停车
engine shutdown procedure 发动机关车程序
engine speed 发动机转速
engine start procedure 发动机开车程序；发动机启动程序
engine starter 发动机起动器
engine starting gear 发动机起动装置
engine status 发动机状态
engine stopped 发动机停车
engine surge 发动机喘振
engine test inspection 发动机试车检查
engine throttle bellcrank 发动机节气门操纵臂；发动机油门杠杆
engine throttled back 发动机节气门关闭
engine torque 发动机扭拒
engine trimming 发动机调整
engine unit 发动机组；动力机组
engine vibration indicator 发动机震动指示表
engine vibration monitoring 发动机振动监控
engine vibration monitoring unit 发动机振动监控系统
engine wear 发动机磨损
engine-aircraft alignment 发动机与飞机同轴性
engine-driven pump 发动机驱动泵
engine-driven supercharger 发动机驱动增压器
engineer brigade 工兵旅
engineer corps 工兵部队
engineering analysis 工程分析
engineering mock-up 全尺寸实体模型（飞机或其部件在工程设计中，为检查结构、系统、设备的三维几何尺寸而创建的金

属模型）

engineering test pilot 技术试飞员

engineering trouble 机械故障

engine-indicating and crew-alerting system 发动机指示和机组告警系统

engine-out 单发停车

enhanced configuration 加强配置

enhanced fire control system 增强型火控系统；增强型射击控制系统

enhanced ground proximity warning system 增强型近地警告系统

enhanced self-propelled artillery weapon system 增强型自行炮武器系统

enhanced target generator 增强的显示目标产生器

enhanced vision system 增强视景系统

enroute 在航线上；飞行中；在途中；沿航线；航线飞行

en-route absorption of terminal delay 航站延误在航线飞行中补救

enroute aircraft 航线飞行的飞机；航线上的飞机；在途中的飞机

enroute automated radar tracking system 航路雷达自动跟踪系统

enroute high altitude 高空航线航行图表

enroute high-altitude chart 航线高空图

enroute hospital（位于飞行航线沿线的）伤病员待运医院

enroute low altitude 低空航线航行图表

enroute personnel 途中护送人员

enroute point 航线点；中途站

enroute reliability 航行可靠率（未发生任何导致改变飞行计划的问题而顺利完成飞行计划的概率）

enroute repair kit 随机成套修理工具；随机工具箱

enroute traffic control service 途中空中交通管制勤务（各空中交通管制中心为从离场到目的地航站区之间按仪表飞行规则飞行的飞机提供的服务）

entering air defense identification zone 进入防空识别区

entering the state of turning to base 三转弯进入

entering vortex ring 进入涡环

entomopter 扑翼机；仿昆虫飞行

器；振翼机

entrance slope 进水角，首尖角

entrance taxiway 入口滑行道

entrance velocity 进口速度

entrance-exit light 跑道出入口灯

entrench 构筑堑壕；挖战壕

entrucking area 搭车地区；上车地区；装载地区

entrucking formation 乘车前部队依次排列的队形

entry airspeed 进入（机动时）空速

entry altitude 进入（机动）高度

entry authority list 进入权限清单

entry beacon 入口信标

entry controls 进入（禁区）控制

entry fix（管制区）入口定位点

entry gate 进入点

entry into service 投入使用；更装

entry point 飞行入境点；进入点

envelope 外壳；封套；包围；（折叠式车顶的）蒙布；（武器的）有效使用范围；蒙皮

enveloping attack 包围攻击

enveloping flank 包围的侧翼

environmental chamber 环境舱；环境试验室；环境室

environmental control system 环境控制系统

EOS（electric optical system）观瞄装置

EOS（electro-optical system）昼夜观瞄装置；昼夜观瞄系统

equal dew-point lines 等露点线

equalizer circuit 均衡电路（在飞机多发电机系统中使各发电机的负载均衡的电路）

equation of time 时差

equator 赤道

equidistant chart 等距航图

equilibrium 平衡；均衡

equilibrium airspeed 平衡空速；空速稳定值

equiphase zone 等相区

equipment 设备

equipment capsule 设备舱；仪器舱

equipment list 设备清单；设备一览表

equipment maintenance facility 装备保养设施

equipment maintenance log 设备（装备）保养记录

equipment maintenance management program 设备保养管理计划

equipment maintenance office 装备保养处

equipped empty weight 带设备空重

equisignal beacon 等强信号导航信标；区域指向标；等强信号无线电信标

equisignal sector 等信号区

equisignal station 等信号台

equisignal zone 等信号区

equivalent 当量；等效；等值；相当的；等效的；等价的

equivalent air speed（EAS）等效空速

equivalent airspeed 相应空速，等效空速

equivalent altitude 等值高度；当量高度

equivalent brake horsepower 等效制动马力；当量功率

equivalent flat-plate area 相当平板面积

equivalent flight hours 当量飞行小时

equivalent headwind 等效逆风；等值顶风；等值逆风

equivalent height 等效高度

equivalent hinge 等效铰；当量铰

equivalent monoplane 等效单翼机（单翼机的机翼升阻特性与两个以上机翼的综合特性相同）

equivalent shaft horsepower（ESHP）等效轴马力

equivalent shaft power of turboprop 涡轮螺旋桨发动机的当量轴功率

equivalent signal generator 等效信号发生器

equivalent single wheel load 当量单轮载荷；等效单轮载荷；等效单轮负载

equivalent temperature 当量温度；等值温度；绝热平衡温度

equivalent transformation 等效变换

equivalent weight 当量

equivalent wing 当量机翼；等效机翼

erecting 安装；装配；建立

erection 架设；安装；竖直；(导弹的）竖起；(陀螺的）修正

erosion（金属直接接触的）腐蚀；磨损；烧蚀；销蚀；消融

error signal 误差信号

ESA（emergency safe altitude）

紧急安全高度

escape capsule 逃逸舱；救生舱

escape hatch 应急离机口；应急离机舱口盖；脱险舱口

escape slide 紧急离机滑道

escape system 应急离机系统；应急脱离系统

escort 护航；护送；护航部队；护航飞机；护送卫兵

escort helicopter 护航直升机

ESHP（equivalent shaft horsepower）等效轴马力

ESM（electronic support measure）电子保障措施；电子支援措施

ESM（electronic warfare support measures）电子战支援措施

essential repair parts stockade list 主要修理零件存储表

essential subjects training 基本科目训练

essential supply cargo 必需的补给品（粮食、石油、医药等战时必需物资）

essential system characteristics, air-borne 机载主要系统特性

essential target 主要目标

establish the flight track and land 建立航线着陆

established 确定的

estimated 概算的；预测的

estimated airspeed 估计空速

estimated approach control 预计进场控制

estimated approach time 估计进近时间；估计进场时间

estimated ceiling 预计升限

estimated coordinate of target 估计目标坐标位置

estimated elapsed time（EET）预计经过时间；预计已飞时间；估计的航程时间

estimated ground time（EGT）估计地面停留时间

estimated mean time between failures 估计故障间隔平均时间

estimated month of loss 预计损耗月份

estimated off-block time 预计滑出时间

estimated time enroute（ETE）预计航路时间；预算航线飞行时间；预计途中时间

estimated time of arrival（ETA）预计到达时间

estimated time of berthing 预计停泊时间

estimated time of block 预计降落时间

estimated time of crew's return 预计机组返回时间

estimated time of departure（ETD）预计离场时间；预计起飞时间；预计出航时间

estimated time of en route（ETE）预计途中时间

estimated time of entry 预计进入时间

estimated time off（ETO）预计起飞时间

estimated time over（ETO）预计通过时间；预计飞越时间

estimated time over target（ETOT）预计飞临目标上空时间

estimating 估计；预计

estimating distance 目测距离；估计距离

ETA（estimated time of arrival）预计到达时间

ETD（estimated time of departure）预计离场时间；预计起飞时间；预计出航时间

ETE（estimated time enroute）预计航路时间；预算航线飞行时间；预计途中时间

ETE（estimated time of en route）预计途中时间

ETO（estimated time off）预计起飞时间

ETO（estimated time over）预计通过时间；预计飞越时间

ETOT（estimated time over target）预计飞临目标上空时间

Eureka 尤里卡信标；地面应答信标

evacuate 搬空；撤退；转移；疏散；排泄

evacuation 后送；撤退；撤运；疏散；撤离；排空

evaluation 估算；评价；评定；鉴定（美军指对情报的可信性、可靠性、适用性与准确性所作的评价）

evaluation flight 评定性能的（试验）飞行

evaporative ice 蒸发结冰；蒸发冰（发动机进气系统表面因蒸发冷却而形成的冰）

evaporator 蒸发器；汽化器

evasion 躲避；脱险；规避动作；规避机动

evasion action 规避动作（回避敌机、导弹、高射火力的机动）

evasive course 规避航向

evasive flight 规避飞行

evasive path 规避机动航迹

evasive tactics 规避战术；机动战术

eventual maintenance 长期维护；最终维护

EW（electronic warfare）电子战

EWH（eye-to-wheel height） 轮视角高度

EWS（electronic warfare support）电子战支援

exceedance 超限记录器（用于记录飞机或发动机使用中每次超出使用限度的有关参数）

exceeding speed limit 超速限制

exceptional circumstances 特殊情况

excess fuel 剩余燃油

excessive force 过度使用武力

excessive maintenance 加强保养维修

exchange line 交换线

exchange of fire 交火

execute a missed approach 进行复飞

executive and support 行政与保障（空中交通管制勤务）

exercise 演习；战斗演习

exercise code name 演习代名

exercise code word 演习代字

exercise on the map 图上作业

exercise readiness condition 演习戒备状态

exercise reconnaissance in force 进行军事侦察；进行威力侦察

exercise route 训练航路

exhaust 尾喷管；排气管；排气装置；放出的；排出的

exhaust back pressure 排气回压；排气背压

exhaust blanking device 尾喷管堵盖

exhaust collector ring 环形热排气收集器；环形排气总管

exhaust cone 尾喷管整流锥；排气锥（排气管内涡轮后的锥体）；喷管内锥；尾喷管

exhaust device 排气装置

exhaust duct（后风扇发动机的）

排气管

exhaust equipment 排气装置

exhaust flame damper 排气管阻焰器；排气阻焰器

exhaust gas 尾气；废气

exhaust gas temperature 排气温度

exhaust gas temperature gauge 废气温度表；排气温度表

exhaust manifold 排气集气管；排气歧管；排气管（总成）

exhaust nozzle 排气喷管；排气喷嘴

exhaust pipe 排气管；尾喷管

exhaust snubber 消声器

exhaust stator blades 排气导流叶片

exhaust stroke 排气冲程

exhaust system 排气系统

exhaust trails 排气尾迹；排气凝结尾迹

exhaust turbine 排气涡轮

exhaust unit 排气装置

exhaust valve 排气阀；排气门

exit fix（管制区）出口定位点（飞离管制区的最后一个报告点）

exit nozzle 喷嘴；喷管；喷口

exit of the nozzle 喷管出口

exit taxiway 滑出道；出口滑行道

exit the area 退出空域

exit time 空投物品离舱时间

exo-atmospheric 大气层外的

exosphere 外大气层；散逸层

exospheric 外大气层的

exotic fuel 特种燃油

expanding search 扩张式搜索

expansion cam 膨胀凸轮；张开凸轮

expansion stroke（内燃机）爆发冲程

expansion turbine 膨胀涡轮

expansion wave 膨胀波

expect 预计

expect approach clearance 预计进近许可

expect departure clearance 预计离场放行许可

expect further clearance 预计进一步许可；等待再次许可

expect further clearance via（airways, routes or fixes）（某跑道、航路或固定点）的预计进一步航路许可

expect higher altitude 等待更高高度

expected approach clearance 预期进场许可；预计进近许可

expected approach time（EAT）预计进近时间；预期到达时间

expected departure clearance time（EDCT）预计离场批准时间

expedite 发送；派遣；加速

expendable light markers 一次性灯光信标

expendable lubrication system 消耗性润滑系统

expendables 消耗件；一次使用件；投放物（可以在飞行中投掉的，包括外挂物、燃油、弹药等）

experimental duty 试验任务

experimental firing 试验射击

experimental mean pitch 实验平均桨距

explode 爆炸；爆发；（座舱玻璃）破裂；炸裂；（飞机编队）散开

explosion 爆炸；爆发

explosive 炸药；爆炸（性）的；爆发（性）的

explosive accident（in the air）爆炸事故（空中）

explosive assembly 爆炸装置；（导弹的）战斗部

explosive decompression 爆炸（性）减（失）压；突然减（失）压

exposure 曝光量；曝光；照射；辐照

exposure interval 曝光时间间隔

exposure time 曝光时间

extend the flaps 襟翼放下

extend/retract（起落架等的）收放

extended range 增大的射程；增程（的）；远程的

extended range operations 远程飞行

extended-root blade 伸根桨叶；伸根叶片

extension 放下（起降架、襟翼等）；延伸；延续；电话分机

extension of window（导弹或航天器）最佳发射时间延伸

extension phrase 延长下滑线飞行阶段

extensive damage 大面积破坏

extensive field maintenance 广泛的野战保养

extensive repair 大修

exterior angle 外角

exterior light control 外部照明控制

exterior lighting 外部照明
external 外部的
external aileron 外副翼
external air 外部空气
external air pipe 外部空气管路
external check 外部检查
external cue 外部线索
external dimension 外形尺寸
external electric supply 外接电源
external force 外力
external fuel duct 外部燃油管路
external intercom jack 机内通话外部插口
external leak 外部泄漏
external lighting 外部照明灯
external load 外挂载荷；外部载荷
external loudspeaker 外部扬声器
external marker 外部标记；外部标线；外部信标
external oil pipe 外部滑油管
external pipe 外部管路
external power 外部电源
external power breaker 外接电源断路器
external power contactor 外部电源接触器；外接电源触点；外部电源接头
external power ready relay 预外接电源继电器
external power receptacle 外部电源插座
external power socket 外部电源插座
external power source 外部电源
external public address system 外部扩音系统
external reference 外部参照物
external thermometer system 外部温度计系统
external view 外部视图；外视图
externally blown flap 外吹式襟翼
externals（飞行员进行的）飞行前外部检查
extinguisher 灭火瓶；灭火器
extinguishing agent 灭火剂
extra air tapped 过量引气
extra high pressure 超高压
extra high tension 超高压
extraction parachute 牵引伞
extraction zone 低空伞投地带
extreme low-altitude air supremacy 超低空制空权
extreme speed 最高限速；极限

速度

extreme turbulence 极度湍流

eye 孔眼；风眼

eye base 眼基线

eye of storm 风暴眼

eye of the cyclone 气旋眼

eye reference point 视力基准点

eye relief 眼睛间隙

eyeball/shooter 搜索 / 攻击机动（长机飞越并目力识别目标，停机在视距外发射导弹）

eyelids（可调喷口的）调节片

eye-to-wheel height（EWH）轮视角高度

F

F layer F 层

F region F 区域

FAA（federal aviation administration）（美国）联邦航空局

face 螺旋桨表面

face alignment 叶面校准矩（螺旋桨叶或直升机旋翼的叶弦中心线与桨叶面的距离）

face blind 防护面罩；（弹射座椅的）防护布帘

faceplate 面板；（电灯开关等上的）护板

facilities and equipment 设施和设备

factor of safety 安全系数；安全因数；保险系数

factored field lengths 机场安全使用长度

FAD（field ammunition dump）战地弹药库

FAD（forward ammunition depot）前进弹药库

fade 衰减；失效

FADEC（full-authority digital electronic control）全权限数字电子控制；全权限数字电子控制装置

fadeout（电视图像）淡出；渐隐；信号衰落；消失

fading 信号衰落；消失（飞机航迹在视界或雷达荧光屏上消失）；制动逐渐失灵

fading period 衰落周期；衰落时期

FAF（final approach fix）最终进近定位点

Fahrenheit 华氏度

fail monitor 故障检测仪

failing load 失效载荷

failsafe 破损安全；失效保护

fail-safe 故障自动防护的；可靠

的；保险的；瞎火保险
fail-safe circuit 故障自动防护电路
fail-safe control 故障自动防护控制
fail-safe system 故障自动防护系统
failure 故障
failure code 故障代码
failure correction 故障排除
failure correction by item replacement 换件排故
failure cycle 故障周期
failure detection 故障检测
failure diagnosis 故障诊断
failure identification 故障标识
failure information 故障信息
failure interval 故障间隔
failure isolation 故障隔离
failure lights 故障灯
failure mode effects analysis 故障模式效果分析
failure record 故障记录
failure state 故障状态
failure test 失效试验；故障检测
failure warning 故障告警
failure-survival autopilot 高生存力的自动驾驶仪；故障生存自动驾驶仪
fairing 蒙皮；整流罩
fairing jettison motor 抛（整流）罩发动机
fairing wire（飞艇蒙皮的）整形张线
FAK（first aid kit）急救箱
faker 假设敌机（防空演习用）
fall off on a wing 机翼下倾；失速转弯（跃升失速倒转）
falling leaf 飘摆滑翔（一种特技飞行动作，飞机如落叶般左右侧滑下降）
fallout 放射性微粒
fallout contamination 放射性沾染
fallout mission 备份战斗任务
fallout-free 无放射性沉降
false cirrus 伪卷云
false glide slope 假下滑道
false horizon 假视地平
false oblique 假倾斜
false ribs 假肋（机翼前缘主肋之间用以支撑布质蒙皮外形用的肋）
false spar 副梁；纵墙（与机身不连接的第二梁，用于安装活动翼面）
false start（发）假启动；假起动；假开车

false target（primary radar）假目标（初级雷达）

fan【口】螺旋桨；风扇

fan air 扇流；冷凝风量

fan beam 扇形波束

fan blade 风扇叶片

fan cameras 多摄影机组；多相机（以不同角度同时对同一目标拍摄的三个或三个以上的照相机）

fan cowl door 风扇整流罩舱门

fan discharge pressure 风扇出口压力

fan frame 风扇框架

fan inlet static air pressure 风扇进口空气静压

fan inlet temperature 风扇进口温度

fan inlet total air temperature 风扇进口空气总温

fan lift 风扇升力

fan mapping 扇形航测制图

fan marker 扇形无线电信标；扇形辐射点标

fan noise 风扇噪声

fan out（飞机编队的）扇形展开；（飞机编队的）扇形解散

fan outlet static air pressure 风扇出口空气静压

fan pressure ratio 风扇压力比

fan reverser 风扇反推装置

fan stream burning 风扇函道气流燃烧

fan-camera photography 多摄影机摄影术；多相机照片（由三个或三个以上相机在不同角度上同时拍摄同一目标，然后拼在一起的宽视角照片）

fanjet 涡轮风扇发动机；装有涡轮风扇发动机的飞机

fanned-beam antenna 扇形波束天线

fanning beam 扇形扫探波束；扇扫波束；扇形射（波）束

FANS（future air navigation system）未来空中导航系统

fan-type marker 扇形标志

FAP（final approach point）最终进近点

far boundary 最远边界

far infrared 远红外

FARP（forward-arming-and-refueling point）前方武器装备及燃料补给站

fast erection 快速架设

fastener 紧固件

fastening system 紧固系统
fast-exit taxiway 高速滑出道
fatal accident 一等事故（机毁人亡事故）；死亡事故
fatigue 疲劳
fatigue level 疲劳强度
fatigue life 疲劳寿命
fatigue rod 疲劳杆
fatigue strength 疲劳强度
FATO（final approach and takeoff area）最终进近和起飞区域
FAU（fault annunciation unit）故障报警装置；故障报警器；故障信号器
fault 故障
fault analysis 故障分析
fault and alarm channel 故障和警告电路
fault annunciation unit（FAU）故障报警装置；故障报警器；故障信号器
fault code 故障代码
fault code diagram 故障代码图；故障代码表
fault code index 故障代码索引
fault correction time 故障修复时间
fault detection and isolation 故障检测和隔离
fault finding 查找故障
fault indication 故障指示
fault isolation 故障隔离
fault lamp 故障指示灯
fault location 故障定位
fault management 故障处理
fault message history 故障信息记录
fault report manual 故障报告手册
fault signal 故障信号
fault tolerance 故障容限；失效容差
faulty component 故障件
favorable air situation 空中有利态势；有利空中形势
FAWS（flight advisory weather service）飞行天气咨询服务；飞行天气咨询勤务
faying surface 搭接面（蒙皮与飞机外露边缘的搭接部分）
FBO（fixed-base operator）固定基地运营商
FCC（flight control computer）飞行控制计算机
FCLT（freeze calculated landing time）固定计算着陆时间

FCOC（fuel-cooled oil cooler）燃油冷却的滑油散热器；燃油冷却式滑油冷却器
FCU（fire control unit）发控电子箱；火控装置；射击控制组
FCU（fuel control unit）燃油调节器；燃调；燃油控制组件
FCU cut-away 燃调剖视
F-day（fire day）发射日
FDC（flight data center）飞行数据中心
F-display F 型显示器
FDR（flight data recorder）飞行数据记录仪（俗称“黑匣子”）
feather 变距；顺桨
feathered propeller 顺桨的螺旋桨
feathering 顺桨；周期变距
feathering angle 顺桨桨叶角；顺桨桨距
feathering axis 顺桨轴；变距轴
feathering button 顺桨按钮
feathering hinge 顺桨铰链；变距铰链
feathering motion 变矩运动；顺桨运动
feathering pitch 顺桨桨距
feathering propeller 顺位螺旋桨
feathering pump 顺桨泵
feature vector 特征向量
FEBA（forward edge of the battle area）战区前沿
federal aviation administration（FAA）（美国）联邦航空局
feed 供给；进弹；馈电器；输入信号；弹带；雷达天线的馈电部分
feed cable 馈电电缆
feed line 馈电线；供气管路；供油管路
feed tank 消耗油箱；给水箱
feedback 反馈；回输；回传
feedback control loop 反馈控制回路
feedback path 反馈路径
feeder 进弹机；托弹盘；馈电线；航空（公路）支线
feeder line 补给线；空中支线；地方航线；馈电线
feeder liner 支线航班；支线航运公司
feeder route 辅助航线；支线
feeder unit 补给单位；进料装置
feeding line 进料管路
feel simulator 操纵感力模拟器；

载荷感觉器

feel system 人工感力系统；载荷感觉装置

feet per minute（fpm）英尺 / 分钟

feint 佯攻；佯动

felt 毡；毡垫片

fence 雷达预警线；雷达预警网；防地物干扰篱笆栅栏；跑道进口；翼刀；导流片；隔流片

fenestron 函道式尾桨

fenestron antitorque rotor 函道式反扭矩尾桨

Ferret【美】“雪貂”电子侦察平台（具有特种装备，可用以探测、定位、记录与分析电磁辐射的飞机、船只、车辆或卫星）

ferry 转场（飞机）；渡运

ferry flight 转场飞行

ferry pilot 转场飞行员；飞机渡运飞行员（负责把飞机送到部队的飞行员）

ferry range 空中转场距离；空载转场航程（无业载时的极限安全航程）

ferry tank 辅助油箱

fiber distributed data interface 光纤分布数据接口

fiber glass reinforced plastic（FRP）玻璃钢；纤维加强塑料

fiber-optics gyro 光纤陀螺

fibreglass cartridge 玻璃纤维滤芯

FIC（flight information center）飞行情报中心；飞行信息中心

fiducial axis 基准轴

fiducial mark 基准标记

field 外场；野战（的）；战地（的）

field alignment error 机场对准误差

field ammunition dump（FAD）战地弹药库

field elevation 机场标高；场高

field helicopter base 野战直升机基地

field helipad 直升机野外起降场

field heliport 野战直升机场

field heliport communications support 野战直升机场通信保障

field heliport navigation support 野战直升机导航保障

field investigation 野战调查；野外调查

field length 机场长度

field maintenance 外场维护

field manual（FM）现场作业手

册；战地参考便览；野战手册

field of view 视场；视域

field performance（飞机）起飞着陆性能

field takeoff and landing 野外场地起降

fielding 野外部署

fighter 战斗机；歼击机

fighter affiliation 战斗机参与（战斗机与轰炸机、其他类型飞机及陆海军部队联合进行训练和演习）

fighter aviation 战斗航空兵；战斗机；战斗机飞行员；战斗机驾驶术

fighter controller 战斗机指挥员；战斗机引导员

fighter cover 战斗机掩护

fighter escort 战斗机护航

fighter sweep 战斗机战斗巡弋；战斗机战斗搜索；战斗机扫荡（战斗机在敌领域内特定地区上空所执行的攻击性任务，其目的是搜索并攻击敌机或临时目标）

fighter-bomber 战斗轰炸机

figure of merit 灵敏值；（对装甲的）侵彻系数

filler block 衬块；垫块；填块

filler cap 油箱盖

filler cap adapter 加油口盖接口

filler depot 补充仓库

filler neck 加油口；漏斗颈；加油口颈

filler unit 注油装置

filler valve 加注阀

fillet 整流片；整流带；（两个面接合处的）整流包皮

filling 加注；装填

film cooling 气膜冷却

film-cooled blades 气膜冷却涡轮叶片

filter 过滤器（油滤、气滤）

filter cap 油滤盖；滤清器盖

filter catridge 滤芯

filter clogging 过滤器堵塞

filter core 滤芯

filter cover 油滤盖

filter downstream pressure inlet 油滤下游压力进口

filter element 滤芯；过滤元件；滤清设备

filter fineness 过滤精度

filter housing 滤网壳体

filter supply pressure 油滤供油压力

filter upstream pressure inlet 油滤上游压力进口
filtering ability 过滤能力
filtering element 滤清设备；滤芯
filtration 过滤；筛选
fin 尾翼；垂直安定面；垂直尾翼；腹鳍；背鳍；散热片
final 最终边；（起落航线的）第五边
final approach 最终进近；最终进场着陆；飞机在接地前的航向飞行；在第五边上着陆；起落航线上的第五边
final approach altitude 最终进近高度；（起落航线的）第五边入口
final approach and takeoff area（FATO）最终进近和起飞区域
final approach area 最终进近区域
final approach course 最终进近航向
final approach fix（FAF）最终进近定位点；最终进近场点
final approach gate 最终进近通道
final approach IFR（instrument flight rules）最终进近仪表飞行规则；第五边仪表飞行规则
final approach leg（直升机起落航线的）第五边
final approach mode 最终进近模式
final approach point（FAP）最终进近点
final approach segment 最后进近段
final approach slope 最后进近斜坡
final assembly 总装；最终装配；输出装置；最末组件
final condition 最终状态
final controller 进场最终阶段控制员；着陆调度员
final heading 最终进场航向；最终航向
final leg（直升机起落航线的）第五边
final procedure turn 最后程序转弯
final turn 四转弯；进入目标的最后一个转弯
fine pitch 小桨距；低距；细螺距
fine sand 细砂
fineness ratio 长细比（流线型物体如机身、导弹的长度与截面最大直径之比）；瘦长比（包括长度排水体积比、长宽比、长度吃水比、（潜艇）长度直径比）；展弦比

fine-pitch latch 小桨距销

finger（航站大楼上下飞机的）走廊桥

finger contact 按钮接触；按钮接点

finger control 手控制；手调节；手调

finger formation 指尖队形（即楔形队形）

finger four 四机指尖队形（美国空军和海军飞机排列的一种队形，形似手掌伸开时的四个手指）

finger-tight 用手拧紧的；松散装配的

fingertip formation 指尖编队；指尖队形

finite wing 有限翼展机翼；有翼尖的机翼

Finmeccanica Company 芬梅卡尼卡集团公司

finned-head fire detector 头部带翅片的火警探测器

FIR（flight information region）飞行情报区；飞行情报保证区；飞行通报区；飞行报告区

fire adjustment 射击修正

fire alarm 火警

fire bottle 灭火瓶

fire bottle discharge indicator 灭火瓶泄漏指示器

fire bucket switch 消防水桶开关

fire control equipment 射击控制器材；火力控制设备；射击控制设备

fire control panel 火控面板

fire control system 火控系统

fire control unit（FCU）发控电子箱；火控装置；射击控制组

fire defect 火警系统不良

fire detect box 火警探测盒

fire detection system 火警探测系统；火力探测系统

fire detector 火警探测器；火力探测器

fire extinguisher 灭火器

fire hazard 火险；易燃性

fire point 射击目标；起火点；着火点；燃点

fire proof switch 防火开关

fire protection 消防；防火

fire sleeve 防火套

fire truck 消防车；救火车

fire unit 射击单元；射击分队；火力分队

fire wall 防火墙；防火壁；防火隔框

fire wall shut off valve 防火墙关断活门

fire zone 火警区；防火区；射击区域

fire-and-forget anti-tank missile system 发射后自动寻的反坦克导弹系统；发射后不管的反坦克导弹系统

fire-and-forget missile 发射后不管的导弹

fire-bomb 火焰炸弹（燃烧炸弹和凝固汽油炸弹的总称）

fire-control radar 火控雷达

fire-extinguishing 灭火

fire-extinguishing system 灭火系统

fire-fighting 灭火；消防

firepower 火力

fireproof bulkhead 防火隔框

fire-pull handle 灭火拉杆

firewire element 火线元件

firing latch 发射扣机

firing mode 发射模式

firing order 发射命令；发射指令；发射口令；射击顺序；（发动机）点火次序

firing pass 实施射击的进入（飞机对目标进入攻击前的机动占位）

firm snow 实雪

first aid kit（FAK）急救箱

first engine to test 首台测试发动机

first front and back, then left and right 先前后后左右

first hop（新型飞机的）首次（地面）滑行试验；短距试飞

first in first out 先入先出；先装先卸

first line maintenance 一级维护；一级维修保障

first moment 静力矩

first officer 副驾驶

first phase of the ground war 第一轮地面战

first pilot 正驾驶员；机长

first-echelon maintenance 一级维护

first-line servicing 一线维护

first-tier center 一线中心

FIS（flight information service）飞行情报勤务；飞行报告勤务（为安全有效地实施飞行而提供建议与情报信息的一种服务）

fishbone antenna 鱼骨天线

fishtail nozzle 鱼尾喷嘴

fit 配合；装备；配备；适合的；合格的

fitting 配件；装配；接头

five by five（收讯情况）很好；讯号强度与清晰度均好

five-digit display 五位数显示屏

five-minute point 五分钟点（起飞功率）

five-way selector switch 五位选择开关

fix 定位；定点；测量点；方位点；标定点；坐标位置；整理；修理；装备安装后的调整；将敌人固定于原地，使其既不能前进，又不能后退

fixed ballast 固定压载

fixed cowling 固定整流罩

fixed geometry 固定几何形状

fixed guide cylinder 固定导筒

fixed landing gear 固定式起落架

fixed landing light 固定着陆灯

fixed light 稳光灯；定光灯

fixed slat 固定式前开缝翼；固定式前缘缝翼

fixed slot 固定翼缝

fixed surface 固定翼面

fixed tab 固定调整片

fixed-area exhaust nozzle 固定面积排气口；固定范围排放嘴

fixed-area nozzle 定面积喷管；面积不可调喷管；不可调喷管

fixed-base operator（FBO）固定基地运营商

fixed-base simulator 固定模拟机；固定基底模拟器

fixed-displacement pump 固定排量泵

fixed-distance marker 定距标记

fixed-gun mode 固定航炮射击模式

fixed-pitch propeller 定距螺旋桨

fixed-shaft turboprop engine 定轴式涡轮螺桨发动机

fixed-spindle 定轴式

fixed-wing aircraft 固定翼飞机

fixing bolt 固定螺栓

fixing force 阻击部队（占领坚固支撑点，阻击并逼使敌人进入歼敌地域的部队）；牵制兵力

fixing nut 固定螺帽

fixture 夹具；（小型）型架；定位器；附属装置

FL（flight level）飞行高度

flag 标识；标志位；故障警告牌

（仪表或其他装置上用以显示发生故障的小牌）

flag alarm 报警信号（空中导航仪表中一个信号指示器，当仪表读数不可靠时即发出警报）

flag tracking 标志物跟踪；（直升机旋翼）打旗法同锥度调整

flag value 特征值

flak 高射炮；高射炮火

flak suit（飞机驾驶员）防弹衣；防弹服（防高射炮弹破片的杀伤）

flak suppression fire 对高射炮火压制射击；压制高射炮的火力

flak vest 防弹背心

flame front 火焰前锋；火焰头

flame holder 稳焰器

flame out（发动机）熄火

flame rate 火焰速率

flame stability 火焰稳定性

flame temperature 火焰温度

flame tube 火焰筒；发火管

flame tube interconnector 联焰管

flameout 熄火；中断燃烧

flameout pattern（发动机）停车着陆航线；（发动机）停车着陆模式

flame-resistant 耐火材料；耐火的

flammable liquid fire 易燃液体火灾

flange 法兰盘；凸缘；弹带

flanged union 法兰连接

flank 翼侧；侧面；侧翼；齿侧；从侧翼攻击；从侧翼包围；侧面的

flap 襟翼；襟板；减速板；阻力板；水舵；火箭舵；（直升机旋翼）挥舞；折翼；瓣状物

flap angle 襟翼角

flap blowing 襟翼区附面层吹除（从高压压缩器中引出的空气吹向下垂的襟翼，以防止气流分离）

flap controls 襟翼控制系统

flap extension 放下襟翼

flare pod 照明弹夹

flap position 襟翼位置；襟翼偏度

flap position indicator 襟翼位置指示器

flap restrainer 挥舞抑制器

flap setting 襟翼调整；襟翼设定

flap station（FS）襟翼位置

flaperon 襟副翼

flap-extended speed 襟翼放下时的飞行速度

flapless landing 无襟翼着陆；不放襟翼着陆；不放襟翼降落

flapless takeoff 不放襟翼起飞

flaplet 小翼片；有圆弧形前缘；窄襟翼

flap-overload valve 襟翼过载阀

flapper valve 瓣阀；止回阀；舌阀

flapping 扑翼；（直升机旋翼桨叶的）挥舞

flapping angle（旋翼的）挥舞角；翼动角

flapping coefficient 挥舞系数

flapping deformation 挥舞变形

flapping distortion 挥舞变形

flapping equality 挥舞平衡

flapping hinge 挥舞铰链；水平铰

flapping motion 挥舞运动

flapping plane 挥舞平面（垂直于旋翼每个水平铰轴的平面）

flare 照明弹；曳光弹；信号弹；拉平（飞机改变飞行轨迹以降低接地时的下降率）；起飞离地后一段平飞；（直升机）瞬时增距

flare effect 光晕效果

flare path 有照明设备的跑道；着陆拉平轨迹；着陆探照灯的光束

flareout（飞机着陆前）拉平

flash 闪光；炮口焰；（火器的）火光；闪现；（飞机）掠过；（简短的）电讯；徽章；肩章；亮度快速的，特急的

flash point 闪点；起爆温度

flash report（执行任务的）初步报告（航空情报用语，就执行某一任务提出初步报告，其中主要包括任务执行过程中飞机或人员的损失）

flash unit 闪光组件

flashback 逆火；回火

flashing beacon 闪光灯标；闪光信标

flashing light 闪光；信号光；闪光灯；信号灯；明暗灯；闪光灯塔

flashlight 手电筒；闪光灯

flashover 跳火；飞弧

flash-resistant 抗闪燃的；阻闪燃的

flat glide 小角度下滑

flat head pin 平头销

flat head reinforcing frame 平封头加强框

flat nose pliers 平口钳；平头钳

flat riser 平放式垂直起降飞机
flat shape 平面形状
flat spin 水平螺旋（飞机在失速状态下依其垂直轴旋转，并以45° 以下的较小的俯冲角下降）
flat tyre 瘪胎；漏胎
flathead screwdriver 一字解刀；一字改锥；一字螺丝刀
flat-rated engine 固定推力发动机；固定输出功率发动机
flattening 扁率；扁平率
fleet 大机群；（飞机的）总数；滑过；飞过；舰队
fleet air arm【英】海军航空兵
fleet scheduling 机队调度
flexibility 灵活性；适应性；柔性，韧性
flexible 柔软的；有挠性的；有弹性的；灵活的；柔性的
flexible ammunition delivery pipe 软输弹导
flexible beam 柔性梁
flexible blade（旋翼的）柔性桨叶（尤指后缘带有补偿片的）
flexible carcass construction 软连接；挠性连接
flexible coupling 弹性联轴节；柔性联轴节；挠性连接
flexible element 挠性元件
flexible pavement 弹性铺面；柔性路面
flexible pipe 软管
flexible rule 软尺；卷尺
flexible runway 柔性跑道；软（地面）跑道
flexible support 弹性支撑
flexible takeoff 弹性起飞（起飞重量小于最大起飞重量时，采用小于最大推力的推力起飞，以节约发动机和降低噪声）
flexible wing 柔性翼；可折翼；挠性翼；软翼；可变翼
flexing 挠曲
flexural aileron flutter（机翼的）挠曲副翼颤振
flexural vibration 弯曲振动
flick roll 快滚；螺旋式横滚
flicker control 脉动操纵
flicker effect 闪烁效应
flicker navigation light 闪烁航行灯；闪烁导航灯
flicker position light 闪烁航行灯
flight 飞行；一次飞行（任务）；班机；定期客机；空军小队（四

机以上的机群）；火箭的齐发；航程；射程；（识别飞机时观察者所见的）飞机投影姿态

flight accident 飞行事故；空中失事

flight advisory message 飞行通报信息

flight advisory weather service（FAWS）飞行天气咨询服务；飞行天气咨询勤务

flight at VNE（velocity never to exceed）最高限速飞行

flight attitude 飞行姿态

flight characteristics 飞行特性

flight check 飞行检查（在实际飞行中对驾驶员或其他乘务员的工作熟练程度所进行的检查；飞行中对机上某些设备的检查）；飞机呼号的前缀

flight clearance 飞行许可；飞行许可证

flight communication center 飞行通信中心

flight control 飞行指挥勤务；航行调度勤务；飞行操纵；飞行控制；飞行管制

flight control check 飞行控制检查

flight control computer（FCC）飞行控制计算机

flight control electronic power supply 飞行控制电子电源

flight control electronic system 飞行控制电子系统

flight control electronics unit 飞行操纵电子组件

flight control line 飞行控制线

flight control panel 飞行控制面板

flight control system 飞行操纵系统；飞行控制系统

flight control unit 飞行控制装置（驾驶舱与自动飞控系统的接口，能提供自动驾驶工作方式、速度、航向、高度选择等功能）

flight controller 飞行自动控制系统；自动驾驶仪；飞行控制员；飞机增稳系统控制台

flight controls 飞行操纵机构；飞行操纵系统

flight crew 飞行人员；空勤组；机组人员

flight crew member 飞行机组成员

flight crew operating manual 飞机机组人员操作手册

flight cycle 飞行循环（机体、推

进系统及其他系统的工作程序，这些程序加在一起完成一次飞行）

flight data 飞行数据；飞行特性数据；飞行试验数据

flight data acquisition card 飞行数据采集卡

flight data acquisition unit 飞行数据收集单元

flight data center（FDC）飞行数据中心

flight data entry and printout 飞行数据输入和输出

flight data equipment 飞行参数设备

flight data input 飞行数据输入

flight data interface unit 飞行数据接口装置；飞行数据接口设备

flight data output 飞行数据输出

flight data processing system 飞行数据处理系统

flight data processor 飞行数据处理器

flight data recorder（FDR）飞行数据记录仪（俗称“黑匣子”）

flight deck 飞行甲板（航母）；（大型飞机的）驾驶舱

flight deck component 飞行甲板（航母）部件；（大型飞机的）驾驶舱部件

flight deck effect 驾驶舱效应

flight delay 起飞推迟；飞行推迟；飞行延误

flight deployment 飞行调配

flight directive 飞行指令

flight director 飞行指挥仪；飞行指引仪；机长

flight director indicator 飞行指引指示器；指引地平仪

flight director standby switch 飞行指挥仪备用开关

flight director system 飞行导向系统；飞行测向系统

flight duty period 执勤时段

flight dynamics 飞行动力学

flight engineer 飞行机械师；随航机械师；空勤机械员

flight engineer certificate 随机工程师证书

flight engineer’s panel 随机工程师仪表盘

flight envelope 飞行状态范围；飞行包线

flight examiner 飞行检查员；驾驶术检查员

flight following facility 飞行跟踪台；飞行跟踪设备

flight following service 飞行跟踪服务

flight frequency 航次频数；航班频率

flight hour 飞行小时

flight idle 空中慢车（飞行中发动机最低转速）；慢速飞行

flight in the airspace 空域飞行

flight information 飞行数据；飞行试验数据；飞行资料；飞行情报

flight information center（FIC）飞行情报中心；飞行信息中心

flight information region（FIR）飞行情报区；飞行情报保证区；飞行通报区；飞行报告区

flight information service（FIS）飞行情报勤务；飞行报告勤务（为安全有效地实施飞行而提供建议与情报信息的一种服务）

flight inspection 飞行检查

flight inspection field office 飞行检查现场办事处

flight instruction 飞行训练；飞行教学；飞行指示

flight instructor 飞行教官；飞行教员

flight level（FL）飞行高度

flight level change 飞行高度改变

flight line 外场；飞行路线；航线；起机线；机场保养工作地区（机场中包括机库、停机坪及机库周围的地区，不包括跑道和滑行道）

flight log 飞行记录簿；飞行日志

flight log book 飞行记录本

flight management computer（FMC）飞行管理计算机

flight management computer system 飞行管理计算机系统

flight management guidance computers 飞行管理引导计算机

flight management system（FMS）飞行管理系统（飞机上计算机化的自动飞行控制系统）

flight management system procedure（FMS procedure）飞行管理系统程序

flight management unit 飞行管理组件

flight maneuver 特技飞行；空中机动动作；飞行动作

flight manual 飞行手册；飞行条令；飞行教范

flight manual supplement 飞行手册补充；飞行手册补遗

flight mechanics 飞行力学

flight monitoring 飞行监听（从地面监听航行中的飞机）

flight navigation control panel 导航控制面板；航行管理小组

flight officer【英】空军中尉；飞行军官；空勤准尉

flight operation manual 航务手册

flight page 飞行界面

flight path 航迹；飞机的飞行轨迹；飞行路线；（导弹、弹丸等的）飞行路径；弹道；（以指向标波束标志出的）航路

flight path angle 航迹倾角

flight path deviation 飞行航路偏差；航线偏差

flight path vector 飞行轨迹矢量（在 EFIS 上指示的未来飞行轨迹）

flight performance 飞行性能；飞行表演

flight plan 飞行计划（驾驶员在起飞前交给空中交通管制站的计划，包括机型、驾驶员姓名、起飞降落地点、时间、航线、高度、时速等）

flight plan briefing 飞行计划简报

flight plan data 飞行计划数据

flight plan message 飞行计划书；飞行计划电报

flight planned route 飞行计划航路

flight position 飞机投影姿态；飞行位置；准备飞行状态

flight profile 飞行（轨迹）剖面图

flight progress strip 飞行进程记录条（空中交通管制中心用以记载高度、预计到达时间等数据，插入飞行进程板内）

flight recorder 飞行记录仪

flight recorder indicator light 飞行记录仪指示灯

flight recorder pushbutton 飞行记录按钮；飞行记录仪按钮

flight reference card（FRC）飞行参考卡

flight refueler 空中加油机

flight safety 飞行安全

flight scheduling 飞行调度

flight segment 航段

flight service data processing system

飞行服务数据处理系统

flight service station（FSS）航空服务站；飞行勤务站

flight signaler 飞行信号员

flight simulation 飞行模拟

flight simulator 飞行模拟机

flight spoiler control lever 飞行扰流板操纵杆

flight status 飞行资格；飞行状态；适航状态

flight strip 简易机场；条幅式侦察照片

flight strip printer 飞行进程单打印机

flight surgeon 航空医生；飞行医生；航空军医（经过参加过航空医疗培训专门对机组人员进行体检和治疗的医生）

flight technical error 飞行技术误差

flight test 飞行实验；飞行试验

flight test bed 飞行检查台；飞行试验台

flight time 飞行时间

flight track 飞行航迹

flight training device 飞行训练器

flight training outline 飞行训练提纲

flight trim 飞行平衡调整

flight using equipment 飞行用设备

flight visibility 飞行能见度；空中能见度

flight warning computer 飞行报警计算机

flight watch 飞行观察哨

flight with external sling load 吊挂飞行

flight with night vision device 夜视飞行

flight-duration engine 续航发动机

flight-path angle 航迹倾角；弹道倾角

flight-path deviation 偏航；飞行轨迹偏差（实际轨迹与预计轨迹之差）

flight-plan area 飞行计划区

flight-plan correlation 飞行计划相关法（根据已知飞行计划来识别飞机的方法）

flight-test vehicle 试飞飞机；飞行实验飞行器

flightworthy 适航的；适于飞行的；做好飞行准备的

FLIR（forward looking infrared radar）前视红外雷达

float 浮筒（水上飞机起落装置）；

（接地前）平飘;（着陆）平飘距离;（操纵面在气流中）飘浮

float gear 浮筒式起落架

float switch 浮控开关

floating line（水陆两用车辆的）水线

floating lines（立体照相）浮线；（航空照相制图中用的）浮动线

floating mark（测距仪或测高仪的）立标；浮标

floatplane 浮筒式水上飞机

float-type carburetor 浮子式汽化器

flood light 探照灯；澜光灯

flotation gear 防沉浮筒；浮筒起落架；飘浮装备；浮水装置；飘浮装置

flow 气流；液流；流量；流动；流程

flow chart 流动状况图；流程图；工艺程序图

flow control 流量控制;（数据）流控制；交通量控制

flow control valve 流量控制阀

flow controller 节流器

flow correction 流量修正

flow divider 分流器

flow indicator 流量指示器

flow limitation 流量限制

flow limiter 流量限制器

flow pattern 流型；流态

flow reverse 气流反向

flow stop 流动停止

flow tube 流管；文氏管；流量管

flow-control valve 流量控制阀

flowmeter 流量计；流量表

fluctuation 波动；不稳定

fluid film lubrication 流体薄膜润滑；液膜润滑

fluid four 四机流动队形（僚机双机组与长机组之间有疏散的正面及大高度差，以便于机动、观察和相互支援）

fluorescent 荧光的

fluorescent penetrant inspection 荧光渗透检查

flush 齐平；平光；冲洗

flutter 脉动干扰；声音颤动；影像抖动；震动；扰动；摇摆；颤动；抖动

flutter speed 颤震速度

flux characteristic experiment 流量特性实验

flux detector unit（磁）通量探测组件;（中子）流探测器组件

flux valve 感应式磁传感器；磁闸

fly 驾驶（飞机）；飞行；航行

fly heading 飞行方向；航向

flyback period 返程周期；返回周期；回描时间

fly-by waypoint 飞过航路点

fly-by-light 光（动力）飞行；光传操纵

fly-by-wire 电子飞行（用电子仪器操纵飞机飞行）；遥控自动驾驶；有线制导飞行，电传操纵系统

fly-in 飞行集会；（作战中）进入目标区

flying 飞行；驾驶技术；驾驶

flying accident 飞行事故；空中失事

flying attitude 飞行姿态

flying boat 飞船；飞艇；船身式水上飞机

flying equipment maintenance 飞行设备维护

flying fatigue 飞行疲劳；战斗疲劳

flying height 飞机高度；飞行高度；航高

flying officer 空军中尉【英】；飞行军官

flying operation cost 飞行成本

flying position 飞行位置；飞行姿态

flying range 航程；飞行状态范围

flying spot 飞点；扫描点；扫描射线

flying tail 全动平尾；可操纵安定面；（靠）平尾飞行（以整个平尾作为主操纵面的飞行）

flying test bed 空中试验飞行器；飞行试验台；空中实验室

flying to the navigation station 向导航台飞行

flying trim 飞行平衡调整

flying wing 空军联队【英】；飞翼；飞翼式飞机（也称无尾飞机）

flying wires 机翼张线；升力张线

fly-off 飞离；飞过；起飞；飞机样机试飞；飞行竞争评选；竞争性飞行演示（通过几种同类型飞机的飞行演示，以选定采购对象）

fly-over waypoint 飞跃航路点；飞过航路点

flyweight 配重；平衡重

FM（field manual）现场作业手册；战地参考便览；野战手册

FM（frequency modulation） 频率调制；调频（使载波频率受调制信号控制变化，一般用于超短波、微波等高频段）

FMC（flight management computer）飞行管理计算机

FMEP（friction mean effective pressure）摩擦平均有效压力

FMS（flight management system）飞行管理系统（飞机上计算机化的自动飞行控制系统）

FMS procedure（flight management system procedure）飞行管理系统程序

foam carpet 泡沫毯；泡沫铺地（在飞机迫降地带铺设的一层泡沫灭火剂）

foam path 泡沫机场跑道；喷洒泡沫的地带；泡沫路段（为紧急着陆飞机在跑道上喷洒灭火剂泡沫的路段）

foam plastic 泡沫塑料；塑胶泡绵

foam strip（强迫着陆前在跑道上喷洒的）泡沫喷洒带

foaming space 泡沫空间（油箱内，特别是滑油箱内，在油面以上供自由蒸发和形成泡沫的空间）

focal length 焦距

focal plane 焦平面

FOD（foreign object damage）异物损伤；异物操作；外来物损伤（是导致喷气发动机故障次数最多的原因）

fog 雾；（照片）模糊

fog patches 积雾

fohn 焚风

foil 水翼；翼型；翼片；（水翼艇起落架上的）升力面；薄桁（机翼后缘展向构件，分上下两层）；（高空测风用）反射薄片；箔片；薄金属片

folded dipole antenna 折叠式偶极天线；折合振子天线；折合振子

folding 撤收

folding wing 折叠翼（舰载的或只能在机库内停放的飞机上，为减少占用面积而可以上折的机翼）

follow-me vehicle 滑行引导车（引导着陆的飞机至停机坪的车辆）

follow-through takeoff 连续起飞（接地后立即起飞）

follow-up 跟进；后续；继续；随动；跟踪

follow-up device 跟踪装置；随动装置;（火箭）导引头；目标位标器

follow-up system 跟踪系统；随动系统

food ration 食物配给

foot control system 脚操纵系统

foot switch 脚踏开关；脚踏电门

foot-pound 英尺磅（力矩单位，1 英尺磅 =0.138 公斤 · 米）

footstep control switch 踏板控制开关

foray 突然侵袭；突然空袭

force designation 部队番号

force feel 操纵力的感觉；力感；载荷感觉器；感力器

force feel system 载荷感觉系统

force landing 强迫降落；迫降

force multiplier 战斗力倍增器

force of gravity 重力

force trim clutch 配平离合器

force trim release（FTR）配平解除

force trim release switch 配平解除开关

force trim switch 配平开关

force-bearing part 承力部件

forced air ventilation 强制通气

forced airflow 受迫气流

forced landing 迫降；强迫着陆；强行降落；强行登陆

forced landing area 迫降场地

forced oscillation 强迫振动

forced vibration 强迫振动

force-feel flying 动力感应飞行

force-feel unit 动力感应系统；动力感应装置

forces 部队

fore-and-aft movement 前后运动

fore-and-aft retractable tricycle type 纵向收缩三角式

forebody 头部；前机身；弹体前部

forecast 预报；预测

foreflap 襟翼前段（双开缝或三开缝襟翼的最前部分）

foreign matter 杂质；外来夹杂物

foreign object 外来目标；外物；异物

foreign object damage（FOD）异物损伤；异物操作；外来物损伤（是导致喷气发动机故障次数最多的原因）

foreign substance 杂质；外来夹

杂物

forensic specialist 法医学专家

foreplane 前置翼面

forge 铸造

fork 最小夹叉阔度；叉；分叉；分支；岔路口；做扫射摇摆

fork lift truck 叉车；叉式起重车

forked joint 叉形接头

forked lug 叉形接耳

fork-shaped steel plate 叉形钢板

form drag 形状阻力

form line 地形线；轮廓线；等高线略图；地形线图；【美】拟构等高线；地表形态线

format 格式；形式；【美】幅面；像幅（摄影学像片的大小与形式）；图幅（制图中地图、海图或图表的大小与形式）

formation 编队；队形；飞机编队

formation drill 队列训练

formation light（便于夜间编队飞行的）编队灯

formation light switch 编队灯开关

formation of helicopter 直升机编队

former 隔框；翼肋；维形肋；假肋

fortification 防御工事；筑城

forward air control 前进空中指挥；前进空中控制

forward air controller 前进空中控制员；空军前进引导员；前进空军控制员；前方对空控制员

forward ammunition depot（FAD）前进弹药库

forward edge of the battle area（FEBA）战区前沿

forward error correction 前向纠错

forward extension 前伸部位

forward fan engine 前置风扇发动机

forward flight（直升机的）前飞；（垂直起落飞机的）机头向前飞行；水平直线飞行

forward frame 前框

forward fuselage 机身前部

forward left 前左

forward looking aft 从前往后看

forward looking infra-red detection 前视红外线探测

forward looking infrared radar（FLIR）前视红外雷达

forward oblique 前方倾斜航空照相

forward opening hinged-door 前开铰链门

forward overlap 前后重叠；航线重叠

forward power drive 前驱动

forward right 前右

forward scatter 前向散射

forward slip 前滑（飞机在偏转状态时仍按原飞行方向前进的现象）；直滑（修正侧风用）

forward swept wing 前掠翼

forward thrust reverser controls 前反推装置操纵

forward tilt（直升机旋翼的桨叶截面心迹线与旋转面之间的）前倾角；前倾（飞机飞行时的向前倾斜现象）

forward wing 前翼

forward-arming-and-refueling point（FARP）前方武器装备及燃料补给站

forward-bias 正向偏置；顺向偏压

forward-looking infrared（FLIR）前视红外设备

fouling 烬渣；火药残渣；碰撞；毁坏；污秽；污染；故障；工作不良；（仪表）指示不准；形成积垢

four greens 四个绿灯齐亮（起落架和襟翼都已放好）

four-axis automatic flight control system（AFCS）四轴自动飞行控制系统

four-course radio range 四航路无线电导航台；四航道无线电指向标

fourier series 傅氏级数

four-stroke cycle 四冲程循环

four-stroke cycle principle 四冲程循环原理

fourth line maintenance【英】四线维护；四线保障

four-way selector 四位选择器

foxhole 散兵坑；散兵孔

fracto 碎雨云

fractocumulus 碎积云

fractostratus 碎层云

fractus 碎云

fragment 碎片

frame 画幅；帧；架；框架；机架；构架；车架；装弹槽；环形天线

frangible object 易碎物

fray 擦伤；磨损

FRC（flight reference card）飞行参考卡

free air temperature 大气温度；自由空气温度

free airport 自由机场；自由航空港

free balloon 自由气球（任其自由飘浮的气球，与系留气球相对）

free canopy 不可抛座舱盖；自由舱盖

free drop 自由空投；无伞空投

free fall（起落架）自由放下；（降落伞打开前的）自由落下；伞降中降落到预定高度由跳伞员自行开伞或自动开伞；标准重力加速度；自由投放

free frequency 自然频率；固有频率

free gyro 自由陀螺仪

free of damage 免损伤；无损伤；损失不计在内；损坏不赔

free of obstructions 无阻碍

free path 自由行程

free power turbine 自由动力涡轮

free stream 自由气流；未扰动气流；自由水流；未扰动水流

free stream dynamic pressure 自由来流动压

free stroke 自由行程

free turbine 自由涡轮

free turbine speed 自由涡轮转速

free turbine speed（Nr）自由涡轮转速

free turbine turboprop engine 自由涡轮螺旋桨发动机

free wheel 活轮；自由离合轮

free wheel clutch 自由转动离合器；自由离合轮

free wheeling unit 自由转动装置

free wing 自由翼

freedom of the airspace 空间自由飞行权

free-fall altimeter 自由降落高度表；跳伞高度表

free-fall bomb 自由投放炸弹；自由下落炸弹

free-lance escort 机动护航；机动护航队

free-stream flow 自由流

free-stream Mach number 自由流马赫数

free-vortex compressor 自由涡流式压缩机

freewheel 空转轮；游滑轮；自由轮；自由回转

freewheel unit 自转机构；单向离合器

freewheeling 自由转动（关车后直升机旋翼的自动旋转）；自由回转；空回转；自由轮转动（汽车后轮速率较发动机高时，后轮与发动机联系自动分离）；单向转动

freeze 冻结；结冰；凝固；滞塞；卡住

freeze calculated landing time（FCLT）固定计算着陆时间

freeze speed parameter 冻结速度参数

freeze-proofing system 防冰系统

freezing drizzle 冻细雨

freezing fog 冻雾

freezing level 冻结高度；凝固高度；结冰高度

freezing mist 冰冻雾

freezing rain 冰雨；冰粒；冻雨

freight 货物；货运

French landing【口】法国式着陆；后三点飞机的两点着陆（利用两个主轮进行着陆并尽量使尾部不着地）

frequency 频率

frequency adapter 频率匹配器

frequency agile radar 频率捷变雷达

frequency band 频带；频段；波段

frequency change 频率变化

frequency conversion 频率变换；变频

frequency division multiple access 频分多重存取

frequency division multiplex 频分多路传输

frequency management study group 频率管理研究组

frequency management system 频率管理系统

frequency matcher 频率匹配器

frequency modulated continuous wave 调频等辐波；调频连续波

frequency modulation（FM）频率调制；调频（使载波频率受调制信号控制变化，一般用于超短波、微波等高频段）

frequency range 频率范围；频段；波段；波面

frequency setting 频率设置

frequency shift keying 频移键控

frequency spacing 频率间隔

frequency swapping 频率交换

frequencyhopping 跳频

frequency-modulation radar 调频雷达

friction 摩擦；摩擦力

friction damper 摩擦减摆器

friction disk 摩擦片

friction error 摩擦误差

friction horsepower 摩擦马力

friction lock 摩擦锁；摩擦制动器

friction lock of the cyclic 驾驶杆摩擦锁

friction mean effective pressure（FMEP）摩擦平均有效压力

friction measurement 摩擦力测量

friction plate 摩擦片

frictional resistance 摩擦阻力

friendly fire 误伤

Frise-type aileron 弗利兹副翼；阻力副翼

frog-leaping support 交替保障

front attachment 前支点（发动机）

front bearing 前轴承

front bearing lubrication 前轴承润滑

front course sector（着陆航向信标台的）前航道扇区

front cover 前盖板

front face outlet 前面出口

front firewall 前防火墙

front flange 前凸缘；前环带（次口径弹的）

front hoisting rings 前吊挂

front lower fairing 下前整流罩

front mounting 前座；前架

front mounting plate 前安装板

front panel 前面板；仪器面板

front part 前部

front part scavenge pump 前回油泵；前汲油泵

front section scavenge pump 前部回油泵

front support attachment 前支架连接处

front swirl plate 前涡流板

front upper 前上框

front view 前视图；正视图；主视图；正面图；对景图

frontal area 迎面面积

frontal attack 正面进攻；迎面攻击；前向攻击（截击机与目标的航向差大于 135°）

frontal inversion 锋面逆温
frontal system 锋系
frontal weather 锋面天气
frost point 霜点
FRP（fiber glass reinforced plastic）玻璃钢；纤维加强塑料
FS（flap station）襟翼位置
FSS（flight service station）航空服务站；飞行勤务站
FTR（force trim release）配平解除
fuel 燃料；燃油；加燃料
fuel additive 燃油添加剂
fuel and water drain valve 放油（水）阀
fuel boost pump 燃油升压泵
fuel capacity 燃油容量
fuel capacity probe 油量探针
fuel collector 燃油收集器
fuel consumption 燃油消耗量；耗油量；燃油消耗率（每千克或每升燃油可运行的里程或每千米消耗的油量）
fuel consumption rate 燃油消耗率
fuel contamination 燃油污染
fuel control 燃油调节
fuel control unit（FCU）燃油调节器；燃调；燃油控制组件
fuel crossfeed valve 交叉供油开关（阀）
fuel cutoff switch 燃油切断开关
fuel density 燃油密度
fuel dipper 切油；燃油减少器
fuel distribution 燃油分配
fuel distributor 燃料分配盘
fuel drain switch 放油开关
fuel dump system 应急放油系统
fuel economizer system 燃油节省器系统
fuel feeding line 加油线
fuel filter 燃油过滤器；燃油滤
fuel filter blockage indicator 燃油滤堵塞指示器
fuel flow 燃油流量
fuel flow controller 燃油流量控制器
fuel flow decrease 燃油流量减小
fuel flow governor 燃油流量调节器
fuel flow indicator 燃油油量表；燃油流量指示器
fuel flow limit 燃油流量极限
fuel flow limiter 燃油流量限制器
fuel flow meter 燃油流量表

fuel flow rate 燃油流率
fuel flow valve 燃油流量活门
fuel for gas turbine engine（AVTAG）航空涡轮用汽油
fuel gun 加油枪
fuel heater 燃油加热器
fuel hydrant 加油栓
fuel icing 燃油结冰
fuel indicating 燃油量指示
fuel injection 燃料喷射；燃油喷射；直接注油；燃油注射
fuel injection system 燃油喷射系统
fuel injector 燃料喷射器；燃料喷注器；喷油嘴；燃料喷嘴
fuel inlet 进油口；燃油进口
fuel inlet orifice 燃油进口孔
fuel inlet union 燃油进油接头
fuel jettison control card 抛油控制卡
fuel line 燃油输送管路；油路；燃料管
fuel load 载油量；燃料负载
fuel low level sensor 低燃油面传感器
fuel manifold 燃油总管；供油管；燃油歧管
fuel mass 燃油质量
fuel metering unit 燃油流量表；燃油调节装置
fuel metering valve 燃油计量活门
fuel nozzle 喷油嘴
fuel oil mouth 燃油口
fuel on board 机上载油量；机上剩余油量
fuel orifices 燃油管嘴
fuel outlet 燃油出口
fuel pressure 燃油压力；油压
fuel pressure indicator 油压表；燃油压力指示表
fuel pressurising valve 燃油增压活门
fuel pump 燃油泵；燃料注射泵；燃料输送泵；汽油泵
fuel pump drive 燃油泵传动
fuel quantity and gauging system 燃油量和测量系统
fuel quantity gauge 燃油表
fuel quantity gauging system 燃油计量系统
fuel quantity indicating system 燃油量指示系统
fuel quantity indicator 油量指示器
fuel quantity processor unit 燃油

量处理器组件

fuel quantity transmitter 油量传感器

fuel receptacle 加油接头

fuel remaining 剩余燃油

fuel selector manifold 燃油选择歧管

fuel shut-off lever 燃油油路切断操纵杆

fuel siphoning 燃油虹吸

fuel specific consumption（FSC）燃油消耗率

fuel spraying jet 燃油雾化喷嘴

fuel starvation 燃油不足

fuel storage 燃油存储

fuel system 燃油系统；燃料供应系统；供油系统

fuel system computer unit 燃油系统计算机组件

fuel system control panel 燃油系统控制面板

fuel system inlet union 燃油系统进口接头

fuel system management card 燃油系统管理卡

fuel tank 燃油箱；油箱

fuel tank arrangement 燃油箱分布

fuel tank capacity 油箱容量

fuel tank filler cap 燃油箱加油口盖

fuel tank sump area 油箱沉淀区域；油箱积垢区域

fuel valve assembly 燃油活门组件

fuel vapor 汽油蒸汽

fuel venting system 燃油通气系统

fuel-air explosive 油气炸药

fuel-air heat exchanger 燃油与空气热交换器

fuel-air mixture 油气混合物

fuel-air ratio 油气比

fuel-cooled oil cooler 油冷式滑油散热器；燃油冷却式滑油散热器

fuel-cooled oil cooler（FCOC）燃油冷却的滑油散热器；燃油冷却式滑油冷却器

fueling truck 加油车

fuel-oil cooler 燃油 - 滑油冷却器

fuel-oil heat exchanger 燃滑油换热器

fuel-pressure switch 燃油压力开关

full authority digital engine control 全权限数字引擎控制

full authority electronic engine control 全权限电子引擎控制

full authority fuel control 全权限燃油控制
full flap position 全襟翼位置
full flight simulator 全程飞行模拟器
full flow 全流量
full ice protection system 完全防冰系统
full modular 完全单元体
full operation capability 全运行能力
full power takeoff 全推力（最大功率）起飞
full rich 全浓（混合比）；全富油（混合比）
full rudder 全舵；满舵
full scale deflection 全刻度偏转
full strength 满员；全额
full throttle 全开油门；全开节气门；最大油门
full-authority digital electronic control（FADEC）全权限数字电子控制；全权限数字电子控制装置
full-flap takeoff 全襟翼起飞；襟翼全放起飞
full-gross-weight takeoff 全重起飞；最大飞行重量起飞
full-life cycle 全寿命周期
full-life restoration 全寿命恢复
full-loaded takeoff 满载起飞
full-pressure suit（宇航员用的）全压飞行衣
full-stop landing 全跑道着陆（滑跑到全停的着陆）；全制动着陆
full-throttle engine 全油门发动机
full-throttle height 全油门高度
full-throttle takeoff 全油门起飞
fully articulated 全铰接式
fully functioning dual controls 全功能双重控制装置
fully-articulated rotor 全铰接式旋翼
fully-closed position 全关位置
fully-coupled 全耦合的
fully-equipped 全装备的
fully-factored 全因数载荷
fully-open position 全开位置
fully-retractable landing gear 全伸缩式起落架
function 功能；机能；职能；作用；职务；职责；任务；函数
function unit 操纵部件；功能部件
functional check 功能检查；功能校验
functional component 功能部件

functional item number 功能项目号

functional statement 功能描述

functional test 功能测试

fundamental 基本原理；基本；基本的；基础的

fundamental frequency 基本频率；基频

fundamental vibration mode 基阶振型

fundamentals 基本原理

funnel 安全进近区；漏斗区（飞机可在仪表着陆系统下滑道上下 0.5° 和左右各 2° 的区域着陆）；第五边（1950 年前用）

funnel cloud 漏斗云

furnishing 机内陈设

fuse 导火索；引信；保险丝；熔化；熔合；熔解

fuse panel 保险丝板

fuse safety 保险丝安全性；引信安全性

fuselage（飞机）机身

fuselage centerline 机身中心线

fuselage datum 机身水平基准线；机身纵轴

fuselage exterior 机身外部

fuselage frame 机身框架

fuselage reference line 机身基准线（飞机机身对称面上的线，作为飞机各主要部件、设备布局的基准）

fuselage station 机身站位；机身测量点

fuselage strake 机身导流片

fusible plugs 易熔塞（常用于机轮，当刹车过猛使轮胎温度急剧升高时，此塞熔化，使轮胎放气以防爆胎）；保险器；保险丝塞

future air navigation system（FANS）未来空中导航系统

fuze 导火索；引信；保险丝

G

g（G）过载

G layer G 层

G/M（gallons per minute）加仑 / 分

GA（go-around）复飞

gain 增益；放大；增量；【雷】放大率；放大系数

gale 强风；大风；强烈风（蒲福氏风级 7 到 10 级，时速 39 到 75 英里的大风）

gall 擦伤

gallery 走廊；过道；弹药储藏室；平台；控制台；地道（导弹发射场内躲人及存放仪器设备的地下场所）；室内靶场

gallery practice 室内射击练习

galling 擦伤的

gallons per minute（G/M）加仑 / 分

galloper 轻野炮；轻野炮炮车；传令官

galvanization inspection 通电检查

GAM（guided aircraft missile）机载导弹

gamma plot 伽马图（航迹倾角与空速图）

gang control 同轴联动控制；共轴控制器

gantry 高架；起重架；塔架；龙门起重机；龙门起重塔（用于装配与维护大型导弹）

gantry crane 高架起重机

gantry scaffold 导弹竖立和维护塔架；龙门起重塔；台式构架

gap 间隙；间隔；缝隙；通路（部队能通过雷场的地段）；（双翼机或三翼机的）翼隔；【雷】盲区；空白区；突破口；漏照区（未拍照地区或未满足最小重叠部的地区）

gap-filler radar 补盲雷达；盲区弥补雷达

gaping 豁口；张口（飞机上的起落架舱等舱门关闭不严）

garbage can【口】深水炸弹
garnished net 伪装网；隐蔽网
gas absorption material 气体吸收材料；吸毒材料
gas cylinder 气瓶；气压作动筒
gas flow 气流；气体流量
gas generator 燃气发生器
gas generator rear bearing scavenge pump 燃气发生器后轴承回油泵
gas generator turbine 燃气发生器涡轮
gas helmet 防毒面具
gas leakage 漏气；气体泄漏
gas lift 气体升力
gas mask 防毒面具
gas pedal 油门舵；油门踏板
gas tube 充气管；导气管
gas tube joint 导气管接头
gas turbine 燃气涡轮；燃气轮机；燃气涡轮发动机
gas turbine engine 燃气涡流发动机；燃气涡轮机
gas turbine numerology 燃气涡轮站位标识；燃气涡轮站位编号
gas welding 气焊
gas-oil type 油气式
gate 登机门；起飞点；油门最大行程止档；入口；油气比范围；拦截飞行
gate position 停机位；停机站位
gate time 门时刻（飞机进入管制空域新扇区的规定时刻，飞机经过航路某一点的时刻）
gate-type check valve 闸门型单向阀
gauge 仪表
gauge pressure 压力表指示的压力；计示压力；表压；测量压力
Gazelle“小羚羊”（直升机）
G-break 过载曲线的骤变；飞机突然偏离直线航迹（飞机左右偏或突然向上）
GCA（ground-controlled approach）地面管制进场；地面控制进场
GCI（ground-controlled intercept）地面控制拦截
GCI（ground-controlled interception）地面控制拦截
GCO (ground-communication outlet) 地面通信出口设施
G-display【雷】G 型显示器；误差显示器（用于机载截击瞄准雷达）
gear 齿轮；飞机起落架各类工

具、器材、机械的统称；装备；齿轮传动；传动机构；传动装置

gear accident 起落架故障引起的飞行事故

gear extension 起落架放下；放轮

gear pump 齿轮泵

gear retraction 起落架收起；收轮

gear train 齿轮组；齿轮系

gearbox 减速器；变速箱

gear-driven supercharger 齿轮驱动增压器；齿轮传动增压器

geared engine 齿轮传动发动机；齿轮变速发动机

geared propeller 齿轮传动螺旋桨

geared tab（飞机舵面上的）随动补偿片

geared-fan engine 齿轮风扇发动机

gear-up landing accident 未放起落架着陆事故

gee 雷达信号；中距离无线电导航设备；G 导航系统；奇异电子航空仪

gel-filled bomb 凝固汽油（炸）弹

general aviation 通用航空

general aviation airport 通用航空机场

general aviation area 通用航空飞行区

general controlled airspace 普通管制空域

general diagram 总线路图；总图

general lighting switch 总照明开关

general method 通用方法

General of the Air Force【美】空军五星上将

General of the Army【美】陆军五星上将;(【俄】空军、陆军）大将（四星）

general offensive 总攻

general oil vent 滑油总通气口

general purpose bomb（GP bomb）普通炸弹

general terms agreement 通用项目协议

general utility airport 通用机场

general vent line 总排气管

general visual inspection 一般目视检查

generalized mass 广义质量

generalized stiffness 广义刚性

generator 发电机

generator breaker 发电机断电器

generator circuit breaker 发电机断路器；发电机自动保护开关

generator contactor driver 发电机接触器驱动器

generator control relay 发电机控制继电器

generator control unit 发电机控制组件；发电机控制装置

generator drive 发电机传动

generator line contactor 发电机线路接触器

generator line contactor relay 发电机线路接触器继电器

generator transfer contactor 发电机转换接触器

gentle turn 平稳的转弯

geodesic 短程线；测地线；大地线

geodetic construction 短程线结构；最短线网状结构

geographic coordinates 地理坐标（地图上的经纬线）；地理坐标制（以地图经纬线为基础的一种坐标制度）；经纬线交叉法；图上坐标法

geographic fix 地理定位

geographic mile 地理英里

geographic reference points 地理参考点；地理基准点

geographical meridian 地理子午线

geographical pole 地极

geographical seperation 地理隔离

geometric mean chord 几何平均弦长

geometric pitch 几何桨距；几何螺距

geometric twist（机翼的）几何扭转（翼弦与基准面的夹角沿翼展变化）

geometrical boresighting 冷校靶

geopotential height 重力高度；位势高度

geopotential of the tropopause 对流顶层的重力势

geostrophic force 地转力

geostrophic scale 地转刻度

geostrophic wind 地转风

geostrophic wind speed 地转风速

germ weapon 细菌武器

gerotor type pump 摆线齿轮泵；摆线泵；旋转齿轮泵；回转泵；转子泵；内齿轮泵；常压泵

G-force 重力；重力加速度

ghost 幻象；双重图像；重影；假目标；反常回波

GIB（guys in back）后座驾驶员

gills 导流板；鱼鳞片

gill-type cowl flap 鱼鳞式导流板

gimbal 万向支架；万向接头；平衡环；陀螺框架

gimbal freedom 万向支架自由度

gimbal joint 万向接头；常平接头

gimbal lock（陀螺）框架锁定；常平架锁；万向支架锁

gimbal ring 万向悬架环；万向平衡环

girder 桁材；大梁

given load 规定载荷

given threshold 规定数值

GL section（ground liaison section）地面联络组

glare shield 遮光罩；闪光屏档

glass cloth 玻璃布

glass cockpit【口】全显示驾驶舱（舱内用显示屏代替传统的大量仪表）；玻璃座舱

glass reinforced plastic（GRP）玻璃钢

glaze 薄冰层；冻雨；冰凌；机翼结冰；镶以玻璃

glaze ice 雨凇冰

glazed frost 凝霜

glide 下滑；滑翔；下滑曲线；下滑道

glide angle 下滑角；滑翔角

glide bomb 滑翔炸弹

glide path（通信）斜面；下滑道；下滑航迹；下滑轨道

glide path angle 下滑角

glide path intercept altitude 下滑拦截高度

glide path station 下滑线电台

glide path width 下滑航道宽度

glide ratio 下滑比；下滑率

glide slope 下滑道；下滑轨道；下滑道斜率（坡度）

glide-path beacon 下滑（道）信标；下滑信标发射台

glide-path transmitter 下滑道信标发射机

glider 滑翔机；滑翔器；滑翔卫星

glider train 滑翔（机）队列

glide-slope-intercept altitude 下滑道截获高度

gliding 下滑；滑翔

gliding angle 下滑角

gliding distance 下滑距离；滑翔距离

gliding performance 下滑性能

gliding range 滑翔距离；下滑距离

gliding rate 下滑率

gliding ratio 下滑率（滑翔机在飞行中所经过的水平距离与下降的垂直距离之比）

gliding speed 滑翔速度；下滑速度

gliding turn 下滑转弯

glim lamp 暗光灯；微光灯

glint 闪光；反光（常指飞机金属表面反射阳光）;【雷】回波起伏（由目标反射面的迅速变化引起）

GLO（ground liaison officer）地面联络官

global navigation satellite system（GLONASS）【俄】全球卫星导航系统；格洛纳斯卫星导航系统

global orbiting navigation satellite system 全球轨道导航卫星系统

global positioning system（GPS）全球定位系统

GlobalMaster Ⅲ 环球空中霸王Ⅲ运输机

GLONASS（global navigation satellite system）【俄】全球卫星导航系统；格洛纳斯卫星导航系统

glory 日冕；光环

glove 翼套；翼板前罩；附加翼面;（机翼）前缘突齿

glue failure 开胶

glue line 胶层

g-meter 过载指示器；加速度表；载荷因数指示器

gnomonic map projection 球心地图投影

go on board 上飞机

go out（飞机）脱离危险状态；停止工作

go-around（GA）复飞

go-around mode 复飞模式；复飞方式；复飞状态

go-around switch 复飞开关

goggles 风镜；护目镜；夜视镜

goniometer 测角仪；测向计；天线方向性调整器

gooseneck flare 鹅颈灯；柔性灯（装在细片支杆上的跑道灯）

gosport tube【空】通话管（特指飞行教官与学员之间单向通话管）

gouge flap 古奇襟翼（上表面呈圆柱面的后退式襟翼）

gouging（压缩机叶片）挖伤

governor 调速器；调节器

governor valve 调速器活门；调节阀

GP bomb（general purpose bomb）普通炸弹

GPI（ground-position indicator）地面位置指示器

GPS（global positioning system）全球定位系统

GPS receiver 全球定位系统接收机

GPU（ground power unit）地面发电机；地面动力装置；地面电源

GPWS（ground-proximity warning system）近地告警系统

grade of accident 事故等级

gradient 坡度；梯度

gradient wind 梯度风

gradual deceleration 平缓减速

grain 颗粒

grape 霰弹；葡萄弹

graphic parts catalog 图解零件目录

graphic symbol 图解符号

graphite 石墨

graphite petrolatum 石墨矿脂

grass 噪声带；（A 型显示器荧光上的）杂波干扰

grass minimum thermometer 最低草温表

graticule 格子线；网格，坐标图；刻度线

graticule tick（地图上）方格短线（表示子午线与纬线相交处）

graveyard spin 死亡尾旋

graveyard spiral 死亡盘旋

gravity 重力；（地球、月球或行星的）引力；一个 G 力（由于加速或减速产生）；重力加速度

gravity fuel filler cap 重力燃油滤帽

gravity tank 重力油箱

gravity-feed fuel system 重力供油系统

graze 低掠；瞬发；着发；弹着观察；炸点观察；弹无虚发；全部击中；低伸炮火射击

graze fuze 瞬发引信；着发引信

grease 硝化甘油；甘油炸药；润滑脂；润滑油

grease packing 油封

grease seal 油封

great circle course 大圆圈航线（以地球中心为圆心的弧形航

线，为地球表面上两点间的最短距离）

green airway 绿色航路；向东的航路

green glass globe 绿色玻璃罩

green run 绿色运转（发动机制造出来或大修后的第一次运行）；试车；初次试车；试运转

green zone 绿色区域（绿色航线与交叉航线相交而成的区域）

Greenwich Meridian 本初子午线；格林尼治子午线

grenade-thrower 掷弹筒

grid 坐标方格；坐标方格网；栅极；（电池的）铅板；格栅；格框

grid bearing 坐标角限角；坐标方位；网格方位（角）；纵坐标方位角

grid convergence 坐标幅合角（坐标北与真北之间的夹角）

grid convergence factor 坐标幅合比（坐标幅合角与经度差之比）

grid coordinate system 网格坐标系统

grid coordinates 网格坐标

grid direction 网格方向；坐标方向

grid heading 网格航向；坐标航向

grid interval 格网间距；坐标线之间的距离

grid magnetic angle 网格磁角；真北与坐标北之间的夹角

grid navigation 网格导航；方格坐标导航

grid north 坐标北；方格北（军用地图上南北纵坐标指示的北方，通常并非真北）

grid point numerical form 格点数值格式

grid ticks 方格坐标标记；方格坐标符号

grid variation 方格坐标偏移；坐标磁偏角（导航）；坐标北与磁北之差

grinding 磨；研磨

grip 手柄；虎钳；夹具；紧固；握把；枪颈

grivation 磁差；真北与坐标北之间的夹角

groove 沟；槽；阴膛线；阴来复线；环行沟槽

gross ceiling 总升限

gross dry weight 总干重（无油料和特殊液体）

gross flight path 总体航迹

gross height 总体高度

gross overload accident 超载事故

gross performance 总性能（不考虑使用过程中性能降低的情况）

gross profile 飞行剖面图（尤指起飞后的飞行轨迹图）

gross takeoff weight 总起飞重量

gross thrust 总推力；毛推力

gross weight 总重；毛重

ground accident 地面事故

ground alert 地面待机（飞机和飞行人员在地面上处于接到命令即可起飞的战斗准备状态）；地面待机信号；地面待机的（飞机）

ground assault 地面攻击

ground battle 地面战；地面战斗

ground check 地面检查

ground clearance 离地距离；离地高度；（装轮炮架、汽车最低部等）离地空隙

ground clutter 地面杂波

ground clutter suppression 地面杂波抑制

ground connection 接地；接地线

ground control 按地面指令操纵；地面引导；地面制导；地面引导设备；地面制导设备；地面控制；地面校正图；大地控制网

ground controlled approach（GCA）地面管制进场；地面控制进场

ground controlled approach system 地面管制进场系统；地面控制进场系统

ground crew 地勤人员

ground delay 地面延误；地面延迟

ground effect 地面效应；地效

ground environment 地面环境；地面系统；地面通信和电子设备系统；地面防空系统

ground equipment 地面设备；地面保障设备

ground fine pitch 地面小桨距

ground fog 地面雾；低雾（不超过人的高度的雾）

ground forces 地面部队

ground handling（飞机）地面运动操作（包括滑行、牵引等）；飞机的地面一般性维护；（飞机）滑行操纵性

ground handling relay 地面操作继电器

ground hydraulic teststand 地面液压试车台；地面液压试验台

ground idle 地面慢车

ground inspection 地面检查

ground interpretation of flight data 飞行参数的地面判读

ground level helideck 直升机甲板

ground level heliport 地平面直升机机场

ground liaison officer（GLO）地面联络官

ground liaison section（GL section）地面联络组

ground lighting illusion 地上灯光幻视

ground lock 地面卡锁

ground loop 飞机地转（飞机在滑行时或在起飞与着陆滑跑时因失控或其他原因而做的猛烈旋转）；接地回路

ground map 地面图

ground nadir 地面最低点；地面底点

ground observer team 地面观察组；对空监视组

ground operation 地面作业；地面活动；地面作战

ground plug 地面插头

ground position 地面位置

ground power 地面电源

ground power unit（GPU）地面发电机；地面动力装置；地面电源

ground power vehicle 地面电源车

ground pressure（坦克、车辆等）地面压力

ground proximity flight 贴地飞行

ground proximity warning computer 近地警告计算机

ground proximity warning system 近地警告系统

ground radar 地面雷达

ground readiness 地面战斗准备状态（地面机场的飞机于接获命令后能在指定时间内处于地面待机状态时的状态）

ground resonance 地面共振

ground return 地面回波；地面反射波

ground roll 着陆滑跑；起飞滑跑；滑行；地面试车

ground run 地面滑跑；滑行；地面试车

ground run test 地面试车

ground service 地勤；地面勤务

ground service relay 地面勤务继电器
ground service select relay 地面勤务选择继电器
ground service transfer relay 地面勤务转换继电器
ground signal 地面信号；机场地面信号；信号（自地面发射的信号弹）
ground sliding 滑跑
ground speed 地速（飞机飞行相对于地面的水平分速）
ground speed and drift indicator 地速和偏流显示器
ground speed readout 地速读取值
ground spoiler selector valve 地面扰流板选择活门
ground start 地面起动；地面发射
ground station 地面站；地面电台
ground stop 暂停起飞；地面逗留
ground strip 信号布板；信号布标
ground support equipment 地面保障设备；地面辅助设备
ground support personnel 地面保障人员
ground temperature 地面温度
ground test 地面试验
ground time 地面停留时间
ground track 地面航迹；航迹地面投影；实航迹
ground visibility 地面能见度
ground wave 地波（沿地面或接近地面传播的无线电波）
ground wire 搭铁线；接地线
ground-adjustable propeller 地面可调（桨距）螺旋桨
ground-based engine monitoring 基地发动机监控
ground-boosted engine 地面增压发动机
ground-communication outlet（GCO）地面通信出口设施
ground-contolled approach（GCA）地面控制进近
ground-controlled intercept（GCI）地面控制拦截
ground-controlled interception（GCI）地面控制拦截
grounding 停飞；禁止飞行；（电）接地
grounding terminal stud 接地接线柱
ground-mapping mode 地形测绘模式

ground-performance aircraft 地面自行飞机（可以不使用飞行动力装置而在地面移动的飞机）

ground-position indicator（GPI）地面位置指示器

ground-proximity warning system（GPWS）近地告警系统

groundspeed（GS）地速

ground-speed mode 恒定地速飞行方式；地速模式

ground-support equipment（GSE）地面支援设备

ground-to-air communications 地空通信；地对空通信

group 大队（空军或海军建制单位）;（雷）集群目标显示；机群目标显示

group army 集团军

group commander 大队长；机群长；群长

group control vessel 特混大队指挥舰

growth（压缩机叶片）增长

GRP（glass reinforced plastic）玻璃钢

GS（groundspeed）地速

GSE（ground-support equipment）地面支援设备

G-tolerance 过载

guaranteed rating 保证功率；保证推力

guard 防护装置；防护设备；遇险信号传输电路；应急通信波道；保护；预防；守卫；警卫；哨兵;（枪）保险；枪扳机的护圈

guard frequency 防相互干扰波段的频率

gudgeon 耳轴；轴头

guidance control panel 制导控制面板

guidance controller 制导控制器

guidance electronic box 制导电子箱

guidance launch control unit（空地导弹）制导发控单元

guidance radar 制导雷达

guidance system 制导系统

guide apparatus 导向器；制导装置

guide rail 导轨

guided aircraft missile（GAM）机载导弹

guided bomb 制导炸弹

guided missile 导弹

gull wing（飞机）鸥形翼；海鸥翼

gun 炮；火炮；加农炮；高射炮；枪；手枪；机枪；步枪；

【空】快速直线运动；急速飞跃
gun camera 摄影枪；射击记录摄影机
gunner 炮手；枪手；火炮瞄准手；【美】（野战炮兵）瞄准手
gunnery 射击学；射击技术；枪炮操作
guns/weapon free 武器自由攻击
gunship 武装直升机；重型攻击机
gunship helicopter 战斗直升机；武装直升机
gust 阵风；阵雨
gust envelope 阵风包线
gust front 阵风锋面
gust load 阵风载荷
gust locks（操纵系统的）阵风锁定装置
gust velocity 阵风速度
gusty weather conditions 阵风天气
guys in back（GIB）后座驾驶员
gyro 回转仪；陀螺仪；陀螺罗盘
gyro benchmark pulse amplification circuit 陀螺基准脉冲放大电路
gyro compass 陀螺罗盘；回转罗盘；旋转罗盘
gyro drift 陀螺漂移
gyro horizon 陀螺地平仪；航空地平仪
gyro pilot 自动驾驶仪；回转导航装置
gyro precession 陀螺进动
gyro stabilization platform 陀螺稳定平台
gyro topple 陀螺仪倾倒
gyro tumble 陀螺仪翻倒
gyro wander 陀螺漂移
gyrodyne 旋翼式螺旋桨飞机；螺旋桨拉进式直升机
gyrograph 陀螺图
gyrohorizon 陀螺地平仪；回转水平仪
gyromagnetic compass 陀螺磁罗盘
gyroplane 自转旋翼机
gyroscope 陀螺仪；回转仪
gyroscope theoretical analysis 陀螺仪理论分析
gyroscopic inertia effect 陀螺定轴效应
gyroscopic moment 陀螺力矩；回转力矩
gyrosyn compass 感应式陀螺磁罗盘

H

HAA（height above airport） 机场上空高度

hail 冰雹

hailstorm 雹暴

HAL（height above landing） 着陆区上空高度

half ball valve 半球形阀；半球阀

half cut-away 半剖面图

half roll（飞机）半滚

halo 晕轮，晕圈；反光；光环；亮环

halt zone 休息地区

halting of troops 部队中途休息

hammer 锤子；榔头；锻锤；锤击；（武器的）击铁（枪击发火装置的一部分，用以击动撞针或击发火帽）

hammer stall 跃升失速倒转

hand arms 手提兵器（可以随身携带使用的兵器，如手提机枪、手枪、自动手枪、左轮手枪、剑等）

hand cranking 手摇起动

hand flying 手动操纵飞行

hand pump 手动泵

hand-cranked inertia starter 手摇惯性起动器

hand-held fire extinguisher 手提式灭火器

handing over 移交；交接

handling accident 操作故障

handoff 控制转交（地面雷达操作员将目标交给另一个雷达操纵员）

hand-operated control 人工驾驶；手动操纵

handover（飞机）指挥的移交（把对飞机的指挥从一个指挥机构移交给另一机构）

hands on throttle and stick（HOTAS）握杆操纵技术；手不离油门驾驶杆；手不离杆操纵装置设计（格斗飞机中所有按钮、开关、扳机

等操纵装置都集中在油门杆和驾驶杆上）

hands on throttle, collective & cyclic（HOTCC）（直升机）三杆操纵技术

hands-off 松开飞行（无自动驾驶仪的飞机，在配平良好时，飞行员可不参与操纵）；用自动驾驶仪飞行；（地面人员）松开气球吊篮

hands-off flying 放手飞行；松杆飞行

hands-on flying 手控飞行

hands-on stability augmentation system（SAS）手控增稳系统

handwheel 手轮

hang glider 悬挂式滑翔机

hang up 中止（操作）

hangar 飞机棚；机库

hangarette（只能容纳一架小飞机的）小机库；机棚

hangfire 迟缓发射；延迟点火

HAPI（helicopter approach path indicator）直升机进近下滑道指示器

harass 袭扰；高空骚扰（旨在降低敌战斗效能的空中袭击）；接近骚扰（反复接近和通过另一架飞机，以打乱其计划）

hard base 硬基地（能抗核爆炸的导弹发射基地）

hard landing 硬着陆；重跌着陆

hard point（发动机、导弹及其升降装置等的）支点；承力点；运输牵引接头；地下发射场；地下指挥所

hard port 向左急转弯

hard site 防原子发射场；硬式发射阵地

hard standing 停机坪；停车场；堆积场

hard starboard 向右急转弯

hard target 硬目标（难于击毁的目标）；防护坚固的目标

hard turn 急转弯；大坡度转弯

hard wing 硬前缘翼；前缘无缝翼；非柔性机翼

hardened 有核防护设施的；防原子的；地下的；坚固的；硬化的；淬火的

hardened aircraft shelters 加固的飞机掩体

hardened shelter 加固掩体；防原子掩体

hardstand 停机场；停机坪；堆积场

hardstart 硬起动（指火箭发动机燃烧室压力增加过剧的启动）

hardware modification 硬件修正

harmonic 谐波

harmonic vibration 简谐振动

harmonic wave 谐波

harmonious excitation force 谐波激振力

harmonization 校靶；调谐；调整

harsh left pedal input 左舵操纵过猛

harsh pedal input 操舵过猛

HAT（height above touchdown）着陆点上空高度

hatch 舱口；出口；开口；舱口盖；画阴影线

hatch cover 舱口盖

haven 港；锚泊地；安全地

hawkish civilians 主战平民

hazard alert 危险报警（在对飞机的检查中发现有危及飞行安全的故障或潜在故障时，用广播通告所有的有关人员）

hazard beacon 危险信标；险情信标

hazard material 危险品

hazardous 危险的；有害的

hazardous materials（HAZMAT）危险品

hazardous situation 危险情况

hazardous weather information 危险天气信息

haze 霾；雾霾

HAZMAT（hazardous materials）危险品

HC（high capacity bomb）高爆炸弹；高能炸弹

HDD（head-down display）俯视显示（在仪表板上显示）

H-display H 型显示器

head 先头部队；弹头；磁头；向……飞行

head down 俯视

head shaft 主轴；驱动轴

head wind 逆风；顶风

head-down display（HDD）俯视显示（在仪表板上显示）

heading 航向

heading bug 航向游标；航向指示

heading control stick 航向操纵杆

heading crossing angle 航向交叉角（空中截击时截击机航向与目

标航向间的角差）

heading error 航向误差

heading hold 航向保持

heading hold datum 航向保持基准

heading hold mode 航向保持模式

heading indicator 航向指示器；航向显示器

heading lock 航向锁定

heading select 航向选择

heading select feature 航向选择装置

head-on collision accident（空中）对头相撞事故

headphone 头戴双耳耳机；飞行帽耳机；耳机

headquarters 指挥部；司令部；队部

Headquarters of Training & Doctrine（HQ TRADOC）【美】训练与条令司令部

Headquarters of Training & Doctrine Command（HQ TRADOC）（美国陆军）训练与条令司令部

headset 头戴式受话器；带耳机的飞行帽

headset connection 耳机连接

headset microphone 耳机麦克风

headset-jack 耳机插孔；耳机接口

heads-up display（HUD）平视显示（使驾驶员或领航员看清重叠在他的正常视域内的指令信息或情况信息的方法）；平视显示器；情况显示器（飞机风挡前的透明屏幕，其上显示有关飞行与武器发射需要的数据）

head-up guidance computer 平视引导计算机

head-up guidance system 平视引导系统

headwind 顶风；逆风

headwind landing 顶风着陆

health monitoring 状态监控

heat consumption 热耗；热量消耗

heat effect 热效应

heat exchanger 热交换器；散热器

heat fire detector 感温火灾探测器

heat homer 热辐射导引头；热辐射寻的导弹；热辐射自导引导弹；主动寻的装置

heat load 热负荷

heat radiator 散热器

heat shield 隔热罩；热屏蔽

heat shield attachment 隔热护罩附属件

heat sink 散热片；热汇

heat stroke 中暑

heat treatment 热处理

heat weapon 热辐射武器

heater 加热器；发热器；保暖设备

heating 供暖；加热

heating control box 加热控制盒

heating cord 电热丝

heating switch 加温开关

heating system 暖气系统；供热系统

heat-seeker 红外线自动导引头部

heavier-than-air aircraft 重航空器

Heaviside layer 海维赛层

heavy aircraft 重型飞机

heavy helicopter 重型直升机

heavy landing 重着陆；粗暴着陆

heavy lift helicopter 重型空运直升机

heavy machine gun 重机枪

heavy rain 大雨（25.0~49.9 毫米 / 日）

heavy-duty skid 重负荷尾橇

heavy-weight landing 大载重着陆

heavy-weight takeoff 重载起飞；大起飞重量起飞

hectopascal（hPa）百帕；百帕斯卡（气压单位）

height 高；高度；垂直距离；海拔；高地；高处；顶点

height（HT）高度

height above aerodrome 机场上空高度

height above airport（HAA）机场上空高度

height above landing（HAL）着陆区上空高度

height above the touchdown zone elevation 接地区高程以上的高度

height above touchdown（HAT）着陆点上空高度

height above touchdown zone 接地区上空高度

height delay 高度延迟

height envelope 高度包线

height hole 无信号高度区；高度空白区

height indicator 高度表；高度计

height lock 高度锁定

height of the cabin door 客舱门离地高度

height/velocity curve 高度 / 速度曲线

height-finder radar 测高雷达

heli 直升机

heliborne assemblies 直升机机载设备

heliborne data link 直升机机载数据链；直升机机载数据传输

heliborne flight 直升机机降飞行

heliborne machine gun 直升机机载机枪

heliborne mine-laying system 直升机（运载）布雷系统

heliborne missile 直升机机载导弹

heliborne operations 直升机机降作战

heliborne rocket 直升机机载火箭

helical antenna 螺旋式天线

helical gear 螺旋齿轮；斜齿轮；螺旋齿轮传动

helical scan 螺旋形扫描

helical spring 螺旋弹簧

helicopter 直升机

helicopter aiming device 直升机观瞄装置

helicopter air alert 直升机空中待战；直升机空中警戒

helicopter air communications relay 直升机空中通信中继

helicopter air cover 直升机空中掩护

helicopter air delivery capability 直升机投送能力；直升机投射能力

helicopter air formation 直升机空中编队

helicopter air navigation 直升机空中领航

helicopter air reconnaissance 直升机空中侦察

helicopter airdrop flight 直升机伞降飞行；直升机空投飞行

helicopter ambuscade 直升机伏击

helicopter anti-air attack 直升机对空攻击

helicopter anti-surface assault 直升机对地突击；直升机反水面突击

helicopter antitank operations 直升机反坦克作战

helicopter approach path indicator（HAPI）直升机进近下滑道指示器

helicopter approach route 直升机进场着陆航线

helicopter apron 直升机停机坪

helicopter assault force 直升机突

击部队

helicopter attack 直升机攻击

helicopter attack station 直升机攻击阵位

helicopter avionic engineer 直升机航电员；直升机航电师

helicopter avionic maintenance 直升机航电维护

helicopter basic flying training 直升机驾驶基础训练

helicopter basic tactical training 直升机战术基础训练

helicopter breakup point（直升机解除编队）解散点

helicopter carrier 直升机航母

helicopter clearway 直升机净空道

helicopter combat 直升机作战；直升机战斗

helicopter combat support 直升机作战支援；直升机战斗支援

helicopter combat zone 直升机作战区域

helicopter communication 直升机通信

helicopter compartment 直升机装备舱；直升机货舱

helicopter compass calibrating flat 直升机校罗坪

helicopter concealed site 直升机潜伏场

helicopter continuous operational capability 直升机连续出动能力

helicopter crab 直升机侧飞；航向法偏流修正

helicopter deck 直升机降落甲板

helicopter direction center 直升机指挥中心

helicopter dive-attack 直升机俯冲攻击；直升机俯冲轰炸

helicopter drop point 直升机空投点

helicopter electronic countermeasures 直升机电子对抗；直升机电子对抗措施

helicopter electro-optical countermeasure system 直升机光电对抗系统

helicopter evasive flight 直升机规避飞行

helicopter field operational support shelter 直升机野战保障方舱

helicopter field refueling device 直升机野战加油装置

helicopter field support team 直升

机野外保障队

helicopter fire assault capability 直升机火力突击能力

helicopter flight principle 直升机飞行原理

helicopter fuel and ammunition supply point 直升机油弹补给点

helicopter fuel-transporting system 直升机运油系统

helicopter ground alert 直升机地面待机；直升机地面值班

helicopter ground taxi 直升机地面滑行

helicopter ground taxiway 直升机地面滑行道

helicopter ground-effect flight 直升机地效飞行

helicopter hedge hopping 直升机掠地飞行

helicopter hovering 直升机悬停

helicopter hovering flight at sea 直升机海上悬停飞行

helicopter hovering gyration 直升机悬停回转

helicopter hovering transportation 直升机外挂作业

helicopter interception 直升机截击

helicopter landing point 直升机着陆点

helicopter landing site 直升机着陆场

helicopter landing transport ship 直升机登陆运输舰

helicopter landing zone 直升机着陆区

helicopter lane 直升机空中走廊

helicopter load 直升机装载

helicopter loading zone 直升机装载地域

helicopter lurking area 直升机潜伏场

helicopter maintenance 直升机维修；直升机维护

helicopter maneuvering flight 直升机机动飞行

helicopter marshaling area 直升机集结（待发）地域

helicopter mechanic maintenance 直升机机械维护

helicopter mechanical engineer 直升机机械员；直升机机械师

helicopter mine countermeasures squadron 直升机扫雷中队

helicopter mine laying 直升机布雷

helicopter mine-laying system 直升机布雷系统

helicopter mobile operation 直升机机动作战

helicopter mooring 直升机系留

helicopter operation support 直升机作战保障；直升机作战支援

helicopter operational flight training 直升机作战飞行训练

helicopter operational tactical training 直升机战术应用训练

helicopter ordnance maintenance 直升机军械维护；直升机军械保养

helicopter over-the-obstacle flight 直升机超越障碍物飞行

helicopter parking position 直升机停放位置

helicopter performance 直升机性能

helicopter performance calculation 直升机性能计算

helicopter piloting 直升机驾驶

helicopter platform takeoff and landing 直升机平台起降

helicopter power network 直升机电网

helicopter range 直升机靶场

helicopter reconnaissance 直升机侦察

helicopter refueling unit 直升机加油装置

helicopter repair 直升机维修

helicopter re-supply field 直升机补给场

helicopter retirement route 直升机撤出航线

helicopter rotor blade folding 直升机旋翼折叠

helicopter rotor head 直升机旋翼桨毂

helicopter running landing 直升机滑跑着陆

helicopter running takeoff 直升机滑跑起飞

helicopter search and rescue 直升机搜救

helicopter search and rescue device 直升机搜救设备

helicopter self-protection smoke 直升机自防烟幕

helicopter service support 直升机勤务保障；直升机勤务支援

helicopter ship 直升机母舰

helicopter sneak attack 直升机偷袭

helicopter special operations 直升机特种作战

helicopter stand 直升机停放点

helicopter standby zone 直升机待命区域

helicopter stern attack 直升机尾追攻击；直升机尾后攻击

helicopter support team 直升机支援小队；直升机支援组

helicopter sweeping gear 直升机扫雷具

helicopter tactical training area 直升机战术训练场

helicopter tail oblique beam folding 直升机尾斜梁折叠

helicopter takeoff and landing ramp 直升机起落坪

helicopter team 直升机运载小队

helicopter terrain-following flight 直升机地形跟踪飞行

helicopter valley flight 直升机山谷飞行

helicopter weighing record 直升机称重记录

helicopter zone 直升机升降区域

helicopter zoom attack 直升机跃升攻击

helicopter-guided attack 直升机引导截击

helicopter-launched anti-tank missile 直升机发射的反坦克导弹

helideck 直升机坪

helidrome 直升机机场

helidrop 直升机悬停空投

helipad 直升机起降场；直升机起降坪；直升机起降块

helipad assembly area 直升机起降集结场

helipad loading ground 直升机起降承载场

helipod 直升机吊舱

heliport 直升机机场

heliport lighting 直升机场灯光设备

heliport markings 直升机场标志

heliport operating minima 直升机场最低飞行条件

heliport protection engineering 直升机场防护工程

heliport reference point 直升机场参照点

helitroops 直升机机降部队

helium gas 氦气

helix angle 螺旋线角；螺旋角

Hellfire air-to-surface missile“地狱火”空地导弹

helmet 钢盔；飞行帽；飞行头盔；潜水头盔；防护帽；罩；机罩

helmet sight 头盔瞄准器；头盔瞄准具

helmet-mounted sight 头盔瞄准器；头盔瞄准具

helo（口）直升机

hemi-spherical globe 航行灯保护罩

hemoglobin 血红蛋白

hermetically sealed integrating gyro 气密式积分陀螺仪

Hertz 赫兹（任何周期运动的频率，每秒钟循环的次数用法定计量单位赫（Hz）来表示）

hesitation roll 间歇横滚

heterodyne 外差（由两个电学（或声学）信号混合而形成第三个信号，其频率是原先两个信号频率的和或差）

hex bolt 六角螺栓

hex head 六角头；六角扳手

hexagon 六角形；六边形

hexagon rod 六方杆

HF（high frequency）高频

HF radio（high frequency radio）高频无线电

hierarchy 分层结构；层次

HIGE（hovering in ground effect）有地效悬停

high 高空（2.5 万至 5 万英尺的高度）；高气压；高压

high altitude 高空（1 万米以上高度）；高海拔

high blower 高速增压器

high capacity bomb（HC）高爆炸弹；高能炸弹

high cloud 高云

high cycle fatigue 高周疲劳

high density civil transport 高密度民航交通

high density raid 密集空袭

high energy ignition unit 高能点火装置；高能点火器

high energy plug 高能电嘴

high explosives 高爆炸弹；烈性炸弹

high frequency（HF）高频

high frequency radio（HF radio）高频无线电

high impedance 高阻抗

high intensity radiation field (HIRF) 高强度辐射场
high intensity runway light (HIRL) 高强度跑道灯
high key point 高目测点
high oblique 大角度倾斜航空照相；高倾斜航空照相
high octane fuel 高辛烷油料
high oil temperature 滑油高温
high performance 高性能
high pitch 高螺距
high pressure 高压
high pressure bleed valve 高压引气活门；高压放气阀；高压旋气活门
high pressure chamber 高压腔
high pressure compressor 高压压气机
high pressure compressor inlet air temperature 高压压气机进口空气温度
high pressure compressor inlet total air temperature 高压压气机进口空气总温度
high pressure compressor rotor 高压压气机转子
high pressure compressor stator 高压压气机静子
high pressure ground connection 高压地面接头
high pressure recoup 高压补偿
high pressure rotor speed 高压转子转速
high pressure shut off valve 高压关断活门
high pressure turbine 高压涡轮
high pressure turbine active clearance control 高压涡轮主动间隙控制
high pressure turbine clearance 高压涡轮间隙
high pressure turbine clearance control 高压涡轮间隙控制
high pressure turbine clearance control valve 高压涡轮间隙控制活门
high pressure turbine cooling valve 高压涡轮冷却活门
high pressure turbine rotor 高压涡轮转子
high pressure turibine nozzle 高压涡轮喷管
high pressure valve 高压活门
high rate discharge fire extinguisher 高速释放灭火器

high speed taxiway 快速滑行道
high speed turn coordination 高速转弯协调
high strength weight ratio 高力重比
high workload situation 高强度工作负载
high-altitude bombing 高空轰炸（投弹高度超过 1.5 英尺的轰炸）
high-bypass 高涵道比
high-bypass turbofan 高涵道比涡轮风扇发动机
high-energy igniter 高能点火器
higher harmonic 高次谐波
higher mathematics 高等数学
higher vibration mode 高振型
high-flotation gear 软道面起落架
high-flotation tire 软道面轮胎
high-frequency communications 高频通信
high-gain antenna 高增益天线
high-ground-hitting accident 与高地相撞事故（撞在高地上毁坏）
high-incidence stall 大迎角失速
high-indensity light 高强度灯光
high-level air route 高空航线
high-level airway 高空航路
high-level fixed area navigation route 高空固定区域导航航路
high-lift device 高升力装置
high-Mach flight 大马赫数飞行
high-pressure（HP）cock 高压开关；高压旋塞
high-pressure compressor（HPC）高压压缩机；高压压气机
high-pressure compressor（HPC）exhaust temperature 高压压气机排气温度
high-pressure compressor（HPC）inlet pressure 高压压气机进口压力
high-pressure fuel system 高压燃油系统
high-pressure oxygen system 高压供氧系统
high-pressure turbine 高压涡轮
high-pressure turbine（HPT）exhaust pressure 高压涡轮排气压力
high-speed exit 高速出口
high-speed landing 大速度着陆
high-speed stall 高速失速
high-speed taxiway（HST）快速滑行道；高速滑行道
high-speed turnoff 高速脱离跑道点

high-tailed aircraft 高尾飞机（水平尾翼在垂直尾翼上方的飞机）
high-velocity drop 高速空投
hinge 铰链
hinge pin 铰销；铰链轴销
hinge shaft 铰链轴
hingless rotor 无铰式旋翼
hinterland 腹地
HIRF（high intensity radiation field）高强度辐射场
HIRL（high intensity runway light）高强度跑道灯
hit the silk 弃机跳伞
hitting area 命中区
HIWAS（hazardous inflight weather advisory service）飞行中危险天气咨询服务
HNS（hybrid navigation system）混合导航系统
hoist operator seat 绞车操作员座椅
hold 保持；稳住；掌握；控制；扼守（阵地等）；牵制；牵制性攻击；将飞机保持在目视或用其他方法识别的位置附近；正（仍）在指定位置待命；暂停；中断（指火箭发射前准备）
hold fire 停止发射；不要发射；摧毁已发射的导弹
hold for release 等待放飞；暂停发放；暂停投放
hold short 在跑道线外等待
Hold your position! 原地等待！
holder 托架；夹板
holder plate 固定板
holding 等待；待命
holding action 牵制行动
holding apron 等待机坪
holding area 等待空域；待降空域；待命地域；集中地区
holding bay 等待坪；待机坪（机场上飞机滑行、牵引时用于等待的场地）
holding fix 等待位置；等待点
holding off（飞机着陆前）保持平飘
holding pattern 等待航线；待命区
holding pattern mode 等待航线方式
holding point 等待点
holding position 等待位置；待降区
holding procedure 等待程序
holding stack 分层飞行等待着陆

的飞机；（定高盘旋飞行的飞机）机群；编队；令飞机分层盘旋飞行等待着陆；使飞机做高度层次配置

holding war 僵持战

hold-short point 短等待点

hold-short position lights 短等待位置灯

hold-short position marking 短等待位置标记

hold-short position signs 短等待位置符号

hollow shaft 空心轴

hollow spinner（螺旋桨的）桨毂整流罩

hollow vane 空心叶片

home 归航；飞机向电台（导航台）飞行；引导；（导弹）自导引；寻的；驻地机场

home station 原驻地；原驻机场；常驻地；常驻机场

home station training 驻地训练

homer 导航台；引导站；归航信标；归航仪；（飞机上无线电接收机的）附加归航装置；自导引导弹；寻的导弹；导引头；寻的装置；目标坐标仪

homing 导航；归航；向电台飞行；寻的；自导引

homing adapter 归航仪

homing beacon 归航台；归航信标

homing guidance 自导引；自动制导

homing mechanism 归航设备

homing station 归航台

homogeneous atmosphere 均匀大气

homopause 均质层顶

homosphere 均质层

honeycomb 蜂窝结构；蜂窝材料；（风洞内的）格栅；蜂房线圈

honeycomb core 蜂窝芯

hood 暗舱罩；防护罩；座舱盖；整流罩

Hook's joint effect 胡克接头效应

Hook's law 胡克定律

hoop 卡箍；环

hop 起飞与着陆间的飞行；从空中攻击飞机；短途飞行；短时间飞行；试验滑跑时使飞机离地片刻又重新接地

horizon 地平线；视地平线；视地平圈；雷达地平线；无线电地平线；虚假地平线；人工地平仪

horizon bar 摇摆指示器（一串可

变强度的白光灯，成直线排列，为驾驶员夜间起降时提供地平线基准）;（仪表上的）人工地平线标志

horizon camera 地平照相机；水平照相机

horizon line 地平线；水平线

horizon point 水平点

horizontal axis 横轴；水平轴

horizontal component 水平构件

horizontal coordinates 横坐标；水平坐标系

horizontal cross hairs 水平瞄准十字线

horizontal displacement wind 横风

horizontal distance 水平距离;（飞行）视线距离；水平能见距离

horizontal equivalent 水平距离当量

horizontal error 水平误差

horizontal fin 水平安定面

horizontal flight 平飞

horizontal hinge 水平铰

horizontal inbound altitude 水平进场高度

horizontal metering needle rod 水平计量油针杆

horizontal parallax 水平视差；横视差；左右视差

horizontal scanning 水平扫描

horizontal situation indicator 水平位置指示器；航线罗盘

horizontal stabilizer 水平安定面

horizontal stabilizer buttock line 水平安定面纵剖线

horizontal stabilizer tank 水平安定面油箱

horizontal stabilizer trim actuator 水平安定面配平作动筒

horizontal stabilizer trim control system 水平安定面配平控制系统

horizontal stabilizer trim control unit 水平安定面配平控制组件

horizontal surface 水平面

horizontally opposed engine 水平对置式发动机

horn（舵面的）操纵摇臂；微波天线耦合波导管；声发射管（截面变化以控制声阻抗及方向性）;（舵面）突角补偿；突击部

horn balance（飞机舵面的）突角补偿；突角配重

horn lamp 喇叭灯

horn radiator 喇叭形辐射器；号角天线

horn switch 喇叭开关
Hornet fighter “大黄蜂”战机
horsepower 马力
horsepower available 可用马力
horsepower loading 马力载荷（螺旋桨功率除飞机最大起飞重量）
horsepower-thrust ratio 马力推力比
hot and high operations 高温高海拔作业
hot electrode 热电极
hot end（燃气轮发动机的）热端；热部件
hot gas shutoff valve 燃气关断活门；燃气关断阀
hot junction 热触点；热端
hot junction signal 热端信号
hot mike 机内直接通话器
hot photo interpretation report 空中照片判读报告；紧急照相判读报告
hot pressurized air 高温压缩空气；高温增压空气
hot reverse thrust 热反推力
hot section（燃气轮发动机的）热端
hot section inspection 热端检查
hot start 过热起动（由于燃气轮机有超温现象而放弃起动）
hot streak 热斑；过热条纹
hot zone 热区
HOTAS（hands on throttle and stick）握杆操纵技术；手不离油门驾驶杆；手不离杆操纵装置设计（格斗飞机中所有按钮、开关、扳机等操纵装置都集中在油门杆和驾驶杆上）
HOTCC（hands on throttle, collective & cyclic）（直升机）三杆操纵技术
hotshot ignition 热射点火
hour 小时
hour counting check 小时数计数检查
hour fuel consumption 小时燃油消耗量
hour meter 计时表
hours flown 飞行小时
hours of darkness 黑夜时间
hours of day light 白昼时间
housing 外壳；机匣；壳体
hover 盘旋；悬停
hover ceiling 悬停升限；静升限
hover check 悬停检查
hover mode display 悬停模式显示

hover point 悬停点
hover taxi 悬停滑行
hover taxiing 悬停滑行
hovering 悬停
hovering attack 悬停攻击
hovering ceiling 悬停升限；静升限
hovering flight 悬停飞行
hovering in ground effect（HIGE）有地效悬停
hovering maneuver flight 悬停机动飞行
hovering point 悬停点
hovering rotation 悬停回转
hovering turn 悬停转弯
hoverway 直升机跑道
howitzer 榴弹炮
HP（high-pressure）cock 高压开关；高压旋塞
hPa（hectopascal）百帕；百帕斯卡（气压单位）
HPC（high-pressure compressor）高压压缩机；高压压气机
HPC（high-pressure compressor）exhaust temperature 高压压气机排气温度
HPC（high-pressure compressor）inlet pressure 高压压气机进口压力
HPT（high-pressure turbine）exhaust pressure 高压涡轮排气压力
HQ TRADOC（Headquarters of Training & Doctrine）（美军）训练与条令司令部
HQ TRADOC（Headquarters of Training & Doctrine Command）（美国陆军）训练与条令司令部
HST（high-speed taxiway）快速滑行道；高速滑行道
HT（height）高度
hub 桨毂
hub force 桨毂力
HUD（heads-up display）平视显示（使驾驶员或领航员看清重叠在他的正常视域内的指令信息或情况信息的方法）；平视显示器；情况显示器（飞机风挡前的透明屏幕，其上显示有关飞行与武器发射需要的数据）
human engineering 人体工程学
human error accident 操作错误事故；人为事故
human factor principles 人为因素原则

human remains 人体残骸；遗骸

humanitarian assistance 人道主义援助

humidifier 增湿器

humidity 湿度

humidity mixing ratio 湿度混合比

humilis 扁平云

hung start（涡轮喷气发动机的）起动转速悬挂

hunt 摆动振荡；（飞机、导弹）左右摆头；在平均航迹上做纵向长周期波动；纵向摆动；（旋翼桨叶）摆振；搜索

hunter 搜索装置；搜索飞机；探测飞机

hunter-killer 协同反潜；猎潜飞机；猎潜潜艇

hunting 摆动；（旋翼的）摆振；（飞机的）摆头；（飞机沿航迹的）长周期波动；寻找（故障）

hunting hinge 摆振铰链

hunting motion 摆振运动

hurricane 飓风；台风

hybrid compressor engine 混合式压缩机的发动机

hybrid navigation system（HNS）组合导航系统

hydrant dispenser 管线加油车；固定加油栓

hydraulic 液压

hydraulic accumulator 液压蓄电池

hydraulic actuator 液压传动装置；液压作动筒

hydraulic battery 液压蓄电器

hydraulic booster 液压助力器；液压助推器

hydraulic brake 液压刹车

hydraulic brake source 断开液压源

hydraulic cut-out inspection 液压切断检查

hydraulic damper 液压减震器

hydraulic damping component 液压阻尼机构

hydraulic failure 液压故障

hydraulic fluid 液压油

hydraulic fluid recirculation 液压油再循环

hydraulic interface unit 液压接口组件

hydraulic isolation valve 液压隔离活门

hydraulic lines 液压管路

hydraulic lock 发动机被液压锁定；起落架锁定；液压锁定

hydraulic motor 液压马达

hydraulic power 液压发电；液压动力

hydraulic power control 液压助力操纵；液压助力操纵机构

hydraulic power pack 液压动力装置；液压供电装置

hydraulic pressure 液压压力

hydraulic pressure indicator 液压指示器

hydraulic pressure module 液压增压组件

hydraulic pump 液压泵

hydraulic pump failure 液压泵失效

hydraulic quantity indicator 液压油量表

hydraulic quantity interface unit 液压油量接口组件

hydraulic reservoir 液压油箱

hydraulic shimmy damper 液压减摆器

hydraulic system 液压系统

hydraulic system control panel 液压系统控制面板

hydraulic test 液压试验

hydraulic test bench 液压试验台

hydro mechanical unit 液力机械组件

hydrobooster 液压助力器

hydrocarbons 碳氢化合物；烃

hydrodynamic lubrication 流体动力润滑

hydrodynamic lubrication system 液压动力润滑系统

hydromechanical 液压机械的

hydromechanical governor 液压机械调节器

hydromechanical manual control 机械液压手动控制

hydromechanical part 机械液压件

hydromechanical valve 机械液压活门

hydrometeor 水汽凝结体

hydrometer 液体比重计；流速表

hydroplaning 滑水现象（飞机在有浅层积水的跑道上着陆时，机轮与地面附着力急剧下降，甚至使机轮脱离地面）;（水上飞机或水翼船）在水上滑行

hydropneumatic 液压气动的

hydrospinning 液力旋压

hydrostatic testing 流体静力试验

hygrogram 湿度图

hygrograph 自动湿度记录计

hygrometer 湿度计

hygrometry 湿度测定法

hyperbolic fix 双曲线（测得的）位置；双曲线定位

hyperbolic navigation system 双曲线导航系统

hyperbolic position line 双曲线定位

hypersonic 超音速的（马赫数大于 5 的）；超音频的

hyperventilation 换气过度

hypobaric chamber 低氧训练室；低压舱；低气压室

hypoxia 低氧症；缺氧症

hysteresis 磁滞；滞后

IAPS（interim antenna pointing subsystem）（国家航空航天局）临时天线指向子系统

IAS（indicated air speed）指示空速；仪表空速

IAS limit（indicated airspeed limit）表速极限

IAT（indicated air temperature）指示气温

IATA（International Air Transport Association）国际航空运输协会

I-band I 波段

I-beam I 字樑；工字樑

ICAC（International Civil Aviation Convention）国际民用航空公约

ICAO（International Civil Aviation Organization）国际民用航空组织；国际民航组织

ICBM（intercontinental ballistic missile）洲际弹道导弹

ice and rain protection 防冰防雨装置

ice crystal 冰晶；冰晶体；结晶冰屑

ice detector 结冰指示器；测冰仪

ice detector control switch 结冰检测控制开关

ice fog 冰雾；冻雾

ice ingestion 吸入冰；吸入冰块

ice light 结冰信号灯

ice needle 冰针

ice patch 小面积流冰群；流冰区

ice patches 积冰

ice pellet 冰丸；雪珠

ice protection system 防冰；防冰装置

ice rain 冻雨；雨夹雪；霰

ice-melter 除冰车

icing 积冰；覆冰；冰泉

icing condition 积冰条件

icing intensity 结冰强度

icing level 结冰高度

ICP（intermediate contingency power）中间应急功率

IDCS（integrated display & control system）综合显示控制系统

ideal circuit element 理想电路元件

ideal pitch 理想桨距

ideal rotor 理想旋翼

identification 识别；鉴定；鉴别；认证

identification beacon 识别信标；密码信号发射信标

identification beam 识别信标；识别波束

identification between ground force and helicopter 陆空识别

identification button 识别按钮

identification document 认证文件

identification friend or foe（IFF）敌我识别器；敌我识别系统

identification light 识别标志灯

identification maneuver 识别机动

identification panel code 识别板符号

identification plate 标牌；识别牌；标签

idiot light 警示灯

I-display I 型显示器

idle 慢车的；空转的；空载的；无功的；无效的

idle corrected speed 慢车修正速度；慢车换算速度

idle cutoff 慢车活门；空转活门；慢车断油

idle mixture 怠速混合气；慢车混合气

idle speed 慢车转速；怠速

idle thrust 慢车推力

idling 慢车；空载；无载

idling bypass 慢车旁路；慢车辅助油路

idling revolution per minute（idling RPM）慢车转速；每分钟惰转数

idling RPM（idling revolution per minute）慢车转速；每分钟惰转数

idling speed 空转速度；慢车速度

IETP（interactive electronic technical publication）交互式电子技术文件

IFCP（integrated fire control processor）综合火控任务机；综合火控处理器

IFCS（integrated fire control system）综合火控系统

IFF（identification friend or foe）敌我识别器；敌我识别系统

IFR（instrument flight rules）仪表飞行规则

IFR（instrument flight rules）en route high-altitude charts 仪表飞行规则航路高空航图

IFR（instrument flight rules）over-the-top 仪表飞行规则云上目视；暗舱飞行规则

IFR aircraft（instrument flight rules aircraft）仪表飞行规则飞机

IFR approach（instrument flight rules approach）暗舱进近着陆；仪表进场着陆

IFR clearance（instrument flight rules clearance）仪表飞行许可；仪表飞行许可证

IFR conditions（instrument flight rules conditions）仪表飞行规则条件

IFR control services（instrument flight rules control services）仪表飞行规则管制勤务

IFR flight（instrument flight rules flight）仪表飞行；暗舱飞行

IFR landing minimums（instrument flight rules landing minimums）（仪表飞行规则）着陆最低天候条件

IGB（intermediate gearbox）中部减速器；中间减速器

IGE（in ground effect）有地效

igniter 点火装置；点火器；点火剂；点火药；电嘴；火花塞

igniter failure 点火器故障

igniter plug 点火电嘴；火花塞

igniter plug spark 电嘴火花

ignition 点火；引燃

ignition electrode assembly 点火电极组件

ignition exciter 高能点火器

ignition harness 点火线装置；点火线束；点火线外套

ignition plug 点火电嘴；点火塞

ignition pulse current 点火脉冲电流

ignition system 点火系统

ignition timing 点火时间

ignition unit 点火装置

ignition unit support 点火装置支架

illuminate 燃亮；照亮；照射

illuminating bomb 照明弹

illumination 照明；照度；照射

illustrated parts 图解零件

illustrated parts breakdown 图解零件分解

illustrated parts catalog 图解零件目录

illustrated parts catalogue 图解零件目录

illustrated parts data 图解部件资料

illustrated parts data publication 图解部件资料文件

illustrated tool & equipment manual 工具和设备图解手册

illustrated tool and equipment list 详细工具和设备清单

illustrated tools and equipment 图解工具和设备

illustrated tools and equipment publication 图解工具和设备文件

ILS（instrument landing system）仪表着陆系统；盲降系统

ILS categories（instrument landing system categories）仪表着陆系统分类

ILS critical area（instrument landing system critical area）仪表着陆系统关键区域

ILS glide-path angle（instrument landing system glide-path angle）仪表着陆系统下滑道角

ILS localizer（instrument landing system localizer）仪表着陆系统定位仪；仪表着陆系统定位信标

ILS reference datum（instrument landing system reference datum）仪表着陆系统参考基准

IM（inner marker）近距信标；场界信标；内信标；内界标

image motion 图像运动；图像移动；图像位移

image motion compensation 图像运动补偿；图像移动补偿

image motion-compensating magazine 图像运动补偿胶卷盒

image movement 图像运动；图像移动；图像位移

imagery 图像；成像

imagery collateral 图像辅助资料；图像判读参考资料

imagery correlation 图像相关性

imagery data recording 图像数据记录

imagery exploitation 图像利用

imagery intelligence 图像情报

imagery interpretation 图像判读

imagery pack 图像档案

imagery sortie 图像侦察架次；成像架次；拍摄图像架次

imaginary component 虚部

IMC（instrument meteorological condition）仪表飞行气象条件；复杂气象条件

immediate air support 紧急空中支援

immediate photo interpretation report 立即相片判读报告

immediate takeoff 立即起飞

immediately 立刻；立即

impact 碰撞；冲击；弹着；命中；冲力；硬着陆；（火箭的）降落；（气流）在驻点的滞止；影响；效果

impact area 接地区；着水区；弹着区；命中区；弹着面积；炸点范围

impact cylinder 撞击筒；冲击气缸

impact error 弹着点误差

impact fuse 着发引信；触发引信

impact ice 碰撞冰

impact point 弹着点；命中点

impact pressure 冲击压力；碰撞压强

impedance 阻抗

impeller 叶轮

impregnate 防毒浸染；化学浸染剂

improvement 优化；改进；改进措施

impulse starter 冲击起动机

impurity semiconductor 杂质半导体

in cloud 在云中；入云

in direct ratio 成正比；成正比例的

in flight 飞行中；空中

in ground effect（IGE）有地效

in parallel with 与……平行；并联

in place 就位；原地

inactive power source 失效电源

inadvertent damage 因疏忽导致的损失

inaugural flight 首飞；首航

inboard engine 内侧发动机

inboard facilities 舱内设施

inboard flap 内侧襟翼

inboard spoiler 内侧扰流板

inboard stabilizing float 内侧稳定浮筒

inbound flight 向台飞行；归航飞行

incendiary 燃烧弹；燃烧剂

incidence 入射；入射角

incidence wire（双翼机上下翼之间的）斜罩张线

incipient failure 早期故障

inclination of the wind 风向倾角

inclined frequency 倾斜频率

inclinometer 倾斜仪；倾角仪；磁倾仪

incoming flow 来流

incoming missile 接近目标的导弹；进入目标区的导弹

indicated air speed（IAS）指示空速；仪表空速

indicated air temperature（IAT）指示气温

indicated airspeed limit（IAS limit）表速极限

indicated pressure altimeter 指示气压高度表

indicated speed 指示空速；表速

indicating components 指示部件

indicating lamp 指示灯；信号灯

indicating range 指示范围

indicating system 显示系统；指示系统

indication data 显示数据

indication sensor 指示传感器

indicator 指示器；指示符；指示项；指针

indicator light 指示灯；信号灯

induced drag 上升阻力；诱导阻力

induced flow 诱导气流

induced speed 诱导速度

induced velocity 诱导速度

inducement rotational resistance 诱导旋转阻力

inducing power 诱导功率

inductance 电感；感应系数；电感线圈；感应体

induction compass 感应罗盘

inductive feedback 磁感反馈；电感反馈

inductive reactance 磁感电阻；感应电阻

inductive type 感应式

ineffective 无效的

inertia 惯性；惰性；惯量

inertial force 惯性力

inertial guidance 惯性制导；惯性导航

inertial lead vertical speed indicator 惯性引导垂直速度表

inertial navigation accelerometer

惯性导航加速度表

inertial navigation gyroscope 惯性导航陀螺仪

inertial navigation system（INS）惯性导航系统

inertial platform 惯性平台

inertial reference mode panel 惯性基准状态面板

inertial reference system 惯性基准系统

inertial reference unit 惯性基准组件

inertial system display unit 惯性系统显示组件

infiltration 渗透

infinite life 无限寿命；无限使用寿命

infinitesimal calculus 微积分；微积分学

in-flight emergency 飞行中的紧急情况

in-flight normal operation 空中正常工作；空中正常行动

in-flight refueling 空中加油

in-flight relight（发动机空中熄火后）重新起动；空中再次点火

in-flight shutdown 空中停车

inflow （通过螺旋桨的）内流气流；吸入气流；流入量

inflow angle 内流角

inflow angle of attack 入流攻角；入流迎角

inflow stream 来流

INFO（information）情报资料；情况；消息；新闻；信息；信息量；提供情报资料的书面报告、计划、图表、模型及侦察报告等

information（INFO）情报资料；情况；消息；新闻；信息；信息量；提供情报资料的书面报告、计划、图表、模型及侦察报告等

information display button 信息显示器按钮

information display knob 信息显示器旋钮

infrared seeker 红外线寻的制导导弹；红外导引头

infrastructure 永久性防御设施；永久性基地

ingestion 吸入（燃气轮发动机吸入外物）

ingress 入口

inherent frequency 固有频率

initial adaptation 初始调节

initial approach 起始进近；起始进场着陆

initial approach fix 初始进场定位；初始进场点

initial contact 首次发现；初次接触

initial phase 初始相位；初始阶段；初相

initial pressure 初始压力

initial pressure rise 起始增压

initial TBO（initial time between overhaul）首次返修时间

initial time between overhaul（initial TBO）首次返修时间

initialization phase 初始化阶段

initiated bit 起始位

initiation into service 开始服役；交付使用

injection manifold 喷射油道；喷射歧管

injection pressure chamber 喷射压力腔

injection pump 喷油泵

injection system 喷油系统；喷射系统

injector 喷管；喷嘴；喷射器

injector bleed valve 喷嘴放泄活门

injector drain valve 喷嘴放泄活门

injector electro-valve 喷油电磁活门；喷嘴电磁活门

injectors cut-out 喷嘴停喷

inlet guide vane 进气导流叶片

inlet guide vane actuator 进气导向叶片作动筒

inner beacon 近距导航台；近距信标台

inner marker（IM）近距信标；场界信标；内信标；内界标

inner navigation station 近距导航台

inner rotor 内转子

inoperative 不起作用的；不能工作的；有故障的；不工作的

input 输入

input concentrator 输入集中器

input shaft 输入轴

input signal 输入信号

input stage 输入级

INS（inertial navigation system）惯性导航系统

inside diameter 内直径；内径

insignia 标志；徽章；识别符号；勋章

inspection 检查；检验；审查；

查阅；检阅；视察

inspection after the last flight of the day 每日末次飞行后检查

inspection before the first flight of the day 每日首次飞行前检查

inspection socket 检修插座

inspection window 检查窗；检验窗

installation 装置；设备；设施；军事设施；安装；装配

installation angle 安装角

installation configuration 安装结构

installed component 已安装组件；附加组件；可卸组件

instant energy 瞬时能量

instant signal 瞬时信号

instantaneous value 瞬时值

instantaneous vertical speed indicator 瞬时升降速度指示器；瞬时升降速度表

instructional flight in navigation 航行带飞

instructional hover and traffic pattern 悬停和起落航线带飞

instructional technical publication 指导性技术文件；指令性技术文件

instructor pilot 飞行教员；飞行教官；飞行教练

instrument 仪器；仪表；工具；设备

instrument approach 仪表进近；仪表（控制）进场

instrument approach chart 仪表进近图；仪器进场图

instrument approach procedure 仪表进近程序；仪表进场着陆程序

instrument board 仪表板

instrument bus voltage sense unit 仪表总线电压感应组件

instrument check 仪表飞行考核；仪表读数检查

instrument current transformer assembly 仪表电流互感器组件；仪表变流器组件

instrument flight rules（IFR） 仪表飞行规则

instrument flight rules aircraft（IFR aircraft）仪表飞行规则飞机

instrument flight rules approach（IFR approach）暗舱进近着陆；仪表进场着陆

instrument flight rules clearance（IFR clearance）仪表飞行许可；

仪表飞行许可证

instrument flight rules conditions（IFR conditions）仪表飞行规则条件

instrument flight rules control services（IFR control services）仪表飞行规则管制勤务

instrument flight rules flight（IFR flight）仪表飞行；暗舱飞行

instrument flight rules landing minimums（IFR landing minimums）（仪表飞行规则）着陆最低天候条件

instrument landing 仪表降落

instrument landing system（ILS）仪表着陆系统；盲降系统

instrument landing system categories（ILS categories）仪表着陆系统分类

instrument landing system critical area（ILS critical area）仪表着陆系统关键区域

instrument landing system glide-path angle（ILS glide-path angle）仪表着陆系统下滑道角

instrument landing system reference datum（ILS reference datum）仪表着陆系统参考基准

instrument light 仪表灯；仪器灯光

instrument lighting 仪表照明

instrument marking 仪表着陆跑道标志

instrument meteorological condition（IMC）仪表飞行气象条件；复杂气象条件

instrument panel 仪表面板；仪表板

instrument remote control 仪表遥控

instrument source select panel 仪表源选择面板

instrument takeoff 仪表起飞

instrumentation 检测仪表；仪表设备；测试设备；使用仪器；装备仪器；仪表学

insulation 绝缘性；绝热，隔热；隔音；隔离

insulation check 绝缘检查

insulation resistance meter 高阻表

insulator 绝缘体；绝缘子；隔电子；隔离物；隔振子

intake 进气道；进入；引入；输入；吸入物；吸入量；输入量；进气道；进口

intake and exhaust device 进排气装置

integer 整数

integer multiple 整数倍

integral 整体的；完整的

integral calculus 积分

integral function 整函数

integral number 整数

integral type 整数类型

integrated avionics processor system 综合航电处理系统

integrated avionics system 综合航电系统；集成航电系统

integrated circuit 集成电路

integrated display & control system（IDCS）综合显示控制系统

integrated display system 综合显示系统

integrated display unit 综合显示组件

integrated drive generator 综合驱动发电机

integrated engine instrument system 综合发动机仪表系统

integrated fire control processor（IFCP）综合火控任务机；综合火控处理器

integrated fire control system（IFCS）综合火控系统

integrated microelectronics equipment 集成微电子设备

integrated navigation system 综合导航系统

integrated standby flight instrument 综合备用飞行仪表

integrated standby instrument 集成备用设备；综合集成备用设施

integrated suspension management system（ISMS）综合外挂管理系统

integration 集成；集合；综合；综合化；一体化

integrity 完整性；整体性；综合性

intelligence 情报；情报工作；信息；信号；智力；理解力；侦察；情报机构

intelligence official 情报官员

interaction 相互作用；相互影响；【空】干扰

interactive electronic technical publication（IETP）交互式电子技术文件

intercardinal points 基点中间方位

点；象限基点

interchangeability 互换性；通用性；变通性

intercom 内部通话系统

interconnect 互连

interconnected crashworthy fuel tank 连通式防撞燃油箱

interconnection box 连接盒

interconnection strip 连接条；连接片

interconnection terminal 连线接线柱

intercontinental ballistic missile（ICBM）洲际弹道导弹

intercooler 中间冷却器；中间散热器

interdiction 空中封锁

interface 接口；界面；接触面；分界面；交界面

interface control unit 接口控制组件

interference 干扰

interference bomb 干扰弹

interference drag 干扰阻力

interim 临时的

interim antenna pointing subsystem（IAPS）（国家航空航天局）临时天线指向子系统

interior fittings 机上用具；机上配件

interior light 内部照明灯

interlock 联动；连接；结合；联锁装置

interlock switch 联锁开关

intermediate approach 中间进场；中期进场

intermediate connector 中间接头；中间连接器

intermediate contingency power（ICP）中间应急功率

intermediate frequency 中频

intermediate gear 中间齿轮

intermediate gearbox 中间减速器

intermediate joint 中间接头

intermediate pinion 中间齿轮

intermediate pressure 中压；中间压力

intermediate pressure compressor 中压压气机

intermediate pressure turbine 中压涡轮

intermediate range ballistic missile（IRBM）中程弹道导弹

intermediate stage 中间级；中级

阶段

intermediate structure 中间结构

intermediate union 中间接头

intermesh 使互相结合；使互相啮合

intermeshing rotor 交叉式旋翼

intermittent 断续的；间断的；间歇的；中断的

intermittent failure 间歇性故障；断续出现的故障

intermittent rain 间歇雨；间断雨

internal and external checks 内部检查与外部检查

internal circuit 内部电路；内电路

internal combustion engine 内燃机

internal damping 内阻尼

internal damping coefficient 内阻尼系数

internal damping losses 内阻尼损失

internal defect 内部缺陷；内部故障

internal duct 内部导管

internal failure 内部故障

internal fan 内部风扇

internal filler parts 内部填充件

internal force of cross-section 剖面内力

internal fuel duct 内部燃油管路

internal leak 内部泄漏

internal pressure plate 内压板

internal radial clearance 内部径向间隙

internal resistor 内阻

internal restrictor 内部节流嘴，内部限流器

internal screw 内螺纹

internal sealing 内部密封

internal speech system 机内通话系统

internal structural component 内部构件

internal supply pipe 内部供油管；内部给水管

international aeronautical radio 国际航空无线电通信

international air route 国际航线

International Air Services Transit Agreement 国际航空过境协定

International Air Transport Agreement 国际航空运输协定

International Air Transport Association（IATA）国际航空运输协会

international airport 国际机场；

国际航空港

International Civil Aviation Convention（ICAC）国际民用航空公约

International Civil Aviation Organization（ICAO）国际民用航空组织；国际民航组织

international operation incident 国际航线上的飞行事故

international standard atmosphere（ISA）国际标准大气压

international standard atmosphere conditions at sea level（ISA conditions at sea level）海平面国际标准大气条件

interphone 内部通讯装置；对讲（电话）机

interphone control panel 内话控制面板

interphone electronic unit 内话电子装置

interphone system 内话系统

interrogator 询问机；询问器；询问者；审问者；询问机的发射机

intersection（几条）跑道中心线的交点；航路交叉点；跑道与滑行道的交点

intersection heading 交点航向

interservice 涡扇核心发动机与飞行连接的；军种之间的

interservice logistics support 军种间后勤支援

inter-stage coupling 级间耦合

interturbine temperature 涡轮间温度

intervalometer 投弹间隔调整器；摄影间隔调整器；空中照相定时器；时间间隔定时器；电动投弹器

intrinsic boundary 固有界线

intrinsic semiconductor 本征半导体

inventory 存货；物资清单；货单；现有器材；库存品；武器总数

inverse feedback 逆反馈；倒反馈

inverse resistance 反向电阻

inversion 逆温；逆增；倒转；反向；倒置；转换；（降落伞）翻伞

inversion illusion 倒转错觉；倒飞错觉

inverter 变换器；换流器；变流机；倒相器；反相器；转换开关；变换电路

ion 离子

IRBM（intermediate range ballistic missile）中程弹道导弹
iron clad 包铁
ISA（international standard atmosphere）国际标准大气压
ISA conditions at sea level(international standard atmosphere conditions at sea level）海平面国际标准大气条件
ISMS（integrated suspension management system）综合外挂管理系统
isobar(气压图上的）等压线;（风洞模型或自由飞行飞行器上的）等滞止压力线
isochronous piston 同步活塞
isoefficiency 等效率
isogonic chart 等风向线图；等磁偏线图；等磁差图
isogonic line 等磁偏线；等向线
isolation 隔离；隔绝；绝缘
isolation control valve 隔离控制活门
isolation of the faulty component 故障件隔离
isospeed 等速度

J

jack 千斤顶；插孔；插座；插口

jacket 外壳；弹壳；套筒

jacketed bullet 披甲弹

jamming 无线电干扰；电子干扰；人为干扰；卡住；堵塞；交通堵塞

jamming device 无线电干扰设备；干扰器材

jamming war 电子干扰战

jar 震动；抖动；颤抖；噪音；噪声；电瓶；缸罐

Javelin anti-tank missile system【美】“标枪”反坦克导弹系统

JDAM（joint direct attack munitions）【美】联合直接攻击弹药（空/地精确制导武器）

jeep 小型航母；小型侦察联络飞机

jeopardize 损害；危及

jet 喷射；喷注；喷流；射流；喷口；喷嘴；喷射器；射流器；喷气舵；喷气发动机；喷气式飞机；喷气式

jet current 喷气流；射流；喷流

jet engine 喷气式发动机

jet engine aircraft 喷气引擎飞机

jet flow 喷气流；喷流；射流

jet fuel 喷气发动机燃油

jet pipe 喷射管；喷管；排气管；尾喷管

jet pipe extension 排气管延伸段

jet pump 引射泵；喷射泵；射流泵

jet stream 喷气流；西风急流带；急流；极地附近的旋转气流

jet-assisted 喷气助推

jettison mechanism 抛弃装置；抛放机构；投弃装置

jetway 喷气式飞机跑道；喷气式飞机乘客上下走廊；航空旅客桥

joint 联合的（两个以上军种）；联合；军种联合；接头；接合；

连接；组件

joint direct attack munitions（JDAM）（美）联合直接攻击弹药（空 / 地精确制导武器）

joint operation 联合作战；协同作战

joint task force 联合特遣部队

join-up 飞机空中集合

joule 焦耳（能量单位）

journal 阵中日记；军中日记；轴颈

journal nut 轴颈螺帽

joystick（飞机的）操纵杆；控制杆

jump ring of the collective 总距杆卡簧

jump signal 跃变信号；阶跃信号

jump speed 反潜直升机从一个水下测音位置到另一个水下测音位置的平均速度；跳伞速度

jump-off 旋翼机的垂直起飞；发起攻击

jump-off line 进攻出发线

jump-off point 进攻出发点；跳伞点

junction 接合；连接；接头；接合件；接合点；汇流点；交叉点；（半导体的）结；枢纽站；中继；中继线

junction box 接线盒；接线箱；分电箱；电缆分线盒；联轴器

juncture 会合处；结合处；会合

junior officer【陆】连级军官之简称；与其他军官相比，军衔较低的军官

Kalashnikov assault rifle 卡拉希尼科夫突击步枪；卡拉希尼科夫冲锋枪

Kamikaze attack 神风式袭击；自杀式攻击

keep in contact 保持联络

keep in touch 保持联络

Kennelly-Heaviside layer 肯内利—海维赛层；E 电离层

key the mike 接通送话器；按下送话器按钮

KIAS（knots indicated airspeed）节指示空速

kick ahead 缓速前进

kick astern 缓速后退

killability 杀伤力；摧毁能力

killer boat 反潜潜艇

killer hunt 猎潜搜索

killer satellite 反预警卫星；作战卫星

killer ship 反潜舰

killing area 杀伤区；杀伤面积；杀伤范围；火炮有效射击地段；包围区

killing zone 毁伤范围；毁伤区；杀伤破坏范围；歼敌区

kilohm 千欧；千欧姆

kilometer fuel consumption 公里燃油消耗量

kinematics 运动学

kinetic energy 动能

kinetic stability 动力稳定性

kinetics 动力学

Kiowa 卡俄瓦人（美国陆军 OH58A 型轻型侦察机绰号）

Kirchhoff's law 基尔霍夫定律

kit 桶；箱；工具箱；工具包；背囊；装具袋；成套工具；成套设备；成套备件；士兵个人装具；应用器具（如仪器、探毒器等）

kit bag 提包；工具袋

klystron 速调管；速度调制管

knob 球形把手；旋钮；按钮；把手

knots indicated airspeed（KIAS）节指示空速

knuckle joint 叉形接头；铰链接合；肘节

L

L.H. static port 左侧静压口

L/D（lift-to-drag ratio）升阻比

LAAS（Local Area Augmentation System）局域增强系统

labile region 不稳定区

labyrinth seal 篦齿封严

lack of oil 滑油不足；缺油

lack of ventilation 吹气不足

lacrimator 催泪性毒气；催泪弹

lag 滞后，落后，延迟，迟缓；外套，防护套；(曲轴在参考点与活门开闭之间的）角位移

lag dampeners 滞后减震器

lag damper 延迟阻尼器

laminar flow 层流；片流

laminate elastic bearing 层压弹性轴承

laminate elastomer 层压弹性体；层压弹性材料

laminated rotor 叠片式转子

laminated shim 调整垫；多层填隙垫

lamp driver unit 灯驱动组件

LAN digital bus 局域网数字总线

land advance 陆路进攻

land effect 陆地效应；海岸效应

land in depth 纵深着陆

land operation 地面作战

landing 登陆；着陆；降落

landing accident 着陆事故

landing aid 着陆辅助设备

landing approach 着陆进近；着陆进场—指在第五边上

landing area 降落区域；着陆区域

landing beam 着陆（信标）波束

landing chart 着陆图

landing check 着陆检查

landing clearance 落地许可

landing configuration 着陆形态

landing craft 登陆艇

landing decision point（LDP）着陆判定点

landing direction indicator 着陆方向指示器

landing distance 着陆距离（使用场地长度）

landing field 降落场

landing flare 着陆照明弹；机场着陆照明灯火；着陆拉平

landing gear 起落装置；起落架

landing gear actuator 起落装置作动筒

landing gear cable “A” 起落架电缆 “A”

landing gear control handle 起落架操纵手柄

landing gear control unit 起落架控制组件

landing gear controller 起落架控制器

landing gear extension 放下起落架

landing gear microswitch 起落架微动开关

landing gear warning horn 起落架警报器

landing gross weight 着陆总重

landing interval 降落间隔

landing light 着陆灯

landing light selector 着陆灯选择器

landing light switch 着陆灯开关

landing load 地面载荷（指起飞、着陆接地、滑行时间作用在飞机上和跑道上的载荷）

landing mat（飞机）着陆垫板；拆装式金属（跑）道面板；着陆接地标示段

landing minimum 着陆最低气象条件

landing monitor system 着陆监控系统

landing pad 起降场

landing pattern 着陆航线

landing point 着陆（接地点）；着陆站；着陆机场；登陆点

landing roll 着陆滑跑；着陆滑跑距离

landing run 降落滑跑；降落滑跑距离

landing ship 登陆舰

landing shocks 着陆冲击

landing strip 着陆跑道；可着陆地区

landing strut 着陆支杆

landing wire 落地线

landing wires 着陆张线，反升力

张线

landmark beacon 地形标灯

landmark compass 地标罗盘

landmark navigation 地标领航

landscape target 全景靶（供射击练习用）

landside 对陆面；非机场控制区

lane 飞行航线，空中走廊；航路，航道

tangent conical map projection 切圆锥地图投影

Lanyard【美】“牵索”（卫星计划代号）

lanyard 拉火绳（发射火炮等的绳索）；牵索，拉索

large aeroplane 大型航空器

large helicopter 大型直升机

large vertical aperture 大垂直孔径

laser guided bomb（LGB）激光制导炸弹

laser ranger 激光测距机

last-minute check 最后检查

last-minute order 最后命令

latch 止动销

latent heat 潜热（物理）

lateral action 翼侧推进

lateral axis 横轴线；侧轴

lateral balance 横侧平衡

lateral beam sensing point 横梁感受点

lateral cargo door 侧面货舱门

lateral central control actuator 横向中央操纵作动筒（器）

lateral command 同级部队

lateral control actuator 横向操纵作动筒（器）

lateral drift 偏流

lateral fuel supply 横向供油

lateral guidance（着陆系统）侧向引导，侧向制导

lateral loading 横向负荷

lateral moment 横向力矩

lateral movement 横向运动

lateral navigation（LNAV）侧向导航；水平导航

lateral navigation light 侧航行灯

lateral position light 侧航行灯；侧锚位灯；侧位置灯

lateral position limit light 侧航行灯

lateral separation 横向分隔，横向间隔

lateral stability 横向稳定性；侧向稳定性

lateral steering 横向导引

lateral transparent panel 横向透明面板

lateral trim equation 横向配平方程

lateral unbalanced torque 横侧不平衡力矩

lateral vibration 横向振动

lateral windshied demister 侧挡风除雾器（装置）

lathe dog 卡箍

latitude 纬度

lattice type 桁架式

launch a suicide car-bomb attack 发动自杀式汽车炸弹袭击

launch strikes 发动进攻

launcher 发射器；发射装置

lavatory service unit 厕所勤务组件

law of conservation of energy 能量守恒定律

layer tint（地图、海图的）渲彩；分层着色

layout 示意图；平面图

layout of passenger accommodations 旅客座位布局；客舱示意图

LCD display unit（DU）液晶显示单元

LDG lever 起落架操纵杆

leach 过滤；过滤器；滤波器

lead 提前量（射击）；发射偏差；前置瞄准；目标长度；先头间距离；引线；导线

lead element 先头部队

leading current 超前电流

leading edge 机翼前缘；前缘

leading edge cap 前缘包片

leading edge counter weight 前缘配重

leading edge device 前缘增升装置

leading edge flaps 前缘襟翼

leading edge joint 前缘接头

leading pin 引脚

leading wire 引线

leading edge flake 前缘包片

leading edge stall 前缘失速

leading edge vortex 前缘涡流

leading-out terminal 出线端

lead-lag 摆震

lead-lag coupled motion 摆振运动

lead-lag direction 摆振方向

lead-lag hinge 摆振铰链；摆振铰；垂直铰

lead-lag load 摆振载荷

lead-lag motion 摆振运动

lead-lag rigid type 摆振刚硬式

leak into casing 漏至机匣

leak test 泄漏实验，密封性检查
leak test plug 查漏堵头
leap-frogging tactics 蛙跳战术
least significant digit 最低有效数字
leaving 现在离开（陆空通话用语）
LED（light emitting diode）发光二极管
leeward 背风坡；下风处
leeway 风压角；偏航；偏航角
left buttock line 左纵剖线
left engine area 左发动机区域
left hand engine 左发动机
left lower fairing 左下整流罩
left nose 机头左侧
left rear 左后方（侧）
left side view 左视图
left static port 左侧静压口
left-hand finger-four formation 左手四指编队
left-handed thread 左螺纹，左旋螺纹
legend（图标的）说明；图例
legibility 易读性;【计】清晰度
length overall 全长
lens axis 透镜光轴
lethal agent 致死剂；致命性物剂
lethal area 杀伤面积，杀伤区域，致命范围，杀伤范围
lethality（弹头的）摧毁效能；破坏效果；致命性；杀伤力
letter of transfer 调令；调职书
letter order 函令
level attitude 水平姿态
level change 高度层改变
level flight 平飞
level fuselage attitude 水平机身姿态
level of safety 安全水平
level off 恢复平飞，拉平，改平
level sight glass 玻璃液面计
level turning 平飞转弯；水平转弯
level yaw heading hold 水平航向保持器
lever 手柄；杆；摇臂，臂
lever mechanism 杠杆机构
lever strength 杆量
leverage 杠杆系统；杠杆传动；杠杆作用
levers 杠杆系统
levorotation rotor helicopter 左旋旋翼直升机
LF（low frequency）低频
LG control handle 起落架操纵手柄

LG warning horn 起落架警报器
LGB（laser guided bomb）激光制导炸弹
LH side view 左视图
liaise 保持联络
liaison 联络
life cycle cost 寿命周期费用
life limit 寿命极限
life limit removal 达到寿命极限拆卸
life limitation 寿命极限
life limited part 时控件
life preserver 救生装置；救生囊；救生圈；救生衣；救生罐头食物
life raft 救生筏；充气救生船
life vest 救生衣；救生背心
lift 升力，浮力；垂直推力；空运；飞机、车、船的一次载运量；空运用飞机
lift capacity 空运能力，空运量，载重量;（飞艇的）浮力，升力；起重量，起重能力
lift characteristics 升力特性
lift coefficient 升力系数
lift component 升力分力
lift curve slope 升力曲线斜率
lift efficiency 升举性能，升举能力，提弹能力，升弹能力
lift off 抬（提）起，起飞，升空
lift system 升力系统；升力产生装置
lift wire 升力线；升力张线
lifting bracket 吊架
lifting force 升力，提升力
lifting line 升力线
lifting point 吊装点
lifting sling 吊具
lift-off 升空（尤指垂直升空）
lift-off speed 离地速度
lift-to-drag ratio（L/D）升阻比
light 使燃亮；点亮
light aircraft 轻航机，轻型航空器，小型飞机，轻型飞机
light alloy 轻合金
light beacon 标灯
light control 灯火管制
light emitting diode（LED）发光二极管
light fog 轻雾
light gun 通信灯；交通管制灯
light helicopter 轻型直升机
light liaison aircraft 轻型联络飞机
light rain 小雨
light turbulence 轻微扰动

light weapons 轻武器
light wind 微风；轻风
lighting button 仪表照明按钮
lighting circuit 照明电路
lighting system 照明系统
lightning 闪电
lightning protection 雷击保护，避雷
lightning strike 雷击；电击
light-scattering 光散射
light-transmitting control system 光传操纵系统
lightweight aluminum alloy base 轻铝合金底座
limit 极限
limit of acceleration 加速度极限
limit switch 限位开关
limit value 极限值
limitations of engine installation 发动机安装极限
limited ice protection system 极限防冰系统
limiter 限制器
line ahead 一路纵队
line astern formation 两架飞机一前一后的编队；纵队
line contactor 线路接触器；线路开关；电路接触器
line current transformer 线路电流互感器
line current transformer assembly 线路电流互感器组件
line formation 横队，横队队形，“一”字队形
line inspection（of aircraft）（飞机）外场检查
line item 册列项目；空军拨发品
line maintenance handling manual 航线维护操作手册
line of combat 战线
line of position 位置线；定位线
line of sight（LOS）视线，瞄准线；视距；雷达发射天线波束对准目标的直线
line replaceable module 外场可更换（组）件
line replaceable unit（LRU）外场可更换（组）件
line select key 行选键
line to line 线与线之间
line to neutral 线与中性线之间
line troops 战斗部队
line-abreast formation 并列一字编队，一字飞行队形

linear actuator 往复缸，直线驱动器；线性作动筒
linear approach and landing flight 直线进近着陆飞行
linear parallax 线性视差
linear predicative coding 线性预测编码
linear rigidity 线系刚度
linear variable differential transducer 线性可变差动传感器
linear variable displacement transformer 线性可变位移传感器
line-of-sight transmission 直线传播
line-of-sight wave 视距波
line-oriented flight training 航线飞行训练；商业航班飞行训练
link ejection chute 抛链道
link line 连线
linkage 连杆系，链系，联动装置；连接，联动，联锁；传动机构；操纵线系
link-up force 会合部队
lip seal 唇式密封
liquid cooled engine 液冷式发动机
liquid crystal display 液晶显示器
liquid slug shimmy dumper 液弹减摆器
list of applicable publication 应用出版物清单
live ammunition（cartridge）实弹（弹药筒）
live attack 实弹攻击
live axle 驱动轴
living wire 输电线
LNAV（lateral navigation）侧向导航；水平导航
LNAV（long-range navigation）pushbutton 远距导航按钮，远程导航按钮
load 负载；载荷；装填
load applied 外加载荷
load center of gravity extremes 负载重心极限
load classification number 载荷分类编号
load compressor inlet pressure 负载压气机进口压力
load compressor inlet temperature 负载压气机进口温度
load decrease 负载减小
load detector 载荷检测器
load factor 载荷系数；负载因数
load increase 负载增加
load limiter 负载限制器

load message 装载电报
load of the rotor 转子负荷
load sensation mechanism 载荷感觉机构
load sharing 负载分配；负荷分配；功率分配
load sharing device 载荷分配设备
load variation 负荷变化
load waterline 满载水线
load-carrying parts 承力件
loaded limit 载重限度
loaded operation 有载工作
loader 装填手；装弹机；弹仓装填器
loading 装填
loading and unloading apron 装卸停机坪
loading bridge（旅客的）登机桥，登机走廊
loading chart 装载位置图
loading ramp 装卸（货物）跳板，装卸货场，装卸机坪，（机身下）装卸斜坡
loading rod 加载杆
loading unit 加载机构
loadmaster 货运管理员；武器操作员
loadsheet 货物清单，舱单
lobe 波瓣；（气球的）舵囊；喷气发动机最大噪声区的一个部分
LOC（localizer）定位（航向）信标；定位器
local action 局部行动
local apparent time 地方视时
Local Area Augmentation System（LAAS）局域增强系统
local area controller 局部区域控制器
local area differential GNSS 局域差分全球卫星导航系统
local communications network 局域通信网络
local mean time 当地标准时间；地方平时
local overheating 局部过热
local time 当地时间
local traffic control 地方交通管制
local war 局部战争
localization 定位，定域；局域化
localizer（LOC）定位（航向）信标；定位器
localizer transmitter 航向台发信机
locating pin 定位销；瞄准机构螺钉

location by coordinates 坐标定位
locator 定位器；探测器；雷达
locator beacon 飞机定位信标
locator middle marker 中距导航台
locator outer marker 远距导航台
lock blade 锁片
lock number 洛克数
lock nut 锁紧螺母
locked 锁定的
locking base 锁座
locking bolt 锁紧螺栓，背帽；防松螺母；对开螺母
locking disc 锁闭盘
locking mechanism 闭锁装置
locking plate 止动板，锁板；机枪体
locking wire pliers 绞丝钳
lock-release mechanism 闭锁解脱机构
lockwire 安全锁线
log 飞行记录；记程仪
log book 飞行记录；记事簿，履历簿；记程仪，测程仪；航向记录器
log-card 履历卡
logic and analog signal 逻辑和模拟信号
logic tree 逻辑树
logical relationship 逻辑关系
logistics 后勤学
logographic（公司）标志，航徽
long final 长五边
long moment 长力矩
long nose pliers 尖嘴钳
long radius exit 大半径出口
long range cruise 远程巡航
long range navigation（LORAN）（双曲线）远程导航系统，罗兰系统
long range navigation system 远程无线电导航系统
long range radar 远程雷达
long range weather forecast 长期天气预报
longeron 大梁，机身桁梁
longitudinal attitude 纵向姿态，俯仰姿态；俯仰角
longitudinal axis 纵轴
longitudinal coordinates 纵坐标
longitudinal dihedral 纵向翼差角（机翼与平尾装角的差）；（直升机）旋翼轴纵向倾斜度
longitudinal overlap 航向重叠
longitudinal separation 纵向间隔

longitudinal stability 纵稳性；纵向稳定性
longitudinal trim equation 纵向配平方程
long-range cruising speed 远程巡航速度
long-range patrol aircraft（LRPA）远程巡逻机
long-range strike aircraft 远程打击飞机
lookout 警戒，监视；眺望台，观察所，监视哨；观察员，观测手
loop 回路，循环，环路
loop antenna 环状天线，闭路天线
loop-type radio 环形天线导航台
loose snow 软雪
loose thermocouple 热电偶松动
loosen the starting button 松开起动按钮
loot 战利品
LORAN（long range navigation）（双曲线）远程导航系统，罗兰系统
Loran receiver 罗兰接收机
LOS（line of sight）视线，瞄准线；视距；雷达发射天线波束对准目标的直线
loss of life 死亡
loss of power 功率损失，功率损耗
loss of signal（航天器、卫星等）信号消失
lost wax casting 失蜡铸造
lot number 批号
loud bang 发动机放炮
loudspeaker 扩音器，扬声器，喇叭
lounge 候机大厅，（飞机上的）头等舱，高级客舱
low approach 低空进入；低空进场
low cycle fatigue 低循环疲劳
low density conflict 低强度战争
low earth orbit 近地轨道，低高度轨道
low frequency（LF）低频
low fuel pressure 最低燃油压力；燃油压力低
low fuel pressure switch 低燃油压力开关
low fuel pressure warning light 燃油油压过低警告灯
low gain antenna 低增益天线

low idle 低慢车

low impedance 低阻抗

low intensity runway light（LIRL）低亮度跑道灯

low intensity warfare 低强度战争

low key point 低目测点

low level wind shear alert system 低空风切变报警系统

low noise amplifier 低噪声放大器

low oil pressure 最低滑油压力；滑油压力低

low oil pressure switch 低滑油压力开关

low order type of vibration 低阶振型

low pass 低空通场

low pitch 低螺距；小角度螺距

low pressure（LP）低气压，低压，气旋

low pressure chamber 低压舱

low pressure compressor 低压压气机

low pressure compressor rotor 低压压气机转子

low pressure compressor stator 低压压气机静子

low pressure rotor speed 低压转子转速

low pressure turbine（LPT）低压涡轮

low pressure turbine active clearance control 低压涡轮主动间隙控制

low pressure turbine clearance 低压涡轮间隙

low pressure turbine cooling valve 低压涡轮冷却活门

low pressure turbine nozzle 低压涡轮喷管

low pressure turbine rotor 低压涡轮转子

low pressure turbine stator 低压涡轮静子

low range radio altimeter 低高度无线电高度表

low range radio altimeter system 低高度无线电高度表系统

low rate discharge fire extinguisher 低速释放灭火瓶

low rumble 发动机发出低沉的响声

low speed flight envelope 低速飞行包线

lower boundary 下边界

lower center 下部中心

lower centre cowling 中下整流罩
lower centre fairing 下部中央整流罩；中下整流罩
lower display 下部显示器
lower index 下标
lower side band 下边带
lower sideband data 下边带数据
lower sideband voice 下边带音频
lower transparent panel 低处透明面板
low-level wind shear 低空风切变
low-level wind shear alert system（LLWAS）低空风切警告系统
low-pressure fuel system 低压燃油系统
low-speed airflow 低速气流
LP（low pressure）低气压，低压，气旋
LPT（low pressure turbine）低压涡轮
LPT (low pressure turbine) discharge temperature 低压涡轮排气温度
LRPA（long-range patrol aircraft）远程巡逻机
LRU（line replaceable unit）外场可更换（组）件
lubricant 润滑剂（油）
lubricant ejector 滑油喷射器，滑油引射器
lubricated component 润滑部件
lubricating duct 润滑导管
lubricating jet 滑油喷嘴
lubricating system 润滑系统
lubrication 润滑
lubrication oil 润滑油
lubrication splasher 滑油泼溅装置
lubrication system 润滑系统
lug 凸耳，接耳；接线片
lull（风波等）暂息
lumped mass 集中质量
lunar day 太阴日
lung irritant 窒息性毒气
Lynx helicopter“山猫”直升机

M

M（mobilize）-Day 动员开始日
MAC（mean aerodynamic chord）平均空气动力弦
Mach 马赫数；M 数
Mach angle cone 马赫角锥
Mach number 马赫数
machine gun【美】机枪
machine gun pod 机枪吊舱
Machmeter 马赫数指示器；马赫表；M 表
mackerel sky【象】鱼鳞天（漫布卷积云的天）
magazine 弹药库；弹匣
magazine fire 连发射击
magnaflux test 磁流检验
magnesium alloy 镁合金
magnet 磁铁；磁石
magnetic azimuth 磁方位
magnetic bearing（MB）磁方位
magnetic bomb 磁性炸弹
magnetic carbon seal 磁性石墨密封环；磁性石墨封严
magnetic clutch 磁性离合器
magnetic compass 磁罗盘
magnetic dip 磁倾角，磁倾
magnetic field 磁场
magnetic flow 磁通
magnetic heading 磁航向
magnetic inclination 磁倾（角）；磁纬度
magnetic latitude 磁纬度
magnetic level indicator 磁油面指示器
magnetic particle 磁性颗粒；磁性屑末
magnetic pick up 电磁式拾音器；变磁阻拾音器
magnetic plug（MP）磁堵；磁塞
magnetic pole 磁极
magnetic seal 磁性密封
magnetic sealring 磁性密封圈
magnetic variation 磁差

magneto 磁电机，永磁发电机，磁化发电机

magnetoscopy check 磁粉检查

magnifier 放大器，扩大器，放大镜

magnitude 震级；光度；星等

magnitude of burst 爆炸强度

Magnus effect 马格纳斯效应

maiden flight 首航，初次飞行

main air bucket 主空气戽斗

main airscrew hub 主桨毂

main base 主基地（提供作战所需之所有舰船、飞机、人员和物资的基地）

main battery 主电池

main battle tank（MBT）主战坦克（用作主要突破兵器的重型坦克）

main body 本队；主力（战术部队的主要部分）；主体，机身

main bus 总线

main cabin air condition 主舱空调

main column 主力纵队

main component 主要部件

main compressed air bottle 主压缩空气瓶

main electric plug 总电气插头

main engine control 主发动机控制

main engine start 主发动机起动

main equipment center 主设备舱

main float 主浮筒

main force（s）主力

main fuel drain valve 主油路放油活门

main gear emergency extension system 主起落架紧急放下系统

main gearbox（MGB）主减速器

main injection 主喷

main injection system 主喷油系统

main landing gear 主起落架

main mast 主轴

main pipe 总管

main points concerning the maintenance 维护要点

main power drive 主动力传动（装置）

main probe 主探头

main rotor axis 主旋翼轴

main rotor damper indicator 主旋翼减震显示器

main rotor drive 主旋翼驱动；主旋翼传动装置

main rotor drive system 主旋翼驱动系统

main rotor head 主旋翼头
main rotor mast 主旋翼轴
main rotor RPM 主旋翼转速
main rotor shaft 主旋翼轴
main shaft 主轴
main shaft drive pinion 主轴驱动小齿轮
main stop 主停止
main tank 主油箱
main transmission 主传动；主传动系
maintenance 机务工作；维护工作；保养；维护
maintenance aid 维修辅助设备
maintenance company 维修连
maintenance concept 维护概念
maintenance control center 维护控制中心
maintenance control display panel 维护控制显示面板
maintenance data terminal 维护数据终端
maintenance day 车场日
maintenance diagnostic computer 维护诊断计算机
maintenance documentation 维护手册
maintenance engineering manual 维修工程手册
maintenance facility and equipment planning document 维修设施和设备计划文件
maintenance general 维护概要
maintenance job card 维护工作卡
maintenance job order 维护工作单
maintenance management system 维护管理系统
maintenance personnel 维修人员
maintenance planning 维修计划
maintenance planning data 维修计划数据
maintenance planning data documents 维护计划数据文件
maintenance planning information 维护计划信息
maintenance processor subsystem 维护处理机分系统，维护处理子系统
maintenance publication 维护文件
maintenance schedule 维修计划；维修作业表
maintenance significant item 重要维修项目
maintenance squadron 维修中队

maintenance support 维护支援
maintenance task card 维护工卡
maintenance team 维修小组
maintenance terminal display 维护终端显示器
maintenance troops 地勤人员；维修部队
maintenance work 保养工作
maintaining 正保持（陆空通话用语）
major 少校（美陆、海、空、海军陆战队）
major combat 主力战
major damage 严重损坏
major failure 主要故障；严重故障
major general 少将（美陆、空、海军陆战队）
major overhaul 大翻修
major world air route area 世界主要航路区
malfunction 故障；失灵
malfunction analysis 故障分析
malfunction procedure 故障程序
management procedure 管理程序
management unit 管理组件，管理单元
mandatory check 强制检查
mandatory maintenance inspection 强制维护检查
maneuver 机动；军事演习
maneuver ability 机动能力
maneuverability 机动性；可操作性，驾驶性能
maneuvering area 操作区
manifest 运（舱）单
manifold 歧管，多支管，排管；总管，集流环
manifold pressure gauge 歧管压力表
man-machine interface 人机接口
man-made accident 人为事故
manning drill 炮手班就定位练习；射击操作练习
manning level 人员配备标准
manning table 人员职务表
manoeuvre 调动部队；机动动作；机动；演习
manoeuvre helicopter in ground effect 有地效机动直升机
manometer 压力计，压力表；流体动力学
man-portable weapon 便携式武器
manual activation 手动激活
manual control 人工操作；人工

控制
manual control knob 手控旋钮
manual control of engine power 发动机功率手动控制
manual control system 人工操纵系统，人力操纵系统
manual deployment unit 人工放出组件
manual entry device 人工输入器
manual fuel flow control 手动燃油流量控制器
manual mode 手控方式；人工操纵工作方式
manual regulator 手动调试器
manual reset 手工复位
manual speed 人工速度
manual start sequence 手动开车程序
manufacture's empty weight 制造空重
manufacturer 生产厂家
manufacturer's supply 制造商提供
manufacturing material 制造材料
manufacturing tolerance 制造公差
MAP（missed approach point）复飞（起始）点
map overlay 覆盖地区
MAR（minimally attended radar）最低程度人力控制雷达
march-past 分列式（阅兵）
marginal takeoff 安全边缘起飞（没有足够安全保证的起飞）
marginal visual flight rule（MVFR）边缘视觉飞行规则
marine climate 海洋性气候
marine radio 航海用无线电台
mark 标记
marker 信标台
marker beacon 信标台；无线电信号发射台
marshaling 集结待发
martial rule 军事管制
martyr 殉难者
martyrdom 殉难
mass addition 加质量
mass media 大众传播媒体
mass point 质点
mast 桁；桅；轴
master caution light（MCL）主警告灯；主提示灯
master control station（全球定位系统）主控站
master minimum equipment list（MMEL）最低设备总目录

master plan 总计划
master sergeant（M.Sgt.）军士长（美陆、空、海军陆战队）
master switch 主控开关；总开关
master warning light（MWL）主警告灯
matchmark 零件（配合）记号，（零件）相配记号
match-up training 配组训练
material data 材料数据
material failure accident 设备故障飞行事故；装备制造缺陷引起的故障
material point 质点
MAU（modular avionic unit）单元体航电组件；可互换航电组件
max emergency rating 最大应急状态
max air tapping flow 最大引气流量
max contingency 最大应急
max contingency power 最大应急功率
max contingency rating 最大应急状态
max continous power（MCP）最大连续功率
max continuity 最大连续性
max continuous 最大连续（推力）
max dry crank time 最大冷转时间
max emergency 最大应急；最紧急
max flow taken 最大引气量
max operating time 最大工作时间
max over-temperature 最大超温
max power available 最大可用功率
max starting 最大起动
max takeoff power 最大起飞功率
max takeoff thrust 最大起飞推力
max torque 最大转矩
max torque limit 最大扭矩限制
max torque limitation 最大扭矩限制
max tropical zone 热带极限区域
max voltage drop during start 最大起动电压下降
maximum authorized IFR altitude 最大允许仪表飞行高度
maximum climb 最大爬升
maximum climbing 最大爬升率
maximum continuous thrust 最大连续推力
maximum crossing altitude（MCA）最高穿越高度
maximum cruising rating 最大巡航功率额定值

maximum cruising speed 最大巡航速度

maximum diameter 最大直径

maximum endurance 最大续航时间

maximum glide distance 最大滑翔距离，最大下滑距离

maximum landing gear extended speed 最大起落架放出速度

maximum landing gear operating speed 最大起落架操作速度

maximum landing weight 最大着陆重量

maximum level speed 最大平飞速度

maximum loiter time 最长巡逻时间

maximum Mach operating speed 最大飞行马赫数

maximum operating speed 最大飞行速度，最大使用空速

maximum payload 最大业载

maximum power 最大功率；最大动力

maximum quantity 最大量

maximum ramp weight 最大滑行重量，最大停机坪重量

maximum range 最大航程

maximum range ability 最大射程

maximum rate of climbing 最大上升率

maximum rate of descending 最大下降率

maximum routing range 最大航程

maximum speed 最大速度

maximum speed for stability characteristics 最大稳定特性速度

maximum stock level 最高存量；最高库存标准

maximum takeoff power 最大起飞功率（推力）

maximum taxi weight 最大滑行重量

maximum zero fuel weight 最大无燃油重量

maximum-performance takeoff 最大性能起飞

maximun power assurance 最大功率保证

mayday 求救信号；无线电话求救信号；国际遇险无线电呼救信号

MBT（main battle tank）主战坦克

MCDU（multi combining and distri-

bution unit）多功能控制显示组件
MCDU（multifunction control display unit）多功能控制显示单元
MCL（master caution light）主警告灯；主提示灯
MCP（max continous power）最大连续功率
Mcrit（critical Mach number）临界马赫数
MD（minimum descent）最小下降
MDA（minimum descent altitude）最低下降高度；最小下降海拔
MDH（minimum descent height）仪表下降的最底高度；最低下降高度
mean aerodynamic chord（MAC）平均空气动力弦
mean line 翼型中线
mean sea level（MSL）平均海平面
mean solar day 平太阳日
mean temperature 平均温度
mean time between alarm（warning）平均告警间隔时间
mean time between failure 平均故障间隔时间；出现故障平均间隔时间
mean time between overhauls 平均翻修间隔时间，平均翻修时限
mean time between removals 平均更换间隔时间；平均移动间隔时间
mean time between unit replacement 平均更换故障单元间隔时间
mean time to failure 平均故障时间；故障发生平均时间
mean time to maintenance 平均维护时间；平均维修时间
mean time to restore 平均恢复时间
mean time to scrap 平均报废时间
message understood 明白（空中交通管制术语）
measuring tool 量具
mechanic 机修工；机械员
mechanical advantage 机械利益；（飞机操纵系统）传动比
mechanical brake system 机械制动系统，机力制动系统
mechanical damper 机械阻尼器
mechanical drives 机械传动装置
mechanical energy 机械能
mechanical failure 机械故障；机械破坏
mechanical fuel control 机械燃油

调节器（机械燃调）
mechanical inspection 机械检查
mechanical linkage 机械联动；连杆机构
mechanical lock 机械锁
mechanical magnetic plug 机械式磁堵；机械磁性堵头
mechanical malfunction 机械故障
mechanical preparation 机务准备
mechanical pump 机械泵
mechanical reeling 机械卷轴
mechanical stop 机械止点
mechanical transmission 机械传动
mechanisation 机理
mechanism 机械装置
mediator 介质；调停人，调停国
medical attendant seat 医护人员座椅
medical equipment rack 医疗设备架
medical kit 医疗箱
medium frequency（MF）中频
medium intensity runway light（MIRL）适中强度跑道灯
medium turn bank 中坡度转弯（坡度 25°~45°）
medium-intensity approach lighting system 中级亮度进近照明系统
medium-range weather forecast 中期天气预报
medium-size helicopter 中型直升机
megohm 兆欧
MEL（minimum equipment list）最低设备单（列出飞机上为执行任务所必不可少的设备）
melting snow 在融化的雪
memo 备忘录
memorandum of understanding 备忘录
memory module 储存模件
mercury 水银（柱）
meridian 子午线；子午圈
meridian of celestial sphere 天体子午线
meshing 啮合（齿轮）齿
mesosphere 中间层
message block identifier 信息块指示器
message number 电报号数
message priority 电报等级
message tape 电报纸条
messages in real time 实时信息
MET（meteorology）气象
metal articulated 金属铰接式

metal cartridge 金属筒；金属壳
metal particle 金属屑，金属微粒
metallic film 金属膜
metal-oxide semiconductor 金属 - 氧化物半导体
METAR（aviation routine weather report）航空例行天气报告
meteorology（MET）气象
meteorological ceiling 气象升限
meteorological information for aircraft in flight（VOLMET）对空天气广播
meteorological instability 气象不稳定性
meteorological knowledge 气象知识
meteorological message 气象电报；气象通报
meteorological report of aerodrome conditions 机场条件气象报告
meteorological satellite 气象卫星
meteorological watch office 气象值班室
meteorology 气象学
metering 计量
metering needle 计量油针
metering needle actuator 计量油针调节器
metering unit 计量装置
metering valve 计量阀，节流阀
MF（medium frequency）中频
MFCD（multi-function colour display）综合彩色显示器
MFD（multi-function display）多功能显示器
MFK（multi-function keyboard）多功能键盘
MGB（main gearbox）主减速器
MGB cowling 主减整流罩
MGB fairing 主减整流罩
MGB oil pressure 主减滑油压力
MGB power input 主减功率输入
MGB skin 主减整流罩
MGB suspension 主减速器悬挂件
MH（minimum height）最小高度
MIA（minimum instrument altitude）仪表飞行的最低安全高度
mica 云母
microbar 微巴（物理单位中用厘米 - 克 - 秒（CCS）表示的压力单位）
microburst 微型下冲气流
micrometer 测微表；测微计；测微器

micron 微米
microphone 话筒；麦克风；扩音器
microprocessor 微处理器
microscopic inspection 显微检验
microswitch 微动开关；微动电门
microwave 微波
microwave landing system（MLS）微波着陆系统
mid density conflict 中强度战争
mid-air collision accident 空中相撞事故
middle assembly piece 装配中段
middle cabin door 中舱门
middle cloud 中云
middle compass locator 罗盘中距指点标
middle marker（MM）中指点信标
mike 话筒；麦克风
mild hypoxia（血液的）轻度缺氧
miles to go 待飞英里数
military action 军事行动
military aircraft 军用飞机
military alliance 军事联盟
Military Area Command 军区司令部
military buildup 军事集结
military capability 军事能力
military checkpoint 军事关卡，军事检查站
military exercise 军事演习
military expenditure 军费，军事支出
military helicopter 军用直升机
military law 军法
military maneuver 军事演习
military operation 军事行动
military operation area（MOA）军事作战区域；军事行动区域
Military Region Command 军事地区司令部
military reserve 军队后备军
military retaliation 军事报复
military science 军事科学
military spending 军费，军事开支
millibar 毫巴
millimeter wave 毫米波
millivolt 毫伏（特）
millivoltmeter 毫伏计，毫伏表
mine clearance 排雷
mini guaranteed power 最小保证功率
mini guaranteed value 最小保证值
mini oil pressure switch 最小油压

开关

mini rotation speed 最小转速

mini voltage to start 最低起动电压；最小起动电压

miniature 缩图，微型

minimal flight path 最短时间飞行路线

minimally attended radar（MAR）最低程度人力控制雷达

minimum aeronautical system standard 最低航空系统标准

minimum descent height（MDH）最小下降高度

minimum control speed 最小操纵速度

minimum control speed with the critical engine inoperative 关键发动机不工作时的最小操纵速度

minimum crossing altitude 最低穿越高度

minimum descent（MD）最小下降

minimum descent altitude（MDA）仪表下降的最底高度；最低下降高度

minimum diameter 最小直径

minimum discernible signal 可辨别的最小信号

minimum en-route altitude 最低航路高度

minimum equipment list（MEL）最低限度装备清单；最低装备需求手册

minimum fuel 最低油量

minimum ground control speed 最小地面操纵速度

minimum ground turning radius 地面最小转弯半径

minimum guaranteed power 最小保证功率

minimum guaranteed value 最小保证值

minimum height（MH）最小高度

minimum instrument altitude（MIA）仪表飞行的最低安全高度

minimum interval takeoff 最小起飞间隔时间

minimum navigation performance specification（MNPS）最低导航功能规范

minimum obstruction clearance altitude（MOCA）最低无障碍高度

minimum of the steering radius 最小转弯半径

minimum of the turning radius 最小转弯半径

minimum oil pressure switch 最小油压开关

minimum operational performance requirement 最低运行性能要求

minimum operational performance standard 最低运行性能标准；最低运行性能规范

minimum rate of descent 最小下降率

minimum reception altitude（MRA）最低接收高度

minimum rotation speed 最小转速

minimum safe altitude（MSA）最小安全高度，安全飞行最低高度，最低安全高度

minimum safe altitude warning 最低安全高度警告

minimum sector altitude 最小搜索区高度, 最低防区高度

minimum shift keying 最小移频键控；最小移位键控

minimum steady flight speed 最小稳定飞行速度

minimum takeoff safety speed 最小起飞安全速度

minimum unstick speed 最低离地速度

minimum use specification 最低应用规格

minimum vectoring altitude（MVA）最低引导高度

minimum voltage to start 最低起动电压

mini-type electrical appliances 小型电器

minor accident 三等事故（一般事故）

minor failure 小故障；次要故障；轻度故障

minor repair 小修

minus（温度）零下的；负数；负号

minute hand 分针

mirage【象】海市蜃楼；蜃景；幻景

Mirage 幻影飞机；幻影直升机

Mirage 2000-5 fighter 幻影 2000-5 战机

MIRL（medium intensity runway light）适中强度跑道灯

misaligned contact 错开了的接触点

misalignment 不对准（直线）；不同心度；不同轴度；不对中；角（度）偏差；失调；校准不当

miscellaneous control panel 多功能控制板

miscellaneous indication 综合显示

mishandling 处理不当；错误操纵

misplacement 错位

missed approach 进场失败；中断进场并复飞

missed approach point（MAP）复飞（起始）点

missile carrier 导弹运载机（由大型客机如波音 747F 等改装而成，可载四枚洲际导弹或八枚中型导弹）

missile launch set 导弹发射器

missile launcher 导弹发射架；导弹发射装置

missile locker 锁弹器

missile rack 弹架

missile retainer 挡弹器

mission flight 任务飞行

mist 薄雾，很差能见度（能见度 4 级，距离 1 000~2 000 米）

mixed mode 混合模式

mixer 混合器

mixer unit 混合器壳体；混合器组件

mixing unit 混合装置

MLS（microwave landing system）微波着陆系统

MLS azimuth 微波着陆系统方位

MLS glideslope 微波着陆系统下滑道

MM（middle marker）中指点信标

MMEL（master minimum equipment list）最低设备总目录

MNPS（minimum navigation performance specification）最低导航功能规范

MNPS airspace 最低导航性能规范空域

MOA（military operation area）军事作战区域；军事行动区域

mobile satellite service provider 移动卫星业务提供者

MOCA（minimum obstruction clearance altitude）越障最低安全高度；最低障碍准许高度

mock-up 实体模型；样机；模型飞机

modal 模态的，模式，情态

modality 模态

mode 模式
mode C intruder 装有 C 模式应答器的入侵飞机
mode check 模式检查
mode control 模式控制
mode control panel 模式控制面板
mode function 振型函数
mode of action 作用方式
mode select unit 模式选择单元
mode switch 方式电门
moderate turbulence 中度颠簸
moderate wind 中速风
moderate windshear 中度风切变
modification 修改，改进；改型，改装
modification index 修正指标
modular avionic unit（MAU）单元体航电组件；可互换航电组件
modular avionics warning electronics assembly 模块化航电警告电子组件
modular concept unit 单元体概念（的）组件
modular construction 单元式结构
modular layout 单元体设计
modular radio cabinet（MRC）模块无线电舱
modular radio system 单元体无线电系统
modularity 单元体性；模块性
modulated pressure 调制压力；调定压力；制动压力
modulated wave 调幅波，已调波
modulator 调制器，调幅器
modulator-demodulator 调制解调器
module 组件；模件；微型组件；模块；单元体
module removal and installation 单元体拆卸和装配
module repairs 单元体修理
module replacement 更换单元体
module test and maintenance 模块测试和维护
module test and maintenance bus interface unit 模块测试和维护总线接口单元
moisture-proof plug 防潮塞
moisture-proof sand 防潮沙
moisturization 保湿
molybdenum bisulfide 二硫化钼
moment arm 力臂
moment characteristics 力矩特性
moment of force 力矩
moment of inertia 转动惯量；惯

性矩
momentary position 瞬时位置
momentary purge 瞬间放油
momentum 冲力；动量
momentum theory 动量理论
monitor 监控，监视器
monitor warning function（MWF）监视警告功能
monitor warning system（MWS）监控警告系统
monitoring parameter 监控参数
monitoring software 监控软件
monitoring station 监听台；监控站
mooring 系留；抛锚；碇泊
Morse circuit 莫尔斯电报电路
Morse code 莫尔斯电码
mortar 迫击炮；深水炸弹发射炮
most significant bit 最高有效位
most significant digit 最高有效位数
mother board 母板
motion tracing 运动追踪
motivation 动力，激励
motor-operated valve 马达作动活门
motto 格言；箴言
mount 安装
mountain and valley breeze 山谷风
mountainous terrain 山地
mounted blades 安装叶片
mounted march 乘车行军
mounting 安装；挂装
mounting base 安装座
mounting flange 安装凸缘，定位凸缘
mounting hole 安装孔
mounting pad（发动机或附件传动机匣的）环形安装凸台
movable joint 活接头
movement area 机场飞行区域；行动地区
movement demonstration test 运动演示实验
movement form 运动形态
movement system 运动系统
moving ring 动环
moving target indicator 活动目标显示器，运动目标显示器
MRA（minimum reception altitude）最低接收高度
MRC（modular radio cabinet）模块无线电舱
MSA（minimum safe altitude）最小安全高度，安全飞行最低高度，最低安全高度

MSL（mean sea level）平均海平面，平均海面，平均海拔

muff coupling 套筒联轴节；联轴套

multi combining and distribution unit（MCDU）多功能控制显示组件

multi-crew operations 多机组作业

multi-engine aircraft 多引擎飞机

multi-engine helicopter 多发动机直升机

multifunction colour display（MFCD）综合彩色显示器

multifunction control display panel 多功能控制显示面板

multifunction control display unit（MCDU）多功能控制显示器

multifunction display（MFD）多功能显示器

multifunction display unit 多功能显示单元

multifunction equipment 多功能设备

multifunction keyboard（MFK）多功能键盘

multifunction spoiler 多功能扰流板；多功能扰流器

multileg itinerary 多航段旅程

multimedia courseware 多媒体课程软件

multi-mode approach and landing system 多模式进近和着陆系统

multi-mode landing system 多模式着陆系统

multi-mode receiver 多模式接收机

multiple address message 多址电报；多地址报文

multiple indicator 多点指示器；多功能指示器；多台指示器

multiple-can combustor 管型燃烧室

multiplexer 复用器

multipurpose equipment 多功能设备，多用途设备

multi-purpose flight data recorder（MPFDR）多用途（功能）飞行数据记录仪

multi-purpose helicopter 多用途直升机

multi-radar tracking 多雷达跟踪

multi-radar tracking using variable update 采用变化更新的多雷达跟踪

multisensor navigation 多传感器

导航

munitions 弹药；军火；军需品

mustard gas 芥子毒气

mutation 突变

MVA（minimum vectoring altitude）最低引导高度

MVFR（marginal visual flight rule）边缘视觉飞行规则

MWF（monitor warning function）监视警告功能

MWL（master warning light）主警告灯

MWS（monitor warning system）监控警告系统

N

nacelle 发动机短舱；短舱；吊舱

NACG（national aeronautical charting group）国家航空制图组

NAND gate 与非门

napalm 凝固汽油；凝固汽油弹

napalm bomb 凝固汽油弹

narrow angle 窄角

NAS（national airspace system）国家空域系统；国家空间系统

NASA（National Aeronautics and Space Administration）美国宇航局

national aeronautical charting group（NACG）国家航空制图组

National Aeronautics and Space Administration（NASA）美国宇航局

national airspace system（NAS）国家空域系统；国家空间系统

national climatic data center（NCDC）国家气候数据中心

National Missile Defense（NMD）国家导弹防御（系统）

National Oceanic and Atmospheric Administration（NOAA）美国国家海洋和大气局

National Transportation Safety Board（NTSB）美国国家运输安全委员会

National Weather Service（NWS）美国国家气象局

natural flapping 自然挥舞

natural frequency 自然频率；固有频率

natural mode of vibration 固有振型

natural vibration type 固有振型

nautical mile（NM）海里

NAV（navigation）航行；导航

navaglobe 远程无线电导航系统

NAVAID（navigational aid）助航设备，航行设备，导航设备；助航；航行法

NAVASCOPE（Airborne Radarscope

Used In Navar（Air Force）导航雷达使用的机载雷达显示器
navascreen 导航雷达地面屏幕
navigation（NAV）航行；导航
navigation database 导航数据库
navigation display 导航显示
navigation facilities 航行设备
navigation instrument 航行仪表
navigation light 闪烁航行灯；航行灯
navigation management system 导航管理系统
navigation service for helicopter mine sweeping 直升机扫雷航海勤务
navigation source 导航资源
navigation source annunciator 导航资源信号器
navigation source key 导航资源键
navigation system 导航系统
navigation track 巡航轨迹
navigational aid（NAVAID）助航设备，航行设备，导航设备；助航；航行法
navigational control 航向操纵
NBC（nuclear, biological, chemical）gear 防核生化武装备
NCDC（National Climatic Data Center）国家气候数据中心
NCO（non-commissioned officer）军士（非委任军官）；士官
ND（network director）网络指挥器；网路指挥器
NDB（nondirectional radio beacon）无方向信标
near collision 有相撞的危险
near-miss 近距脱靶；近失弹；靠近弹
neck of axle 轴颈
neck-shaped part 颈型件
needle nose pliers 尖嘴钳
needle roller bearing 滚针轴承
negative 反对（陆空通话用语）
negative dihedral 下反角
negative focal plane 负焦平面
negative lift 负升力
negative pole 负极
negative torsion 负扭转
neoprene gasket 氯丁橡胶垫片
neoprene gloves 氯丁橡胶手套
nephoscope 测云器
nerve gas antidote atropine 神经毒气解毒剂阿托品
net effect 净效应

net weight 净重

network control processor 网络控制处理器

network coordinating station 网络协调站

network director（ND）网络指挥器；网路指挥器

network interface controller（NIC）网络界面控制器

network interface module 网络接口模块

network interface system 网络接口系统

network protocol data unit 网络协议数据单元

neutral notch 中间槽

neutral position 中立位置；中间位置；空挡

neutral position switch 中间位置开关

neutralize and lock the cyclic 驾驶杆中立锁紧

neutron bomb 中子（炸）弹

never-exceed speed 最高限速；极限速度；不可超越的速度

Newton's third law of motion 牛顿第三运动定律

NEXRAD communications interface unit 下一代气象雷达通信接口单元

NEXRAD radar（next-generation radar）下一代气象雷达

next position and time over 下一位置及预达时间

next-generation radar（NEXRAD）radar 下一代气象雷达

Nf 自由涡轮转速

Ng 燃气发生器转速

Ng datum lever 发动机转速（Ng）调整杆

Ng dismatch（twin engines）双发转速（Ng）的不协调

NGV（nozzle guide vane）涡轮喷管导向叶片

NIC（network interface controller）网络界面控制器

night effect 夜间效应；夜间误差

Night Hawk stealth fighter 夜鹰隐形战机

night vision goggles（NVG）夜视镜

nightscope 夜视镜

nighttime operations 夜间作战

nil 无，零

nimbostratus 雨层云

nimbus 雨云

nipper pliers 尖嘴钳

nipple 接嘴；连接螺帽；连接套管，螺纹接头

nitrogen 氮

NM（nautical mile）海里

NMD（National Missile Defense）国家导弹防御（系统）

no ignition 不点火

no lighting on the warning light 警告灯不亮

no risk of collision 无碰撞危险

no start 不起动

NOAA（National Oceanic and Atmospheric Administration） 美国国家海洋和大气局

nocturnal bombardment 夜间轰炸

no-fly zone 禁飞区

noise abatement 噪音抑制；消音（尤指潜艇、飞机等）

noise level 噪声水平

noise silencer 消声器，噪声防除器

nominal rotor speed 额定转速

nominal shaft power 标称轴功率

nominal speed control 名义转速控制

nominal value 标称值；额定值；名义价值

nominal working pressure 额定工作压力；公称工作压力

non illumination on test 校验时灯不亮

non-active connection 无效连接

non-amphibian operations 非两栖作战

non-break power transfer 无间断电源转换

non-commissioned officer（NCO）军士（非委任军官）；士官

nonconductor 非导体；绝缘体

non-conventional attack 非常规袭击

non-destructive 非破坏性的

non-destructive inspection 非破坏性检验；无损检查

non-destructive testing manual 无损探伤手册

non-dimensional Ng 无因次发动机转速

non-directional radio beacon（NDB）导航台；不定向无线电信标；全向无线电信标

non-firm power 不恒定功率

non-flammable gas 不易燃气体
non-modular part 非单元体件
non-precision approach 非精密进近
non-precision instrument runway 非配备精密仪表着陆设备的跑道
nonprocurable 非采购件
non-proliferation 不扩散
non-retractable landing gear 非收放式起落架；固定式起落架
non-return valve 止回活门；单向活门；单向阀；止回阀
non-reversible gear 不可逆齿轮
non-rotary swash plate 固定斜板凸轮；静盘
non-stable region 不稳定区
nonstop flight 不着陆飞行；连续飞行
nonsymmetrical airfoil 非对称型桨叶；非对称翼型
non-torque loading 非扭矩载荷
non-toxic gas 无毒气体
non-vectored interrupt 非向量中断
nonvolatile memory 非易失性存储器
NOR gate 或非门
normal angle 常规视角
normal course of weather 常规天气
normal engine running 发动机正常工作
normal illumination 正常灯亮
normal inspection 正常检查
normal operating zone（NOZ）正常作业区
normal operation 正常工作，正常运行
normal procedure 常规程序
normal rate of climbing 正常上升率
normal rate of descending 正常下降率
normal rundown time 正常惯性转动时间，正常停转时间
normal running 正常运转
normal start 正常起动
normal turn 正常转弯；标准转弯
northerly turning error 北转误差
northern hemisphere 北半球
Norton's theorem 诺顿定理
nose 机头（直升机）
nose avionic bay 前航电舱
nose compartment 机头舱；前舱
nose compartment access door 前舱检修门；机头舱检修门
nose down 俯冲（直升机）；推机头

nose exterior 机头外部
nose gear emergency extension system 前起落架紧急放下系统
nose landing gear 前起落架
nose LG（landing gear）emergency extension system 前起落架紧急放下系统
nose over 拿大顶
nose ribs 机头肋
nose taxi lamp 前滑行灯
nose tricycle landing gear 前三点式起落架
nose tricycle type 前三角式
nose up 抬机头
nose wheel 起落架前轮；前轮
nose wheel lock 起落架前轮锁；前轮锁
nose wheel status 起落架前轮状态
nose wheel steering control 前轮转弯操纵
nose wheel steering hydraulic system 前轮转弯液压系统
nose-down 机头下俯
nose-down moment 低头力矩
nose-up moment 抬头力矩
nose-wheeled device 前三点式装置
NOTAM（notice to airman）航空通告；航行通告
NOTAR 无尾桨直升机
notch effect 冲孔效应；刻槽影响；切口效应；缺口效应
notes on maintenance 维护说明
notice to airman（NOTAM）航空通告；航行通告
nozzle 尾喷管（嘴）；喷管（嘴）；排气管
nozzle guide vane（NGV）涡轮喷管导向叶片
nozzle orifice 喷口
Nr（free turbine speed）自由涡轮转速
NR selector 旋翼转速选择器
N-sector N 区
NTL（non-torque loading）非扭矩载荷
NTSB（National Transportation Safety Board）美国国家运输安全委员会
NTZ（no-transgression zone）非海侵入区
nuclear, biological and chemical reconnaissance 核生化勘测
nuclear，biological，chemical（NBC）gear 防核生化武装备

number of consecutive starts 连续起动次数

number representation system 数制

numerator 分子

numeric keypad 数字小键盘

numerical indicator 数字指示器

numerical map 数字地图

NVG（night vision goggles）夜视镜

NWS（National Weather Service）美国国家气象局

O

O level maintenance（first line maintenance）第一级维护

OAT（outside air temperature）外界温度（大气温度）

object of interest（radar）重要目标（雷达）；有价值的目标（雷达示波器上出现的有意义的目标回波）

objective of training 培训目的

oblate and long-tube type 扁圆形长筒状

observation and sighting turret 观瞄球

obstacle awareness alerting and display 障碍物提醒警告与显示

obstacle clearance altitude（OCA）障碍许可高度；越障安全高度

obstacle clearance height（OCH）越障高度

obstruction 遮障地物，障碍物；（妨碍飞机低空飞行的）低空障碍物

obstruction free 无障碍

obstruction light 障碍物标示灯

obturating membrane 闭气膜

OC（airport obstruction charts）机场障碍图

OCA（obstacle clearance altitude）障碍许可高度；越障安全高度

occasional shower 阵雨

occluded front 锢囚锋

occult light 明暗灯；歇光

occupant 乘员

OCH（obstacle clearance height）越障高度

octane number 辛烷数;【象】辛烷值

OEI（one engine inoperative）一台发动机停车

OEI（one engine invalidation）应急状态，一台发动机停车

OEI 2 min. rating 2 分钟 OEI 状态

OEI 30 sec. rating 30 秒 OEI 状态

OEI continuous OEI 持续

OEI continuous rating 连续 OEI 状态

OEI rating（one engine inoperative rating）OEI 状态；超应急状态（一台发动机不工作）

off-airway area 航路外空域

off-field landing（outside landing）场外着陆

office of foreign disaster assistance 国外灾害救助处

officer commanding 指挥官

officer on duty 值班军官；值日军官

official documentation 正式文件

offset 偏置，补偿

offshore wind 离岸风

OGE（out of ground effect）无地效

OGE hover 无地效悬停

Ohm's law 欧姆定律

oil 滑油

oil circuit 滑油路

oil circulation 滑油循环

oil contamination 滑油污染

oil cooler 滑油散热器

oil cooler air intake cowling 滑油散热器进气整流罩

oil dilution 滑油稀释

oil dilution check 滑油稀释度检查

oil duct 油管

oil filter 滑油滤

oil filter clogging 滑油过滤器堵塞

oil filter pre-blockage indicator 滑油滤堵塞指示器

oil filter replacement 更换滑油滤

oil filtering unit 滑油滤组件

oil leak 滑油泄露

oil leakage 滑油泄露

oil level 滑油油面；滑油压力

oil line 滑油管路

oil lines 滑油管路

oil mist 油雾，油气混合物

oil outlet line 滑油出口管路

oil pressure 滑油压力

oil pressure indicator 滑油压力表

oil pressure probe 滑油压力探头

oil pressure transmitter 滑油压力传感器

oil pump 滑油泵

oil pump pack 滑油泵组件

oil pump unit 滑油泵组件

oil quantity 油量

oil quantity indicating system 滑油油量指示系统
oil replenishment 滑油补充
oil sample 滑油取样
oil sampling for analysis 滑油取样分析
oil specification 滑油规格
oil splash lubricated part 喷油润滑部件
oil strainer 滑油过滤器
oil sump 收油池
oil supply line 滑油供油管
oil system 滑油系统
oil tank 滑油箱
oil tank vent line 油箱通气管
oil temperature indicator 滑油温度表
oil temperature indicator system 滑油温度指示系统
oil temperature probe 滑油温度探针
oil temperature switch 滑油温度开关
oil valve assembly 滑油活门组件
oil viscosity 滑油黏度
oil well 油井
oil-gas 油气式
oiling venting nozzle 加油通气嘴
OM（outer marker）远（外）指标点
omni-bearing 全方位
omni-bearing indicator 多方位指示器
omni-direction 全方位；全方向
omni-directional 全向的，无方向性的
omni-directional airspeed transmitter 全向空速传感器
omni-directional approach lighting system 全向进场照明系统
on call 待命（准备执行特定的任务）
on duty 值勤
on station 进入阵地；空中待命（升空飞机已做好攻击目标准备）
on top of cloud 在云上
on-board launch control circuit 机上发射控制线路
on-board recorder 机载录音机；机载记录仪
oncoming air 迎面（而来的）空气
oncoming stream direction 迎流方向
on-condition maintenance 视情维护

on-condition monitoring 视情监控；视情检测；视情维护
on-course signal 正航信号
one engine inoperative（OEI）一台发动机停车
one engine invalidation（OEI）应急状态，一台发动机停车
one-piece propeller 整片螺旋桨
one-way clutch 单向离合器
one-way communications 单向通讯
one-way valve 单向阀；单向活门；止回活门
one-wheel landing 单轮着陆
onshore wind 向岸风
on-the-spot investigation 事故现场调查
open circuit 断路；开路
open formation 疏开编队
Open Skies Agreement《开放天空条约》
open sky 开放天空
open wrench 开口扳手
opening 距离或范围的增大
open-loop state 开环状态
operability bleed valve 可用引气活门
operating base 现用基地；作战基地
operating ceiling 实际上升限度（系留气球的最大有效上升限度）
operating check 操作检查
operating cycle 工作循环
operating empty weight 使用空机重量，使用空重
operating envelope 工作包线
operating frequency 使用频率
operating hours 工作时间（小时）
operating limitation 工作限度
operating limitations not exceeded 不超过工作极限
operating line 工作线
operating manual 使用手册
operating on reduced power 使用低电力工作
operating parameter 工作参数
operating phase 工作状态
operating phases 工作阶段
operating power 工作功率
operating procedures 工作过程
operating rocker arm 操纵摇臂
operation 作战；行动
operation and maintenance 运行和维修
operation check 工作检查

operation flight 作战飞行
operation load 操作载荷
operation of free turbine blocked forbidden 禁止锁住自由涡轮
operation specification 操作规范
operational amplifier 运算放大器
operational area 作战区域
operational deployment of army aviation 陆军航空兵作战部署
operational flight plan 操作飞行计划；运营飞行计划
operational readiness demonstration 战备示范
operational requirement 作战需求，作战要求
operations 操作；行动；作战行动
operations manual 操作手册
optical axis 光轴（线）
optical equipment 光学设备
optical illusion 光幻视
optical unit 光学组件
optimised thermodynamic cycle 热力优化循环
optimization 优化
optimized airfoil profile 优化翼型
optimum speed 最佳速度
optional chip pulse system 选装碎片脉冲系统
optional equipment 备选设备；附加设备；选装件
optional equipment supplement 补充备选设备
optional fuel flow meter 选装燃油流量计
optional kit 备用工具箱
OR gate 或门
ORBAT（order of battle）作战序列
orbital flight 轨道飞行
order of battle（ORBAT）作战序列
ordnance 武器，军械
ordnance compressed air bottle 军械压缩空气瓶
ordnance equipment 军械设备
organic light aircraft 建制轻型飞机
orient 定向
origin of the coordinate 坐标原点
original equipment manufacturer 原始设备制造商
O-ring seal O 型密封圈
orographic rain 地形雨
orthogonality 正交性
orthomorphic map projection 正形

地图投影
oscillation 震动，摆动
oscillation damper 减震器
oscillation of the angle of attack 迎角振荡
oscillation reducer 减震器
oscillator 振荡器
oscillatory bending stress 振荡弯曲应力
oscillatory differential equation 振动微分方程
oscilloscope 示波器
out of ground effect（OGE）无地效
outboard engine or outboards 外侧发动机
outboard flap 外侧襟翼
outboard float 外侧（安定）浮筒
outboard spoiler 外侧扰流板
outbound flight 背台飞行
outer beacon 远距导航台
outer casing 外机匣
outer compass locator 罗盘远距指点标
outer east marker 东远台
outer fix 远距定位点
outer gimbal 外万向架
outer marker（OM）外信标台
outer navigation station 远距导航台
outer rotor 外转子
outer shell 外壳；机匣
outfight 战斗优势；作战优势
outfit 全套设备
outflank 翼侧包围；翼侧延伸
outflow 流出量，流出口，溢出物
outflow valve 出流活门，放泄活门
outlet 出口
outlet guide vane 出口导向叶片
outlet union 出口接头
outmaneuver 智取
outmaneuver the missile 实施机动以规避导弹的攻击
outnumber 数目超过
out-phase 反相
outpost 警戒部队
output 输出
output command 输出指令
output concentrator 输出集中器
output drive 输出轴
output gear 输出齿轮
output pinion 输出齿轮
output shaft 输出轴

output shaft speed 输出轴转速

output signal 输出信号

outrigger 弦外支架；翼下起落架支架；外伸架

outside air pressure 外界空气压力

outside air temperature（OAT）外界温度（大气温度）

outside diameter 外径

outside landing 场外着陆

outside retaining cable 外部防护锁链

outward opening emergency exit hatch 外开式紧急出口门

over run the R/W 冲出跑道（滑跑冲出跑道）

overall compression ratio 总压缩比

overall height 全高；总高

overall length 全长

overcast 阴天的，满云的；满天云

overcurrent 过载电流

overcurrent protector 过流保护

overfrequency 过频

overfuelling 燃油供给过量；供油过量；过量供油

overhang 外伸量；航尾凸出部；多翼机的上翼伸出下翼的长度；悬于……之上

overhaul 翻修；大修

overhaul manual 大修手册；翻修手册

overhead console 顶部操纵板

overhead electronics unit 头顶电子组件

overhead panel 顶部仪表盘；顶棚控制板；上仪表板；舱顶板

overhead panel lighting 顶部仪表盘照明；顶棚控制板照明

overhead projection 高射投影

overheat 过热

overheat protector 过热保护

overlay 覆盖，叠加；重复占位；透明图

overload 过载，超载;【炮】使用强装药

overload accident 超载事故

overload detecter 过载探测器

overload takeoff 超载起飞

overpressure 超压

override 超控，超载控制（超载或绕过自动驾驶仪的手控）

overrunning clutch 超转离合器，超速离合器

overshielding 超屏蔽

overshoot 复飞；超越机场；超越

跑道；着陆目测高;（轰炸、射击时）超越目标；远弹

overshooting 进场过高;【炮】远弹；高于目标的炸点

overspeed 超转

overspeed arming 超转复位

overspeed circuit 超转线路

overspeed control box 超转控制盒

overspeed drain valve 超转放油活门

overspeed electro-valve 超转电磁活门

overspeed electro-valve jammed in opened position 超转电磁活门在打开位置阻滞

overspeed governor 超速调节器；超转控制器

overspeed protection 超转保护

overspeed rearming light 超转复位指示灯

overspeed safety circuit 超转安全电路

overspeed setting 超转调定

overspeed shut-down 超转保护停车

overspeed solenoid valve 超转电磁活门

overspeed valve 超转活门

overstrength 超编制（人员），超编，超额

overtaking 超越

overtemp（overtemperature）超温，过热

overtemperature（overtemp）过热，超温

overvoltage 过压

overwater 在水域上空，水上，水面

overwater emergency equipment 水上救生装备，海上应急装备

overwater operations 水面行动；水上作战

oxide 氧化物

oxygen mask 氧气面罩

P

P3 capsule stop P3 膜盒止动点

pack discharge pressure sensor 组件排气口压力传感器

pack discharge temperature sensor 组件排气口温度传感器

pack of oil pumps 滑油泵组件

pack temperature controller 组件温度控制器

pack temperature sensor 组件温度传感器

packed encoding rule 分组编码规则

packup 成套件;（能装箱运输的）部件

pad 底座，台，场地;【箭】发射坪；发射台（指承载和发射火箭飞行器的基地或平台）;（小型飞机）降落场;（直升机的）起降平台

paddle 桨，短桨；桨叶，叶片；（降落伞的）桨形包装尺;（速率陀螺的）阻尼片

paddle airfoil profile lift drag test 桨叶翼型升阻力实验

page header 页眉；页头信息；页标题

paint 涂料;（显示器）图像；回波图像；识别图像；用激光目标指示器照亮目标

paint thinners 涂料稀释剂

painting off 掉漆

paired system 配对系统

pallet（飞机货舱内标准尺寸的）货盘，集装箱货盘;（美国防部专用的）463L 型货盘;（空投货物的）底座;【箭】运输架;（航天实验室用的）标准尺寸底座；标准尺寸台架

pane 窗玻璃片；窗格

panel lighting 面板照明；面板照明系统

panoramic sight 全景瞄准镜

panzer（德语）装甲车；坦克
PAOAS（parallel approach obstacle assessment surface）平行进近障碍物评估面
PAPI（precision approach path indicator）精密进近航道指示器
PAR（precision approach radar）精密进场（着陆）雷达
parabola 抛物线
parachute 降落伞
parachute fragmentation bomb（parafrag）带伞杀伤弹；伞投杀伤炸弹
parachute regiment 伞兵团
parachutist 跳伞者；伞兵
parade 分列式（阅兵）
paradrop 空投；伞投
parafrag（parachute fragmentation bomb）带伞杀伤弹；伞投杀伤炸弹
paraglider 翼伞飞行器；滑翔伞，翼伞
parallax 视差
parallel approach obstacle assessment surface（PAOAS）平行进近障碍物评估面
parallel planform 平行俯视图
paralleled 并联作动器，并联舵机
parameter（PRMTR）参数
paramilitary fighter 辅助军人；准军人
paramilitary force 准军事力量，准军事部队
parasite drag 废阻力；寄生阻力
paratrooper 伞兵
paratroops 伞兵部队（复）
parent aircraft 母机；运载飞机
parent base 主补给基地；主航空基地
parent bomb rack 主炸弹架
parent bomber 运载火箭（导弹）的轰炸机
parent radar 制导雷达（导弹）
parent station 主补给站
park 滑进停机坪，停机坪;（枪炮、军需品等的）放置场；车场，炮场
parking apron 停机坪
parking brake 停车闸；停车制动器
parking brake shutoff valve 停放刹车关断活门
parking brake status 停放刹车状态
parking brake switch 停机刹车

开关

parking brake valve 停放制动活门

parley 休战谈判

part condition assessment 零（部）件状况评估

part number 件号，零件编号

part repalcement code 零件更换代码

part task trainer 局部任务训练器

partial clogging of a jet 喷嘴局部阻塞

partial failure 局部故障

partial obstruction 局部堵塞

partial shock wave 局部激波

particle 质点；颗粒；颗粒物；碎粒

particles detection 粒子探测

partition 隔板，区隔，分区，分块

parts replacement 更换零件

passage of gas exhaust pipe 排气管通道

passenger address system 旅客广播系统

passenger addressing 旅客广播

passenger addressing function 旅客广播功能

passenger cabin 客舱

passenger cabin door 客舱门

passenger control unit 旅客控制组件

passenger door’s height and width 客舱门尺寸

passenger gate 登机口

passenger light 乘客灯

passenger seat 乘员座席

passenger service unit（PSU）乘客服务组件

passenger speaker amplifier 乘客扩音喇叭

passing 现在通过（陆空通话用语）

passing shower 过境阵雨

passivation 钝化；钝化处理

passive component 无源设备，无源部件

passive countermeasures 消极对抗措施；消极干扰

passive device 无源器件

passive two-terminal network 无源二端网络

password 密码

patchboard 接线板；配线盘

patches of water 积水

path 航迹，轨迹，通道

path following error 路径跟踪误差

Pathfinder “探路者” 卫星（美第三代军事卫星第一颗卫星）；“开拓者号”（美国航天飞机的全尺寸模型）；“开拓者” 计划（美国未来月球和火星载人航天活动的研究和开发计划）；“开拓者”（马丁 · 玛丽埃塔公司研制的前视红外系统）

pathfinder 导航飞机；机载导航雷达；（空降）目标引导组；空降信号员；（着陆）下滑道指示器；目标引导信标；航向指示器

pathway 路径

patrol 巡逻

patrol ship 巡逻舰

Pave Hawk【美】“铺路鹰” HH/MH — 60G 直升机

pavement classification number（PCN）道面等级号（表示机场跑道能承受飞机着陆负荷的特征数）

pavement strength 道面强度

pawl 掣子；锁子；掣爪；【炮】驻钣

payload 净载重；有效载荷

PCM（power control module）动力控制模块；功率控制模块；电源控制模组

PCN（pavement classification number）道面等级号（表示机场跑道能承受飞机着陆负荷的特征数）

PDP（power distribution panel）配电板

peace process 和平进程

peak envelope power 峰值包络功率

peak value 峰值

pedal 脚蹬

pedal application 蹬脚蹬

pedal force trim 脚蹬力量配平

pedal trim 脚蹬配平

pendular action 钟摆运动

penetrating bomb 钻地弹；穿甲弹

penetration adjustment 插入量调整

pentropic organisation 五群热带组织

percussion cap 雷管；火帽

percussion fuse 着发引信

perfectly readable 非常清楚（通话质量）

performance assessment 性能评估

performance data computer (system)

性能数据计算机（系统）

performance navigation computer system 导航计算机系统性能

performance engineer's manual 性能工程师手册

performance data 工作特性；性能数据

periodic check 定期检查

periodic function 周期函数

periodic inspection 定期检查

periodic maintenance 定期维护；定期维修；定期保养

periodical quantity 周期量

periodical test 定期测试

periodicity 维护时限

permanent grounding 永久停飞

permanent heliport 永备直升机场

permanent magnet 永久磁铁；永磁体

permanent magnet alternator（PMA）永磁交流发电机

permanent magnet rotor 永久磁铁转子

permanent magnetic generator 永磁发电机

permanent virtual circuit 永久性虚拟电路，常设虚拟电路

permeability check 渗透性检查

permeability injection manifold check 喷射油道渗透检查

peroxide 过氧化物

perpendicularity 正交性

persistency of vision 视觉持续

person on board 机上人员

personal injury 人身伤害

PFD（primary flight display）主飞行显示器（常指飞机上显示飞行姿态的显示器）；初级飞行表演

phase detector 鉴相器

phase difference 相位差

phase displacement 相位移

phase displacement type 相位型

phase inverter 反相器

phase localizer 相位定位器，相定位仪

phase meter 相位计

phase sequence 相序

phase shift keying 移相键控

phase-shift omnidirectional radio range 相移全向无线电指向标

phasometer 相位仪

phasor diagram 相量图

phasor expression 相量式

Philips screw 十字螺钉
Philips screwdriver 十字解刀；十字改锥；十字螺丝刀
phonic wheel 音轮
phosphor screen 荧光屏，发光板
photoelectric cell 光电管；光电池；光电元件
photoelectric device 光电器件
photoelectric turret 光电转塔
physical motion 物理运动
physical parameter 物理参数
physical quantity 物理量
physiological reaction 生理反应
PIC（pilot in command）机长；正驾驶员
picked men 精兵
picked troops 精锐部队
pier 候机楼登机走廊；旅客桥
pierced plank runway 跑道钢板
pigmented dope 着色涂料
pillow block 轴台
pilot balloon 测风气球
pilot call panel 飞行员呼叫面板
pilot cockpit door 飞行员座舱门
pilot collective lever 驾驶员总距杆
pilot door 飞行员舱门
pilot error 飞行员驾驶误差，驾驶员失误
pilot error accident 飞行员责任事故（驾驶错误事故）
pilot glareshield 飞行员遮光板；飞行员防眩板
pilot headset connection 飞行员耳机连接
pilot in command（PIC）机长
pilot instrument panel lighting rheostat 正驾驶仪表面板照明变阻器
pilot interface 驾驶员界面；飞行员系统接口
pilot interphone 驾驶员机内通话机
pilot proficiency check 飞行员熟练水平检查，飞行员技术考核
pilot report（PIREP）飞行员报告
pilot ventilation 驾驶员通风口
pilot's Mach airspeed indicator 驾驶员的马赫数空速表
pilotage 领港，领航，地标领航，目视领航；领航术，目视飞行术
pilotage chart 地标航行图
pilot-caused accident（pilot factor accident/pilot error accident）飞行员责任事故（飞行员因素事故/飞行员错误事故）

piloting 地标领航，目视领航；驾驶；引水

piloting parameter 飞行参数

pilotless reconnaissance drone 无人驾驶侦察机

pilot's control unit 飞行员操纵装置

pilot's DCP（digitally controlled potentiometer）驾驶员数控电位计

pilot's ND（network director）正驾驶网路指挥器

pilot's PFD（primary flight display）and its controls 正驾驶主飞行显示及其操纵系统

pin 销，钉，栓

pinion 链齿；小齿轮；前翼

pinpoint 精确测定的位置

pip【雷】（阴极射线管上的）信号标志；脉冲，尖头信号；反射点；（瞄准具上的）光点

pipage（水、气、油等的）管道输送；管道；管道系统

pipe cap 管帽

pipe pliers 管钳

pipe rupture 管子开裂

pipe wrench 管钳

pipeline 补给线；补给系统；输油管；管道，输送管

pipeline aircraft 补给运输飞机

pipeline status 补给线状态

pipeline time 补给时间，申请至领到补给品的时间

pirate raid 对非军事目标空袭；海盗式空袭

PIREP（pilot report）飞行员报告

piston engine 活塞式发动机

piston engine aircraft 活塞发动机航空器

piston valve 活塞式活门

piston-engined helicopter 活塞式直升机

pitch 俯仰

pitch angle 桨距角；螺旋角；俯仰角；仰角

pitch attitude 俯仰姿态；俯仰角

pitch attitude beep rate 俯仰姿态“滴”音速率

pitch attitude target line 俯仰姿态目标线

pitch axis 俯仰轴

pitch balance 俯仰平衡

pitch control 变距；变距操纵，俯仰操纵

pitch control horn 俯仰控制臂

pitch control rod 俯仰控制杆；节距控制杆
pitch control wheel steering 驾驶盘俯仰操纵
pitch damping moment 俯仰阻尼力矩
pitch decrease 变小距
pitch down the aircraft 推机头
pitch feel simulator 俯仰感力模拟器
pitch increase 变大距
pitch inoperative 俯仰操纵系统故障
pitch linear actuator 俯仰线性作动筒
pitch rotary inertia 俯仰转动惯量
pitch stability 俯仰稳定性
pitch trim actuator 俯仰配平作动筒；俯仰配平舵机
pitch up the aircraft 抬机头
pitch/roll beep trim selector switch 俯仰 / 滚转蜂鸣器配平选择开关
pitch-control axis 俯仰控制轴
pitching moment 俯仰力矩
pitching oscillation 俯仰摆动，俯仰振动
pitot head 皮托头
pitot heater 空速管加热器
pitot source-select valve 全压源选择活门
pitot tube 空速管；皮托管
pitot tube anti-icing 空速管防冰
pitot tube heating switch 空速管加温电门
pitot tube warming system 空速管加温系统
pitot tube warning system 空速管告警系统
pitot-static pressure system 全静压系统
pitot-static probe 皮托管；皮托静压管；空速管
pitting（压缩机叶片）锈斑
pivot 支点，中枢；中心点，枢轴点
pivot point【舰船】回转枢心；【舰船】旋转点；支点；中心点
pivoting turn 原地转弯（指飞机在地面上）
placard 标语
plaited sleeve 编织护套
plan coordination 规划协调
plan position indicator（PPI）平面位置指示器

plan view display 平面图显示器
plane contact type 面接触型
plane of rotation 转动面
plane shape 平面形状
planetary gear 行星齿轮
planetary reduction gear 行星减速齿轮
planform 平面图；俯视图
planned flight level 计划飞行高度
plano-convex aerofoil 平凸形翼型
plasma jet 等离子流
plastic packaging 塑料封装
platform takeoff and landing 平台起降
plier 老虎钳；钳子；手钳
plotter 炮兵标图员；对空标图员；测绘板；绘图器，绘图仪，绘图机
plug screw 螺塞
plummet 铅垂；水砣；测深器
PMA（permanent magnet alternator）永磁交流发电机
pneumatic bleed valve 气动式放气活门
pneumatic brake system 气动制动系统
pneumatic drive unit 气动驱动组件
pneumatic flex shaft unit 气动挠性轴组件
pneumatic ground service air connection 地面空气充气接头
pneumatic system 气压系统；气动系统
pneumatic butterfly valve 气动蝶型节流阀
pod（机身外）吊舱；短舱；分离舱；可拆卸的货舱
point company 尖兵连
point contact type 点接触型
point of effort application 施力点；作用点
point of intersection 交点
point of no return（不能用自带的燃料飞返基地的）航线临界点；无法返航点
point of regulation 调节点，调压点
point of safe return 安全回转点，安全返航点
point of saturation 饱和点
point of support 支点
point on one engine 单发承载点
pointer 指针
points to be checked 检查项目

point-to-point communication 定点通信，点对点通信

polar air 极地气团；极地空气

polar atlantic air mass 大西洋极区气团

polar front 极锋

polar maritime air mass 极地海洋气团

polar pacific air mass 极地太平洋气团

polarity 极性

polarity leading wire 电极引线

polarization effect 极化效应

polyester fiber 涤纶

polyvinyl chloride 聚氯乙烯

pools of water 跑道积水

poor braking action 刹车状况不好

port 端口；左舷（舵）；舱门；接口；射击孔，炮眼，枪眼；（装甲车）展望孔

port engine 左发动机，左舷发动机

portable data loader 便携式数据装载机

portable fire extinguisher 手提灭火机

portable receiver 便携式接收机

portside door 左舱门

position（vertical scroll） 位置（垂直滚动）

position at rest 静止状态

position command 位置指令；位置命令

position dilution of precision 位置精度扩散因子

position feedback 定位反馈；位置反馈

position light 航行灯；锚位灯；位置灯

position light signal 方位灯号志

position limit light 闪烁航行灯

position report 位置报告（飞机用无线电向地面控制站发出关于该机位置、高度、地速及航向等情况）

position sensor 位置传感器

position servo loop 位置伺服回路；位置伺服环路；位置伺服回圈

position transmitter 位置传感器；位移传感器

positioning 定位；引导；安置；配置；驻扎

positive area 正区

positive control area（空中交通）全面管制区；全面管制空域
positive lift 正升力
positive plane 正焦平面
positive pole 正极
post-flight check 飞行后检查
post-flight inspection 飞行后检查
post-flight maintenance preparation 飞行后机务准备
post-flight preparation 飞行后准备
posthumous medal 追颁勋章
post-war reconstruction 战后重建
potential 电位
potential accident（飞行）事故征候
potential energy 势能，位能
potential instability 位势不稳定
potentiometer 电位仪；电位器
pound force 磅力
power approach 动力进场，发动机工作进场着陆；带油门进场着陆
power available 可用功率，可用马力
power balance point 功率平衡点
power check 功率检查
power conditioning system 电力调节系统
power contactor 电力接触器
power control actuator 动力操纵作动筒
power control module（PCM）动力控制模块；功率控制模块；电源控制模组
power control unit 动力控制组件
power difference indicator 功率差指示器
power distribution panel（PDP）配电板
power drive flange 功率传动法兰盘
power drive shaft 传动轴
power drive unit 动力驱动组件
power drop 功率下降
power electromagnetic equipment 电力电磁设备
power evolution 功率变化
power input shaft 功率输入轴
power lever angle 油门杆角度
power loading 动力负载；功率载荷
power loss box 功率损失盒
power loss unit 功率损失单元
power management control 功率管

理控制

power margin 功率余量

power output 动力输出

power output shaft 功率轴出轴

power plant 发动机，动力装置

power plant buildup manual 动力装置手册

power relay 功率继电器，电源继电器

power requirement 需用功率

power reserve 功率储备

power shaft 传动轴

power shaft bearing 传动轴轴承

power shaft seal 传动轴密封圈

power socket 电源插座

power source 电源

power source element 电源元件

power source unit 电源装置

power storage 储能

power supply 电源

power supply element 电源元件

power supply equipment 供电设备

power supply module 电源供应组件

power supply unit 电源装置；供电设备

power train 动力传动机构；动力传动系

power transfer unit 动力传输组件

power turbine 动力涡轮

power turbine assembly 动力涡轮组件

power unit 发动机组；发电设备；动力设备，动力装置；功率单位

power-driven equipment 用电设备

powered flight 动力（推进）飞行

power-enrichment system 功率强化系统

power-on check 通电检查

power-on clear 电源接通清除

power-on mode 上电模式

powerplant 动力装置，推进系统；发电站

powerplant system 动力系统

power-to-weight ration 功重比

power-up 通电

power-up built-in test 通电自检

PPI（plan position indicator）平面位置显示器

PPT（press-to-transmit）一按通

practical ceiling 实用升限（使用升限）

practical pitch 实际桨距

practical work（PW）实际操作

pre-blockage indicator 预堵塞指示器

pre-blockage pressure switch 预堵塞压力开关

precautions 注意事项，预防措施

precipitation 降水

precise positioning service 精密定位服务，精确定位系统

precision approach 精确进场

precision approach path indicator（PAPI）精密进近轨迹指示器

precision approach radar（PAR）精密进场（着陆）雷达

precision bombing 精确轰炸；对小面积目标轰炸

precision distance measuring equipment 精密测距仪，精密测距装置

precision instrument runway 配备精密仪表着陆设备的跑道

precision runway monitoring 精密跑道监视

precision spin 准确螺旋

precision turn 准确转弯，精确转弯

precision very high frequency omnidirection range 精确甚高频全向导航台

precision-guided bomb 精确制导炸弹

precision-guided bunker-penetrating bomb 精确制导破地堡炸弹

precision-guided munitions 精确制导弹药

precooling 预冷

precursor（核爆炸冲击波前的）前驱波；母体;（材料的）前身；原材（制成某种材料所用的材料）

Predator spy drone 掠夺者侦察机

predecessor activities 先行活动

pre-departure clearance 起飞前的放飞指示

prediction angle 提前量修正角；总修正角

predictive windshear system 预测风切变系统

predictor 预测器；预报器；预报员

pre-dip belt 预浸带

preemptive attack 先发制人的进攻

preemptive strike 先发制人的攻击

preferential routes 优选航线（由航线交通管制中心用计算机选出的最有利航线，公布在《公报》上）

preferential runway 优选跑道（为减少当地噪声并使航线远离居民区而指定使用的跑道）
preflight adaptation 飞行前适应
preflight check 飞行前检查
preflight inspection 飞行前检查
preflight maintenance preparation 飞行前维护
preflight preparation 飞行前准备
preflight test 飞行前测试；飞行前检测
pre-landing check 着陆前检查
prelaunch console 发射前控制台，发射前操作程序控制台
preliminary performance estimation 初步性能估计
preliminary weight estimation 初步重量估计
preparation before start 起动前准备
preprogrammed guidance 预置导引
prescribed limit 预定范围
prescriptive technical publication 指令性技术文件
preselect 预选
pre-selected heading 预定航向
present position 当前位置，现在位置
presented target area 显示目标区
preset frequency 预调频率；预定频率
preset guidance 预定程序制导；预置制导
preset limit 预定极限
pre-sortie briefing 开飞前指示
press roller 镇压轮，压辊
press wheel 镇压轮
pressing-belt panel 压链板
press-to-transmit（PPT）一按通
pressurant gas 挤压气体
pressure 压力
pressure after jet 喷嘴后压力
pressure altitude 气压高度
pressure altitude variation 压力高度差
pressure altitude warning 压力高度警报
pressure anemometer 压力风速表
pressure angle 压力角
pressure anomaly 压力异常;【气】气压距平
pressure bias modulation 偏压调节
pressure breathing 加压呼吸，增压呼吸
pressure coefficient 压力系数

pressure contour 等压线
pressure control 压力控制
pressure control valves 压力控制阀
pressure demand oxygen equipment 压力需求式供氧装备
pressure differential 压力差
pressure differential resistance 压差阻力
discharge compressor pressure 压气机出口压力
pressure drag 压差阻力，压阻
pressure drag coefficient 压力阻力系数
pressure drop 压（力）降（低）
pressure face 压力面（螺旋桨或旋翼的下表面）
pressure fall center 降压中心
pressure filler inlet 压力加油口
pressure flap 均压阀；压力活门
pressure fluctuation 压力脉动（波动）
pressure foredrag 头部压差阻力
pressure fueling station 压力加油仓口
pressure gradient 压力梯度（机翼升力面的或大气中的压力变化率）
pressure head 总静压头，全静压管；压力头
pressure increasing valve 增压活门
pressure indicator paper 压力指示器纸
pressure instrument 压力式仪表
pressure lapse rate 气压递减率
pressure line 压力管路
pressure line of position 气压位置线
pressure lubrication system 压力润滑系统
pressure modulating device 压力调制装置
pressure nozzle 压力喷嘴
pressure off switch 断压开关
pressure pattern 气压（分布）型（式）
pressure pattern flying 等压面飞行
pressure plate anemometer 压板式风速表
pressure probe 压力探测器
pressure pump 增压泵
pressure raising shutoff and dump valve 升压关断和安全活门
pressure recovery 压力复原，压力恢复
pressure reducing valve 降压活门

pressure regulating and shut-off valve 调压与断油活门

pressure regulating valve 压力调节活门

pressure regulation 压力调节

pressure regulation valve 调压活门

pressure relief valve（PRV）减压阀，释压阀

pressure rigid airship 充压刚性飞艇

pressure rise center 升压中心

pressure sensor 压力传感器

pressure shut off valve 压力关断活门

pressure suit 加压服；宇宙服；抗荷服

pressure switch 压力开关

pressure system 气压系统，压力系统

pressure tapping 测压管接头

pressure tendency 压力趋势，气压倾向

pressure transducer 压力传感器

pressure transmission 压力传输；传压

pressure transmitter 压力传送器

pressure tube anemometer 压管风速表

pressure warning unit 压力警告器，压力报警器

pressure wave 压力波；激波；冲击波

pressure-bag molding 压力袋成形

pressure-fed 压馈

pressure-pattern flying 等压航线图飞行，等压面飞行

pressure-reducing valve 减压活门

pressure-volume diagram 压容图，压力容积线图

pressurising valve 压力活门；增压活门

pressurization 增压

pressurized cabin 加压舱，增压机舱

pressurized construction 加压结构

pressurized labyrinths sealing 增压篦齿封严

pressurized seal 压力密封圈

pressurized tank 加压箱，增压油箱

prestall buffet 失速前抖震

prestart 起动前

pre-start checks 起动前检查

pretest simulation 试验前模拟

preventable accident 可避免的事故（可防止的事故）

preventive maintenance 预防性维护

preview（PRV）预检；预观；预测

primary aerodynamic characteristics 主要气动特性

primary air stream 主气流

primary airflow 主气流

primary circulation 主环流，一级环流

primary control surface 主要控制面

primary cyclone 主气旋

primary display system 主显示系统

primary flight control 主要飞行控制

primary flight control system 主要飞行控制系统

primary flight display（PFD）主飞行显示器（常指飞机上的显示飞行姿态的显示器）；初级飞行表演

primary flow 主气流

primary flying training 初级飞行训练

primary frequency 主用频率，主要频率；正常使用频率

primary instability 最初失稳

primary instruments 基本飞行仪表

primary modulus of elasticity 基本弹性模数

primary parallax 初级视差

primary radar 初级雷达；主要雷达；一次雷达

primary surface 主平面

primary surveillance radar 一次监视雷达

primary target 主要目标

primary trainer 入门教练机；初级教练机

primary training aircraft 初级教练机

prime 加注；装填；装火药；装雷管；起动注油；涂底漆

prime meridian 原始子午线，本初子午线

priming device 起爆装置；起动注油装置

principal axis 主轴

principal moment of inertia 主惯性矩

principal part 主体机件

principal restraint 主限制

principal strain 主应变
principal stress 主应力
principal user processor 主要用户处理器
principle 原理；原则
principle axis 主轴线
principle of adaptation 匹配原理
printed circuit board 印刷电路板
priority landing 优先降落
prior-to-flare speed 拉平前飞行速率
prisoner exchange 交换战俘
prisoner of war camp 战俘营
PRM（precision runway monitoring）approach 精确跑道监视进近；精密跑道监察进近
PRMTR（parameter）参数
probability of hit 命中概率
probability of kill 击落概率，摧毁概率；杀伤概率
probable cause 可能原因；合理根据
probable error 概率误差；概率偏差
probable maximum precipitation 最大可降水量
probe 探测；探针，探头，探测器；（空中加油的）受油探管；（对一阵地进行）试探性攻击
procedural training 程序训练
procedure 程序；规程；过程，步骤
procedure turn 程序转弯，标准转弯（指一条航道上以等转弯角速度转弯，再按反方向转弯，以便飞机准确地按往复航线做机动飞行）
processing speed 处理速率
processor 被动、主动机载数字处理机；处理机；处理程序
procurement 采购；人员调入；补给品、器材或经费的获得
product 乘积；产品
product support 产品支持
production quality action request 生产质量措施要求
production test requirement 生产测试请求
proficiency 熟练
proficiency flight 熟练飞行（飞机驾驶员或其他空勤人员，为了提高自己的技术所作的飞行）；检查飞行，考核飞行
profile 翼型；剖面；侧面

profile drag 型阻；翼型阻力

profile drawing 剖面图

profile grinding 轮廓轮磨，轮廓磨光

profile line 剖面线

profile map 半剖面图，断面图

profilometer 测面仪，测平仪，表面光度仪

prognostic chart 天气形势预报图，预报图

prograde orbit 顺行轨道

program for regional observing and forecasting service 地区观测和预报服务项目

programmable read only memory 可编程只读存储器

programmed roll 预定滚转

programmed turn【箭】程序控制转弯

progress 进度，进行，进步

progress chart 工作进度表

progress report 进度报告

progressive approximation 逐次近似

progressive burning 增面燃烧

progressive propellant 增面燃烧推进剂

progressive schedule 进度时程

progressive stall 渐进失速，渐进型失速

prohibited area 禁航区，飞行禁区

prohibition of re-injector 防止重新喷油

projected area 投射面积；伸进气流中的面积

projected planform 投影平面形

projected propeller area 螺旋桨投影面积

projectile 射弹，抛射体

projectile charging 装弹

projectile loading 装弹

projectile trajectory 抛射体弹道

projection chart 投影图

projection（map）投影（地图）

proliferation（核）扩散

proliferation of nuclear weapons 核武扩散

promulgate 发布，颁布

prone position 卧姿；俯卧姿势

proof of concept 概念验证，概念论证；方案论证

proof strength 保证强度

proof test 证明测验，验证试验

propellant 发射药；火箭推进剂；

燃料

propellant tanks 推进剂箱

propellant loading ratio 推进剂装载比

propellant utilization system 推进剂输送调节系统

propeller 螺旋桨；推进器

propeller aerodynamic balance test 螺旋桨气动力平衡试验

propeller airplane 螺旋桨飞机

propeller anti-icer 螺旋桨防冰器，螺旋桨防冰装置

propeller axis of rotation 螺桨转轴（线）

propeller balance 螺旋桨平衡

propeller balance stand 螺旋桨平衡台

propeller blade angle（螺旋桨）桨叶角

propeller blade aspect ratio 桨叶展弦比

propeller blade element theory 螺旋桨叶素理论

propeller blade shank 螺旋桨桨叶柄

propeller brake 螺旋桨制动器

propeller camber ratio 螺旋桨（桨叶）厚弦比

propeller cuff 螺旋桨护甲

propeller disc 螺旋桨盘

propeller disk area 螺旋桨盘形面积

propeller effective thrust 螺旋桨有效推力

propeller feathering 螺旋桨顺桨

propeller flutter 螺旋桨颤振

propeller governor 螺旋桨调速器

propeller horsepower 螺旋桨马力

propeller hub 螺旋桨毂

propeller interference 螺旋桨（气流）干扰

propeller left-hand 左转螺旋桨

propeller load 螺旋桨负载

propeller load curve 螺旋桨负载曲线

propeller master blade 螺旋桨主叶

propeller pump 旋叶泵

propeller radius 螺旋桨半径

propeller rake 螺旋桨倾角

propeller reduction gear 螺旋桨减速齿轮

propeller reinforcing girder 螺旋桨增强樑

propeller root 螺旋桨叶根，螺旋

桨桨根
propeller section 螺旋桨剖面
propeller slip function 螺旋桨滑流函数
propeller speed 螺旋桨转速
propeller static balance test 螺旋桨静力平衡试验
propeller thrust 螺旋桨推力
propeller tip speed 桨叶尖端速率
propeller tipping 螺旋桨包梢
propeller torque 螺旋桨转矩，螺旋桨扭力
propeller wash 螺旋桨洗流
propeller width ratio 螺旋桨（桨叶）宽度比
propeller-blade speed 桨叶周围速率
propelling nozzle 推进喷嘴
proper value 特征值
proper vector 特征向量
proportional type 比例式
proportional/integral type controller 比例 / 积分类控制器
propulsion 推进力
propulsion wind tunnel 发动机（试验）风洞
propulsive efficiency 推进效率
propulsive power 推进功率
propulsor 推进器
protection tube 保护管；保护套管；防护管
protective gear 防护装备
protector 保护设备
prototype 原型；原型机，样机
prototype construction 原型机结构
prototype test flight 原型试飞
prototyping 原型机制造
protractor（航图上用的）量角器，分度规
protrude 弹出；出来
proving flight 验证试飞；技术（性）试航
proving ground（新武器）试验场，靶场
provisions 准备，预备，防备；措施，预防措施；设备，装置；供应，保障，供应品
proximate analysis 近似分析
proximity 近程，近位；接近，临近；近似
proximity fuze 近发引信；近爆引信
proximity sensor system 近位传感系统

proximity switch electronic unit 接近电门电子组件
proximity warning instrument 防撞警告器
prudent limit of endurance 安全续航时间极限；续航力可靠极限
prudent limit of patrol 巡逻可靠极限
PRV（pressure relief valve）减压阀，释压阀
PRV（preview）预检；预观；预测
pseudo random number 伪随机噪数字
pseudo-noise 伪噪声
pseudopursuit navigation 假追踪导航
pseudorange correction 伪距校正
pseudo-range error 伪距误差
pseudo-range rate 伪距变化率
PSU（passenger service unit）乘客服务组件
psuedo-range 伪距离
psychrometer 干湿计，冷却湿度计，干湿球湿度计
PTT（push to talk）对讲机；按下发话
public data network 公共数据网
public-address amplifier 扩音机
pull ring（降落伞的）开伞拉环
pull rod 拉杆
pull up 急跃升，（从平飞中）拉起，急剧上仰；停止，刹住，拨起
pullback 撤军
puller 拉拔器
pulley 滑轮，滑车
pulley block 滑车组，滑轮组
pulling force 拉力；推力
pull-off switch 拉断开关
pull-out（拉杆）改出俯冲；中断进场着陆，复飞；撤走；退出攻击；（将拦阻索）拉到头
pull-up turning 滚转
pulsating current 脉动电流，脉动气流
pulsating DC voltage 脉动直流电压
pulsating direct current 脉动直流电
pulsation 脉动
pulse altimeter 脉波高度计，脉冲式无线电高度表
pulse code 脉波密码，脉冲编码
pulse code modulation 脉波电码

调制

pulse duration 脉波宽度，脉冲持续时间

pulse modulation 脉波调制

pulse noise jamming 脉冲噪音干扰

pulse radar 脉冲雷达

pulse repetition frequency 脉冲重复频率

pulse repetition rate 脉波重现率

pulse signal 脉冲信号

pulse spacing 脉波间隔

pulse width modulator 脉冲宽度调制器

pulsed plasma accelerator 脉冲电浆加速器

pulse-frequency modulation 脉冲频率调制

pulsejet 脉动式喷气发动机，间歇式空气喷气发动机

pulsejet missile 脉冲喷射导弹，脉动喷气发动机导弹

pulse-time modulation 脉波时间调制

pulse-width modulation 脉波宽度调制

pultrusion 拉挤成形

pump 泵

pump discharge pressure（齿轮）泵出口压力

pump efficiency 泵效率，泵浦效率

pump pressure 增压泵压力

pump shaft rupture 泵轴损坏

pump speed 泵转速

pump suction efficiency 泵吸入效率

pump supply pressure 泵供应压力

pump-fed 泵馈

pumping 抽运，泵浦（泵压、泵汲）；唧筒作用，（气压指示的）急剧升降

pumping frequency 泵频

pumping friction 排吸摩擦

pumping speed 泵压速率

punch 铳子

punch press 压孔机

pundit 用莫尔斯电码作为识别信号的机场信标台；闪光灯标

pure helicopter 常规直升机；普通直升机

pure pursuit course lead 纯追击航线前置量

pure radial position 径向位置

purge 清洗，吹洗；（油箱）充气

（指向燃油箱内泵入惰性气体，排出油面空间的空气）

purge outlet union 放油活门出口接头

purple【通】可能携带核武器的飞机分队

pursuit 追踪，追击，驱逐

pursuit by fire 火力追击

pursuit course 追击航线；跟踪航线

pursuit missile 沿追踪曲线接近目标的导弹

pursuit navigation 追踪法制导；沿追踪曲线引导

pursuit troops 追击部队

push button annunciator 按压式电门信号器

push button indicator 按钮指示器

push down 推头下降

push forward 前推

push plate conveyor 推板运送机

push rod 顶杆，推杆

push the cyclic forward 向前推杆

push to reset 按压复位（置零）

push to talk（PTT）对讲机；按下发话

push/pull microphone 推拉式传话筒

pushback 推出，后推，拖机作业

pushbutton 按钮

pusher dog 拨爪

pusher propeller 推进式螺旋桨

pusher-type aircraft 推进式飞机

pushing force 推力

pushing over 进入俯冲

push-pull cable 推拉索，推拉式电缆

push-pull rods 推拉杆

pushrod 顶杆，推杆，提升活门顶杆

PW（practical work）实际操作

pylon（飞机场）标塔；塔式建筑物；（架高压输电线的）电缆塔；（机翼下悬挂副油箱或做弹等的）外挂架

pylon eight 标杆 8 字下行（指以标杆作为固定地所作的 8 字飞行）

pyranometer（平面）总日射表，日射强度计，总辐射表

pyrgeometer 大气辐射表

pyrheliometer 日射强度表

pyroelectric detector 热电探测器

pyrogen 致热源

pyrolysis products 解热生成物

pyrometer 高温计，高温表

pyrometric harness plug 高温电缆插头

pyrometric harness test 测温电缆测试

pyrotechnic shock 烟火冲击

pyrotechnics 烟火品，烟火信号弹，烟火施放术，烟火制造术

pyroxylin primer 硝基纤维性底漆

Q

QFE 场面气压

QNE 标准海平面气压；标准海压；标准气压

QNH 修正海平面气压；修正海压

QRH（quick reference handbook）快速参考手册；快速检查单

quadrant 象限；【炮】象限仪，射手象限仪；扇形支架，扇形摇臂；（等信号区之间的）无线电信标区；无线电指向标的 4 个等强信号扇形区

quadrant angle 象限角

quadrant elevation【炮】射角；仰角；水平射角

quadrantal point 中间基本方位

quadrantal altitude 象限高度

quadrantal deviation 象限罗差，象限差

quadrantal error 象限误差

quadrantal heading 象限航向

quadrantal points 象限点，象限基点

quadrantal separation 象限隔离

quadrantal signal 区信号

quadrate amplitude modulation 正交调幅

quadruple phase shift keying 四相相移键控

quadruple register 四用记录器（指启示风速、风向、阴晴、降雨用的记录器）

quaker gun 假炮；模型火炮

quality assurance 品质保证，质量保障

quality control 质量控制，质量管理

quantity of electricity 电量

quartering area 设营地域，设营区域

quartermaster 军需军官；军需主任

quartermaster corps 陆军军需兵

（种），陆军军需部队

quartermaster depot 军需仓库

quartermaster plan 军需勤务计划

quartermaster unit 军需部队

questionable aircraft 可疑航空器，可疑飞机

quick disconnect ring 快速脱开环，快速断开环，快卸环

quick fire 急射

quick fuse 瞬发引信

quick march 齐步行进（口令）

quick reaction force 快速反应部队

quick reference handbook（QRH）快速参考手册；快速检查单

quick release mounting bracket 快脱安装托架

quick response 响应快，快速响应

quick-fire 快速射枪（炮），急速射

quick-release buckle 快卸扣子，快卸扣环

R

R.H side view 右视

R.H static port 右静压口

rack 架，台，（尤指飞机空投货物）载物架，挂物架；炸弹架；发射架，（机载设备的）安装架，支架

rack and pinion mechanism 齿条和齿轮机构

RADAN（radar Doppler automatic navigator）多普勒（自动）导航仪

RADAN（radar navigation）雷达导航

radar acquisition 雷达探测，雷达截获（目标）

radar air navigation chart 雷达航行航图

radar alphanumeric display system 雷达字母和数字显示系统

radar altimeter 无线电高度表

radar altitude hold 雷达高度保持

radar approach control 飞机进场的雷达控制；雷达进场着陆控制

radar approach control（RAPCON）（美国联邦航空局的）雷达控制进近中心

radar babble 雷达干扰

radar beacon 雷达信标，无线电信标

radar bearing 雷达方位

radar check point 雷达校对点，雷达地标

radar clutter 雷达混扰，雷达杂波，雷达地面杂乱回波

radar confusion 雷达干扰

radar control area 雷达控制区域，雷达作用区

radar controller 雷达管制员；（用于）雷达控制（中的）设备

radar countermeasure 反雷达干扰；雷达对抗；反雷达措施

radar coverage 雷达覆盖范围，雷

达（有效）探测范围
radar data function 雷达数据功能
radar data processing 雷达数据处理
radar data processing system 雷达数据处理系统
radar detection 雷达探测
radar Doppler automatic navigator（RADAN）多普勒（自动）导航仪
radar echo 雷达反射波；雷达回波
radar flight-following 雷达飞行追踪
radar hand off 雷达警戒
radar height 雷达高度
radar identification 雷达识别
radar jamming 雷达干扰
radar microwave link 雷达微波链路，雷达微波传输线
radar modernization project 雷达现代化计划
radar monitoring 雷达监视，雷达监控
radar navigation（RADAN）雷达导航
radar navigation chart 雷达航图
radar navigation guidance 雷达导航制导
radar navigator 雷达领航员，雷达导航系统
radar netting 雷达网
radar netting station 雷达网中心站，雷达网络（控制）站
radar netting unit 雷达网装备，雷达网络装置
radar picket 雷达哨（为增大雷达的探测距离而活动于边缘地区的装有警戒雷达的车辆、飞机或舰只）
radar picket escort 雷达哨护航
radar pilotage 雷达地标导航，雷达领航
radar plot 雷达标图，雷达标示的航迹图
radar point 雷达站
radar position symbol 雷达位置符号
radar post 雷达站，雷达哨
radar prediction 雷达预测，雷达预报
radar products generator 雷达分量发生器
radar pulse 雷达脉波，雷达脉冲
radar range finding 雷达测距

radar range marker 雷达距离标尺
radar ranging 雷达测距
radar rating 雷达检定
radar receiver 雷达接收机
radar reconnaissance 雷达侦察，雷达探空
radar reflectivity 雷达反射率
radar repeat-back guidance 雷达反馈制导
radar report 雷达报告
radar resolution 雷达辨别力
radar response 雷达反应
radar return 雷达回波；雷达反射信号
radar scan 雷达扫描；雷达扫描法
radar separation standards 雷达航线间隔标准
radar service 雷达勤务
radar signature 雷达信号特征
radar silence 雷达静默，雷达静止状态；雷达静止时间
radar storm detection 雷达风暴侦测，雷达探测风暴
radar surveillance 雷达监视，雷达观测，雷达侦察
radar target 雷达目标，雷达目标回波
radar target designation control 雷达目标标定控制
radar target marker 雷达目标标示器
radar target simulator 雷达目标模拟器
radar track 雷达跟踪
radar track position 雷达追踪位置
radar track speed 雷达测定的地速
radar tracking 雷达追踪
radar tracking station 雷达追踪站，雷达跟踪站
radar vector 雷达引导
radar wind 雷达测风
radar wind system 雷达测风系统
radar-equipped inertial navigation system 雷达惯性导航系统，装有雷达的惯性导航系统
radar-guided missile 雷达制导导弹
radar-guided weapon system 雷达制导武器系统
radarscope 雷达显示器，雷达示波器
radial 径向的，辐式的
radial（flow）compressor 径流式压缩机，离心有压气机，离心式压缩器

radial acceleration 径向加速度
radial antithrust bearing 径向止推轴承
radial bearing 径向轴承
radial diffuser 径向扩散器
radial displacement 径向位移
radial drill 旋臂钻床
radial drive shaft 径向驱动轴
radial end thrust ball bearing 径向止推滚珠轴承
radial fan 辐流式风扇
radial flow 径向流，辐射流
radial fuel supply 径向供油
radial G 径向重力
radial heading 径向航向，辐射航向
radial jet 横向推进器
radial load 径向载荷
radial position 幅（径）向位置
radial search 辐射式搜索
radial spar 辐射樑
radial stator vane 径向静子叶片
radial stress 径向应力
radial turbine 径流式涡轮
radial unbalance torque 辐向不均衡扭矩
radial velocity 径向速度，视向速度
radial wing configuration 辐射翼外形
radial wire 辐条式钢索
radially expanding clutch 辐张离合器
radian 弧度
radiance 辐射度，面辐射强度；放射亮度，光度
radiant energy 辐射能
radiant heat 辐射热
radiation 辐射；放射；发射；射线；辐射能
radiation aid 无线电设备
radiation belt 辐射带
radiation dosage 辐射剂量
radiation fog 辐射雾，放射雾
radiation heat transfer 辐射热传递
radiation pressure 辐射压力
radiation scattering 辐射散射
radiational cooling 辐射冷却
radiator fan 散热风扇
radiatus 辐状云，辐辏状云
radio 无线电；无线电设备；无线电通信；无线电通信术；用无线电发报（或通信联络）
radio aid 无线电导航设备，无线

电设备
radio altimeter 无线电高度表
radio altimeter computer 无线电高度计算机
radio altimeter indicator 无线电高度计指示器
radio antenna 无线电天线
radio approach aid 进场着陆无线电导航设备;（飞机）无线电助降设备
radio astronomy 射电天文学
radio band 频带（波段）
radio beacon 无线电信标，无线电指向标
radio beacon marker 无线电信标标志，无线电指点信标台
radio beam 无线电波束
radio beam coupler 无线电波束耦合器
radio bearing 无线电方位（角）
radio bearing deviation 无线电方位差，无线电方位偏差
radio broadcast data system 无线电广播数据系统
radio channel 无线电信道，无线电波道
radio communication failure 无线电通信故障，无线电通信中断
radio communication failure message 无线电通信中断通报
radio communication panel 无线电通信面板
radio communications link 无线电通信链路
radio compass 无线电罗盘
radio control equipment 无线电控制设备
radio countermeasure 无线电对抗措施
radio data system 无线电数据系统
radio detector 无线电探测器;雷达
radio determination satellite 无线电定位卫星
radio determination satellite service 无线电测位卫星业务
radio direction finder 无线电定向仪，无线电定位器
radio directional beacon 无线电方向信标
radio direction-finding 无线电定向
radio direction-finding chart 无线电定向航图
radio direction-finding net 无线电

定向网

radio direction-finding station 无线电定向电台

radio directive device 无线电定向装置，无线电定向设备

radio distance magnetic indicator 无线电距离磁指示器

radio distortion 无线电失真

radio facility 无线电设备

radio fadeout 无线电波衰减

radio failure 无线电失效，无线电中断

radio failure procedure 无线电失效程序，无线电中断程序

radio fan-marker beacon 无线电扇形标志信标，无线电扇形指点标

radio fix 无线电定位

radio frequency 无线电频率，射频

radio frequency interference 无线电频率干扰

radio frequency oscillator 无线电频率振荡器，射频振荡器

radio frequency modulator（RFM）射频调制器

radio frequency regulator（RFR）射频调制器

radio guidance system 无线电制导系统

radio height hold reference bug 无线电高度保持基准半自动发报键

radio homing beacon 无线电归向信标

radio humidity test 无线电湿度试验

radio inertial guidance system 无线电惯性制导系统

radio interference 无线电干扰

radio line of position 无线电定位线

radio log 无线电台日志

radio loop 无线电环形天线

radio loop nacelle 无线电环形天线罩

radio magnetic indicator（RMI）无线电磁指示器

radio management panel 无线电管理面板；无线电管理专家组

radio marker 无线电指点信标；无线电信标台；无线电信号显示器

radio marker beacon 无线电标志信标，无线电指点标

radio marker transmitter 无线电标志发射机

radio mast 无线电柱
radio master control 无线电总控制
radio master system 无线电总系统
radio meteorograph 无线电气象计
radio meteorology 无线电气象学
radio navigation 无线电导航
radio navigational aid 无线电助航设备
radio orientation 无线电定向
radio pressure test 无线电压力试验
radio range beacon 无线电导航信标
radio range leg 无线电测距射束
radio range receiver 无线电测距接收机
radio relay 无线电中继
radio sensor system 无线电传感系统
radio telephony network 无线电话通信网
radio telescope 无线电望远镜，射电望远镜
radio teletypewriter 无线电传打字机
radio transmitter 无线电发射机，无线电发信机
radio transmitter receiver 无线电收发报机，无线电收发两用机
radio tuning unit 无线电调谐组件
radio vibration test 无线电振动试验
radio wave 无线电波
radioactivity 放射性
radiobiology 辐射生物学
radiochemistry 放射化学
radioecology 辐射生态学
radiograph 放射线照相，射线照片
radioisotope 放射性同位素
radiolocation 无线电定位，无线电定位学
radiological defense 放射防护，放射性辐射防护，防辐射
radiological weapon 放射性武器
radiometry 辐射度学，辐射测量学
radio-range station 无线电导航台
radiosonde 无线电探空仪，无线电高空测候器
radiosonde balloon 无线电探空气球
radiosonde observation 雷送观测，无线电探空
radiotelegraph（RTG）无线电报
radiotelephone（RTP）无线电话
radiotelephony（RTP）无线通话，

无线电话学
radius 半径，径向，射线
radius of action 活动半径，作战半径；有效破坏半径
radius of damage 损坏半径，破坏半径
radius of gyration 回转半径
radius of navigation 航行半径；导航系统有效半径
radius of operation 作战半径，活动半径
radius of turn 转弯半径
radius vector 径向量，向径
radome 雷达（天线）罩；雷达天线罩；天线屏蔽仪（雷达）
ragged clouds 碎云，飞云
raid 袭击，入侵；空袭
rail type launcher 滑轨式发射架，导轨式发射器
RAIM（receiver autonomous integrity monitoring）接收机自主完好性监测；接收机自主完善性监测；接收器自主合成监视
rain climate 多雨气候
rain cloud 雨云
rain day 雨日
rain drop 雨滴
rain removal system（风挡玻璃）除雨系统
rain repellent 排雨剂
rain shower 阵雨
raininess 雨量强度，雨量特性
raise the collective 上提总距杆
rake of wingtip 翼尖斜削
ram air exit 冲压空气出口
ram air inlet 冲压空气进口
ram air intake 全压空气受感器；全压接收管
ram air temperature（RAT）冲压空气温度
ram air turbine 冲压空气涡轮
ram airflow 冲压气流
ram drag 冲压阻力
ram pressure 冲压力
ram rocket 冲压火箭，冲压火箭复合式发动机
ram wing 冲翼，“冲翼”式
ramjet 冲压式喷气发动机，使用冲压式喷气发动机的飞机（或导弹）
ramjet missile 冲压喷射导弹，冲压式喷气发动机导弹
ramjet speed 冲压喷气发动机打开时之速率

ramming 冲击，冲压，锤击；输弹，送弹；撞击动作

ramp 停机坪;（无人机、靶机或导弹的）发射斜轨，滑轨;（水上飞机的）下水坡道，滑道

ramp weight 最大停机重量

rampart 壁垒，防御物

ramshorn airfoil 翘尾翼形

random access memory 随机存取存储器

random drift rate 随机漂移率

random error 随机误差，偶然误差

random vibration 随机振动

range 靶场，靶区，射击场;（尤指）导弹及无人驾驶飞行器的试验场，武器试验场;（导弹等）射程，航程，距离；测距，距离试射；区域，幅度，范围；无线电指向标;【无】波段

range ability 航程，活动半径；最大射程

range analog bar 距离模拟线（指光学显示器、平视显示器和雷达瞄准显示器出现的水平线。当截获目标时，该线的长度即表示距离）

range angle（轰炸）瞄准角，投弹角

range attenuation 距离衰减

range correction 射程修正，距离修正

range error probable 可能射程误差，距离公算误差

range extended missile 增程弹

range finder 测距仪

range gating 距离栅，距离门

range light 航路指示灯，跑道头灯

range markers 距离标识器

range octagon 射程八角形显示

range of action 活动范围

range of propeller 螺距范围

range panel 靶场标板

range resolution 距离分辨力

range station 无线电指向标，无线电导航台

range to target（离）目标距离

range tracking 距离追踪

range/azimuth display unit 射程/方位显示器

Rankine thermometric scale 兰氏温标

RAPCON（radar approach control）（美国联邦航空局的）雷达控制进近中心

rapid decompression 急速减压
rapid exit taxiway 快速出口滑行道
rapid eye movements 眼球快速移动
rapid-extraction parachute 快速开伞
rapid-reload capability 快速再装载能力
rare gas 稀有气体
rarefaction wave 稀疏波
rarefied gasdynamics 稀薄气体动力学
RAT（ram air temperature）冲压空气温度
ratchet 棘轮
ratchet wrench 棘轮扳手
rate damping 速率阻尼
rate gyro 速率陀螺仪，速率陀螺
rate of ascent 上升率
rate of attitude change 姿态改变速度；姿态改变率
rate of climb 爬升率
rate of climb indicator 升降速率表
rate of closure 接近率，接近速度
rate of descent 垂直下降速度
rate of heat release 释热率
rate of roll 滚转角速度
rate of sink 下沉率
rate receiver 速率接收机
rate station 主控站
rate transmitter 速率发射机
rated altitude 额定高度，计算高度
rated brake horsepower 额定制动马力
rated engine speed 发动机额定速率，额定发动机转速
rated horsepower 额定马力
rated load 额定负载，额定载荷
rated power 额定功率，额定动力
rated power output 额定动力输出，额定功率输出
rated revolution 额定转速，额定转数
rated thrust 额定推力
rated value 额定值
rate-integrating gyro 速率积分陀螺仪
rating 额定值，标称值，定额；参数，特性，工作状态
rating availability 可用状态
rating light 状态灯
rating stop 状态限制；状态止点
ratings 工作状态
ratings table 数据表

ratio changer module 传动比变换组件

ratio of transmission 传动比

ratiometer 比率计，比值计

rationale 基本原理

raw material 原料

rawin 雷达测风器，无线电测风仪

Rayleigh formula 瑞利公式

Rayleigh line 瑞利线

Rayleigh number 瑞利数

Rayleigh process 瑞利过程

Rayleigh scattering 瑞利散射

RB（relative bearing）相对方位；航向角；舷角

RBI（relative bearing indicator）距离方位指示器

RCC（Rescue Coordination Center）搜救协调中心，援救协调中心

RCLS（runway centerline lighting system）跑道中心线灯光系统

RCO（remote communications outlet）远程通信引出线

RCP（reversion control panel）反向控制面板

RCP switch（radar conversion procedure switch）雷达转换程序开关

reaching 现在到达（陆空通话用语）

reaching cruising altitude 到达巡航高度

reacting force 反作用力

reaction 反作用，反作用力；反向辐射

reaction control jet 喷气操纵系统喷管；反作用喷气舵；反作用控制喷嘴；小推力发动机

reaction control system 喷气操纵系统

reaction engine 反作用发动机

reaction propulsion 喷气推进；喷气发动机

reaction time 反应时间；延迟时间;【箭】发射准备时间

reaction torque 反作用力矩

reaction torque force 反扭矩力

reaction wheels 反作用轮

reactionary torque 反作用力矩

reactivation 重新建成（部队）

reactive force 反作用力

readable 能听清（通话质量）

readable but with difficulty 能听清但很困难（通话质量）

readable now and then 可断续听

到（通话质量）
readiness 待机，准备状态；备用状态；戒备，战备
reading 读数，读
read-only memory 只读存储器
readout 读数
readout mode selector 读数模式选择器
ready reserve 第一类预备役
ready room 飞行员待命起飞室，战斗值班室
real basis 真实基准
real fluid 真实流体
real time 即时
real time kinematics carrier tracking 实时动态载波跟踪（技术）
real time reconnaissance 即时侦察
real time reconnaissance cockpit display system 即时侦察座舱显示系统
real-time DGPS 实时差分全球定位系统
real-time interrogate 实时询问器
rear（aft）part 后部
rear access door 后部舱门
rear airframe structure 后机身结构
rear attachment 后支点
rear avionic bay 后航电舱
rear engine mounting 发动机后安装架
rear firewall 后防火墙
rear flange 后法兰盘
rear furrow wheel 尾轮
rear hoisting ring 后吊挂
rear limit stop 后限止器
rear mounting 后支座
rear seat 后座
rear section scavenge pump 后部回油泵
rear stay bar 后撑杆
rear view 后视；后视图
rear wheel drive tractor 后轮驱动式牵引机
rearming 补充弹药；重新装定引信
rearward c.g. stalling speed 后重心失速速率
rearward flight【直】后飞（指机尾朝前的飞行）
rearward speed 后飞速率
reattached flow 再附着流
reattachment point 再附着点
reaumur thermometric scale 列氏温标

rebound stroke 伸张行程
recalescent point 再炽点
receiver 接收机;(空中加油的)受油机
receiver autonomous integrity monitoring（RAIM）接收机自主完好性监测；接收机自主完善性监测；接收器自主合成监视
receiver gain 接收机增益
receiver transmitter 收发机；收发信机
receiving only 仅可收报
receptor 感受器；接收机
reciprocal bearing 反向方位
reciprocal value 倒数值
reciprocating compressor 往复式压缩机
reciprocating engine 活塞式（往复式）发动机
reciprocating motion 往复运动
reciprocating pump 往复泵
reciprocating sieve 往复式筛
reciprocating type supercharger 往复式增压机
recirculation 环流；二次循环；再循环
recognition light 识别灯
recoil 后坐力；后坐距离；后坐，反冲，跳回
recoil reduction unit 减少后坐力装置
recommended check 推荐检查
recommended speed 建议速率
recommit 重标战术行动
recompression 再加压
recon reconnaissance 侦察，搜索，探测
reconditioning 状态恢复；整修，检修
reconfiguration 重整军备；改装
reconfiguration unit 重新确认装置
reconnaissance 侦察，搜索，探测
reconnaissance aircraft 侦察机
reconnaissance detachment 侦察分队
reconnaissance element 侦察分队
reconnaissance missile 侦察导弹
reconnaissance mission 侦察任务
reconnaissance satellite 侦察卫星
reconnaissance troop 侦察部队
reconnaissance unit 侦察部队
reconnoiter 勘察，侦察，搜索，探察，踏勘
record observation 记录观测

recording airspeed indicator recording 自记空速表

recording altimeter 自记高度表，自记高度仪

recording anemomete 自记风速表，自记风速仪

recording barometer 气压计；自记气压表

recording documents 履历文件

recording optical tracking instrument 光学追踪记录仪，记录式光学跟踪仪

recover from stall 从失速中改出

recovery 回航；（作战飞机）安全返回基地（或海上平台）；（遥控飞机、靶机）场外回收；（直升机将坠毁飞机）吊回；（从螺旋俯冲、特技飞行中）改出；救援，救生

recovery capsule 回收舱

recovery factor 回收因数

recovery footprint 回收地点；回收降落场（指回收舱、载人航天器或其他物体预期降落的场所）

recovery net 回收网

recovery percentage 回收率

recovery speed 改出速度

recovery spring 恢复弹簧

recovery temperature 恢复温度

recovery time 恢复时间；救援（持续）时间

recovery voltage 恢复电压

recruit 新兵；征召（新兵）；补充，添足，充实

rectangular aperture 矩形口径；矩形孔径

rectangular coordinate system 直角坐标系

rectangular course 矩形航线

rectangular wave 矩形波

rectenna 校正天线

rectification 整流，检波，校正

rectification error 校正误差

rectified airspeed 校正空速

rectified altitude 校正高度

rectified three phase current 整流的三相电流

rectifier 整流器；检波管；（航摄）纠正仪

rectilinear flow 直线流

rectilinear flying 直线飞行

recurvature 转向

red alert 紧急警戒；红色警报

red bug 设定红标

red gang bar 红色联动手柄；红色共轴手柄

red indicator piston 红色指示活塞

red line 红线标志

red mechanical indicator 红色机械指示器

red snow 红雪

red warning 红色警报

red zone of intersection 红色交叉区

redeploy 重新部署；变更部署

redout 红视（在空中飞行员因向心加速度使头部充血而引起的头痛和视野变红的现象）

reduced comfort boundary 降低舒适度

reduced flight idle 减推力飞行慢车

reduced lateral separation minima 减小的最小侧向间隔

reduced longitudinal separation minima 减小的最小纵向间隔

reduced mass 折合质量

reduced modulus 折合模数

reduced modulus of elasticity 弹性折减模数

reduced pressure 订正压力；折算压力

reduced removal and installation time 拆卸装配时间少

reduced takeoff and landing 缩短起降

reduced temperature 订正温度；折合温度

reduced thrust 减推力

reduced vertical separation minimum（RVSM）缩小最小垂直间隔

reducing gear 减速齿轮

reduction gear bearing 减速器齿轮轴承

reduction gearbox 减速齿轮箱；减速机匣；减速箱；减速器

reduction transformer 降压变压器

redundancy failure 多余度故障；冗余故障

redundancy signal 冗余信号

redundant computer system 冗余计算机系统

redundant electrical supply 冗余供电

redundant structure 超静定结构，静不定结构，冗余结构

reel antenna 卷轴式天线，卷盘式天线

reemergency 再度出现

reentrant angle 凹角；折弯角（指

飞机蒙皮或构件外表面突然改变伸展方向，形成小于 180° 的角，超音速时能引起附体激波）
re-entry 重返大气层，重返
re-entry body 重返物体
re-entry corridor 重返走廊
re-entry plasma 重返电桨；再入等离子体
re-entry profile 重返剖面
re-entry speed 再入速率
re-entry trajectory 再入轨迹；返回轨道，再入弹道，重返弹道
re-entry vehicle 重返载具；再入（大气层）飞行器
reequipping 重新装备
reference area 参考面积；基准面
reference axis 参考轴；基准坐标轴；读数轴
reference channel 基准通道
reference circle 参考圆；分度圆；（机载雷达瞄准具荧光屏上的）瞄准环，基准圈
reference data 参考数据；参考资料；基准数据
reference datum 参考基准面；基准面；基准点
reference datum height 基准高度
reference direction 参考方向；基准方向
reference frame 基准坐标系
reference landing distance 参考着陆距离；基准着陆滑跑距离，基准着陆距离
reference monitoring station 基准监控站
reference orbit 参考轨道
reference parameter 基准参数
reference plane 参考面；基准面
reference point 基准点
reference pressure 参考压力
reference speed 参考速度，基准速度，标准速度
reference station 基准站
reference surface 基准面
reference target parameter 参照目标参数
reference temperature 参考温度，基准温度
reference trajectory 参考弹道；基准轨道，参考轨道，基准弹道
reference value 参考值；参照值
reference voltage 参考电压
refire time 再发射时间；第二次点火；再次发射间隔时间

reflection 反射；反射波，反射信号

reflection rainbow 反射虹

reflection suppression 反射抑制

reflection-interference wave 反射干涉波

reflective strip 反光条

reflectivity 反射性，反射率，反射系数

reflector 反光镜；雷达反射面；反射器；反射物；反射镜

reflex angle 反射角

reflex circuit arrangement 反射电路配置

reflex curvature airfoil 反翘翼形

reflexed trailing edge airfoil 反翘后缘翼形

reflux 回流；反流，逆流

refraction（光线、电波的）折射

refraction dome 折光罩

refractive index 折射率

refrigerant 制冷剂；冷冻剂；冷却液

refrigeration cycle 冷冻循环

refrigeration tunnel 结冰测试风洞；冷却风洞

refuel 加油；空中加油；（再）加燃料

refueler 燃料加注车；加油车；加油（飞）机；加燃料器

refueling 再加油；再加燃料

refueling flight 空中加油飞行

refueling probe 空中受油探管

refueling speed 空中加油飞行速度

refueling stop 补给站

refueling plane 加油机

refurbishment 整修；翻新

refusal speed 中断起飞极限速度

refused takeoff 中断起飞

regelation 再冻（作用）

regenerative cooling 再生冷却

regenerative cycle 再生循环

regenerative detector 再生检波器

regenerative gas turbine cycle 再生气涡轮循环

regenerative reheating cycle 再生重热循环

regime 管理制度；体制

region of influence 影响区

region of normal command 正常飞行状态范围

region of reversed command 反常飞行状态范围

region of reversed flow 反流区

regional air navigation 区域空中导航
regional airport 区域性机场
regional augmentation system 区域增强系统
regional supplementary procedure 地区补充程序
registered horse power 登记马力
registered thrust 登记推力
regression analysis 回归分析
regressive burning 退行性燃烧，减压燃烧
regressive orbit 逆行轨道
regular army 正规军
regulated current 稳定电流
regulated servo pressure 调节伺服压力
regulated takeoff weight 规定起飞重量
regulated voltage 稳定电压
regulation coefficient 调节系数
regulator 调节器
rehearsal 演习；预演；练习；训练
rehearsal day（演习、作战）预演日
reheat 再热
reheat effect 再热效应
reheating 重热
reheating cycle 重热循环
REIL（runway end identifier light）跑道末端标识灯
reinforce 增援；支援；加强；增加强度，加固；炮（枪）身的粗厚尾部
reinforced bulkhead 加强隔框
reinforced concrete 钢筋混凝土
reinforced fiber composite 纤维加强复合材料
reinforced patch 加强片；加强补丁
reinforced plastics 强化塑胶
reinforced plywood 加强层板；加强夹板；加强胶合板
reinforced shell fuselage 硬壳机身
reinforcing 加强，补强；增援
reinforcing bar 钢筋
reinforcing tape 加强带；补强胶带
re-injection 回注；重新喷油
reject the takeoff 停止起飞
rejected item 剔退件
rejected landing 中止着陆
rejected takeoff 中断起飞
rejection 报废；阻碍，抑制，

拒绝

rejector 带阻滤波器；拒收器；抑制器

relapsing fever 回归热

related ground service 地面相关服务

relateral tell 中继同级传送；间接报知；情报中转

relation curve 关系曲线

relational expression 关系表达式

relational position 关系位置

relative airflow 相对气流

relative altitude 相对高度；海拔高度

relative azimuth angle 相对方位角

relative bearing（RB）相对方位；航向角；舷角

relative bearing data 相对方位数据

relative bearing indicator（RBI）距离方位指示器

relative displacement 相对位移

relative efficiency 相对效率

relative humidity 相对湿度

relative inclinometer 相对倾斜仪；相对倾斜计

relative motion 相对运动

relative parallax 相对视差

relative pitch 相对螺矩

relative speed 相对速率

relative target altitude 相对目标高度

relative target bearing 相对目标方位

relative target speed 目标相对速率

relative velocity 相对速度

relative vorticity 相对涡度

relative weight 相对重量；相对重力

relative wind 相对风；相对气流；视风

relativity 相对性，相对论

relaxation of stress 应力松弛

relaxed static stability 缓和静态稳定性

relay 继电器；中继，转播；接力；传递；转运

relay aircraft 中继转信飞机

relay box 继电器盒

relay station 中继站；无线电中继站，转报站，转发台；转运站

release agents 离型剂

release altitude 投射高度；投弹高度；释放（滑翔机等）高度；（导弹、火箭等的）分离或发射

高度

release hand 释放手柄；解脱手柄，分离手柄；投弹杆，投弹手柄

release hook 释放钩

release mechanism 释放装置；分离机构，解脱装置；投弹装置

release pressure 释放压力

release process 释放过程

release speed 投放飞行速率

release spring 释放弹簧；回位弹簧

release time 放行时间，释放时间；投弹时间

reliability 可靠性；可靠度

reliability and maintainability 可靠性和可维护性

reliability control program 可靠性控制方案

reliability evaluation and corrective action program 可靠性评估和修改工作项目；可靠性评估和改正性措施项目

relief map 地势图；立体图

relief valve 泄压活门；释压阀；安全阀

relight 再点火；(发动机空中熄火后)重新起动

relight in flight 空中再点火

reluctance 磁阻

remark 附注；评语，注解

remedial action 补救措施；补救性行动

remedy 补救，改正方法，校正，修理，检查

remote apron 偏远停机坪

remote area precision positioning system 边远地区精密定位系统

remote communication outlet 遥控通讯站

remote communications facility 遥控通信设施

remote communications outlet (RCO) 远程通信引出线

remote compensation unit 远距补偿器；远距补偿单元

remote control 遥控，远程控制

remote control station 遥控站

remote duty sites 遥控任务区

remote electronic unit 远程电子组件；遥控电子组件

remote instrument controller (RIC) 远距离仪表控制器

remote maintenance monitoring

system 远端维护监视系统
remote measurement 远隔测定；遥测
remote monitoring subsystem 远端监视分系统；远端监视子系统
remote monitoring system（RMS）远端监视系统；远端测控系统；远程监控系统
remote sensing 遥控探测，遥测，遥感
remote system 遥控系统
remote transmitter/receiver 远端无线电发 / 收信机
remote velocity 整体速度
remotely monitored battlefield-area surveillance system 遥控战场监视系统
remotely piloted helicopter（RPH）遥控直升机
remotely piloted mini-blimp 遥控小型飞机
remotely piloted research vehicle 遥控研究载具；遥控研究机
removal 拆卸，消除，删除
removal and replacement 拆卸及替换
remove the idling conditions 解除慢车状态
removed component 已拆除组件
removed position 拆开状态
rendezvous 会合
rendezvous maneuver 会合机动
rendezvous point 会合点
repair shop 修理车间
repairability 可修性，适修性，可修度（指系统发生故障时，在规定的有效修理时间内能恢复到工作状态的概率）
repeatable accuracy 重复精准度
repertory 库存；储备；仓库
repetitive flight plan 长期飞行计划
replaceable component 可替换部件；可更换组件
replacement 替换；代替者；补充人员；补充兵员；归还；复位；取代；代替；替换；更换
replacement of accessories 附件的更换；更换附件
replacement of components 部件更换
replenish 添加；补充；补给（兵力）等
replenishment 添加；灌充；装满；加注；加足；补充；补给

reply mode 回答方式；回复模式

reporting point 飞机（在空中的）报告点；飞机相对的地理位置（国际民航组织用语）

representation 表示法，表示，表达式

representative cruising air speed 典型巡航空速

representative fraction 代表分数；比例

representative observation 示例观察

representative scale 图示比例尺

reproduction speed 再现速率；重现速率

request 要求，要求；需要

request descent 请求下降

request homing 要求归航；请求归航

request radar blip identification message 要求雷达回波识别信号

request to send 请求发送

required check 必需的检查

required communication performance 所需通信性能

required horsepower 马力需求；需用功率

required inspection item 必检项目

required navigation（RNAV）所需导航

required navigation performance（RNP）所需导航功能；所需导航性能

required navigation performance capability 所需要导航性能能力；要求的导航能力

required power for hovering 悬停需用功率

required system performance 所需系统性能

required time of arrival 要求的到达时间

requirements 需求，要求；必要条件；需要物；需要量

re-rate engine 固定推力发动机

rescue 救难，救援，搜救

rescue aircraft 救援飞机

Rescue Coordination Center（RCC）搜救协调中心，援救协调中心

rescue operations 救援工作

research aircraft 研究用航空器

reseau 气象台群；观测网；网格；栅，光栅

reserve 预备队；预备役；后备部

队；保留；保存；储备；预定；预备补给品；储（藏）量；伞兵的备用伞
reserve fuel 备份油量；贮备燃料
reserve margin 裕度；储备余额；备用容量
reserve officers training corps 后备军官训练队；预备役军官训练队
reserve power 备用功率
reserve service 预备役
reserve tank 副油箱，备用油箱；储备箱；
reserved controller 预备控制器
reservoir 液压油箱，油箱；储油箱；容器，储存器；水箱；气罐
reset 重启，重置，重调，重新装定；重设；清除；转换，转接；复原，复位，置零；进入预定点，进入作战区域
reset（RST）重置；复位键
reset switch 复位开关
reset the pressure supply 恢复供压
Reset/Set 复位 / 调定；复位 / 置位
resetting 复位；重新设置
residual deviation 剩余偏差
residual fuel 余油；剩余油量
residual gas 剩余气体
residual strength 残余强度；剩余强度
residual thrust 剩余推力；参与推力
resilience energy 弹性能；弹回能量
resin 树脂
resin content 树脂含量
resin rich 多树脂，富树脂
resin selection 树脂选择
resin transfer molding 树脂转注成型；树脂转注模塑
resistance 阻力，电阻，抵抗；阻力；抵抗组织
resistance character 电阻性
resistance characteristics 阻力特性
resistance derivative 阻力导数；空气动力导数（指空气动力或力矩对线速度、角速度或操纵面偏转角的偏导数）
resistance of taxing 滑行阻力
resistance temperature coefficient 电阻温度系数
resistance temperature detector 电阻式温度探测器
resistance thermometer 电阻温度计
resistance-capacitance coupling 阻

容耦合

resistance-coupling 电阻耦合

resisting moment 阻力矩

resisting shear 阻剪力

resistive 应变式

resistive temperature device 电阻温度检测器

resistor 电阻器

resojet engine 谐振喷射发动机；喷气发动机引擎

resolution advisory 决断咨询，决断提示

resolver 求解器；溶剂；分解器；解算器；旋转变压器

resonance 谐振，共振；共鸣

resonance diagram 共振图

resonance frequency 谐振频率，共振频率

resonant burning 谐振燃烧，共振燃烧

resonant pulsejet 谐振脉冲喷射发动机；共鸣间歇式冲压发动机

resource requirement 资源需求，所需资源

respirator 防毒面具；防尘口罩，防尘面罩

responder 应答器，应答机；响应器

responder beacon 应答信标

response 反应，效应，响应；答复；回答；【雷】应答（信号）；响应曲线，特性曲线；灵敏度，敏感性，感扰性

response set 反应趋向；反应定势

response speed 反应速度；响应速度

response time 反应时间，响应时间；过渡过程时间

rest mass 静止质量

restoring moment 回复力矩；恢复力矩

restrained beam 固定端梁

restraining force 约束力

restraint 货物系紧装置；限制器，约束器；限制；为使货物便于运输而进行的捆扎过程

restricted altitude 限制高度

restricted area 限航区，空中禁区；禁区，限制区

restricted propellant 限制推进剂，限燃推进剂

restricted space 限航空域；有限空间

restrictor 节气嘴；节流嘴；气流

限制装置；限定器；节流孔；节流阀

resultant 合力

resultant action 合力作用，合成作用，总的效果

resultant force 合力

resultant force/moment 合力 / 合力矩

resultant velocity 合速度

resultant wind 合成风

resume 重新开始；恢复；恢复到上一次的编队（或阵地、任务等）

retainer 止动器；止动装置

retaliatory strike 反击；报复打击

retardant 防火剂，阻燃剂

retarded bomb 减速炸弹，慢降炸弹

retarding force 制动力（减速力）

retention 保持力；超期服役；延长服役时间

reticle aiming dot 光网瞄准点

reticle ring 光网圈

retract the flaps 襟翼收上；收上襟翼

retract the gear 收上起落架

retractable device 可收放装置

retractable float 伸缩浮筒；收放式浮筒

retractable landing gear 收放式起落架

retractable landing light 收放式着陆灯

retractable launcher 伸缩式发射器；收放式发射装置

retractable tail wheel 收放式尾轮

retractable wing 可收翼

retracting lock 锁上锁

retraction jack 升降机；收放机构

retractor strut 伸缩支柱

retreating blade（旋翼的）后行桨叶

retreating blade stall 后行桨叶失速

retrenchment 内线防御工事（指壁垒、胸墙、堑壕等）

retrofire 制动火箭发射；制动发动机点火；减速点火，制动点火

retrofit 返厂修改；改装；改进；改型

retrograde motion 逆行运动

retrograde orbit 逆行轨道

retrograde wave 后退波，退行波，逆行波

retrogressive wave motion 后退波

动，逆行性波动

retrolaunch 向后发射，逆飞行方向发射

retrorocket 反向火箭，制动火箭，反推火箭；减速火箭；离轨火箭；制动火箭发动机

retrothrust 反向推力，制动推力

return trace 返回跟踪；回迹

return 回油；返回，返航；回程；回路；回路管道;【雷】回波（信号）；反射信号；反射;【雷】应答（信号）；报告

return filter 回油滤

return flow 逆流，回流

return flow tunnel 回流式风洞

return lightning 回闪电

return line 回油管；回油导管;（阴极显象管上的）回扫描的描迹

return streamer 回返闪流

return to base 返回基地；返航

return to shop interval 返厂间隔

return to the factory for overhaul 返厂翻修

reverberatory furnace 反射加热炉

reversal 逆转；反向；回动；倒转;（操纵）反效

reversal error 反置错误；倒转误差

reverse breakdown voltage 反向击穿电压；反射击穿电压

reverse current 逆向电流；反向漏电流

reverse curve 反向曲线

reverse flow combustor 逆流燃烧室；回流燃烧室

reverse flow engine 逆流发动机；回流发动机

reverse lever angle 反推手柄角度

reverse pitch propeller 逆桨螺旋桨

reverse sleeve 反向器衬套

reverse thrust 反推力，逆推力

reversed control 反操纵，反控制

reversed flow region 反流区

reversed rolling moment 反向滚转力矩

reverse-flow annular combustor 回流式环形燃烧室；逆流式环形燃烧室

reverse-flow methods 反流法；逆流法

reverse-flow turbofan-type engine 逆流涡轮风扇型发动机

reversement 倒转，倒转动作；跃升倒转；反转

reverser 换向器；反推力装置
reverser cowl 反推力装置整流罩
reverse-video value 反向显示值
reversible cycle 可逆循环
reversible pitch propeller 反距螺旋桨
reversible process 可逆过程
reversing engine 可逆发动机
reversing propeller 逆桨螺旋桨
reversing pump 换向泵；可逆式水泵
reversion control panel（RCP）反向控制面板
revolutions per minute（RPM）每分钟转数
revolving body 回转体，旋转体
Reynolds analogy 雷诺类比关系
Reynolds experiment 雷诺实验
Reynolds number 雷诺数
Reynolds stress 雷诺应力
RFM（radio frequency modulater）射频调制器
RFM（rotorcraft flight manual）旋翼飞机飞行手册
RFR（radio frequency regulator）射频调制器
RH standard cabin door ASSY 右侧标准舱门组件
rheostat 变阻器，电阻箱
rhombic antenna 菱形天线
rho-theta navigation system 距离角度导航系统
RHT（radar altitude hold）雷达高度保持
RHT（radar height）雷达高度
rhumbatron 空腔谐振管
ribbon antenna 扁形天线，带形天线
ribbon coil 带式线圈
ribbon fuse 带状保险丝
ribbon lightning 带状闪电
ribbon microphone 条带式话筒，带式传声器
ribbon parachute 条带式降落伞；带条伞，带式伞
RIC（remote instrument controller）远距离仪表控制器
ride control 载波控制；行驶控制
ridge 脊，岭，山脉，分水岭；海脊，海岭；狭长隆起部；（气象图上的）狭长空压带，脊
rifle optics 瞄准器
rifle telescope 步枪瞄准望远镜
rigging 传动装置；调整飞行操纵

系统，飞机水平测量

right aft service door 右后使用舱门

right bank 右侧气缸排；右倾斜，右坡度，右坡度转弯

right bundle branch block 右束支传导阻滞

right buttock line 右纵剖线

right cabin door 右舱门

right cargo cabin 右侧货舱

right circular cone 正圆锥体

right circular cylinder 直立圆柱

right engine area 右发动机区域

right engine outboard 右发动机外侧

right engine pylon 右发动机挂架

right horizontal stabilizer 右水平安定面

right inboard wing 右翼内侧

right landing gear well 右起落架

right lower fairing 右下整流罩

right main landing gear 右主起落架

right of way 交通优先权，先行权

right of way rules 路权规则，优先通行权规则

right panel（wind shield）右风挡

right pitot static probes 右皮托管静态探头

right side engine 右发动机

right side separation 右侧隔离

right side view 右视，右侧视图

right sideslip 右侧滑

right static port 右静压口

right wheel well 右轮舱

right wing area 右翼区

right wing gear 右翼起落架

right wing tip 右翼顶部

right wing trailing edge 右翼后缘

right wing underside 右翼下侧

right-hand engine 右转发动机

right-hand finger-four formation 右手四指编队

right-hand propeller 右转螺旋桨；顺时针转螺旋桨

right-hand rotation 右转，向右旋转

right-hand side tool 右向工具

right-hand thread 右旋螺纹

right-hand-circular polarized 右旋圆极化

rigid body 刚体

rigid cleat 夹板

rigid foam 硬塑胶

rigid motion 刚体运动

rigid pipe 硬管，刚性管
rigid pull rod 硬式拉杆
rigid rotor system 无铰式旋翼系统；刚接式旋翼系统
rigid structure 刚性结构
rigidity 刚性，刚度；硬度；稳定性
rigidity of linear system 线系刚度
rigid-rotor helicopter 刚接式旋翼直升机
rim speed 轮缘速度
rime fog 霜雾；雾凇，雾凇雾
rime ice 霜状冰，毛冰
ring flange 法兰盘
ring gage 环规
ring laser gyro 环形激光陀螺
ring shear 环剪
rinsing 冲洗；水洗；冲洗
rinsing procedure 清洗过程
rip cord（降落伞）张伞索，开伞拉绳
rip panel 开幅，裂幅
risk not determined 危险无法确定，不确定风险
risk of collision 碰撞危险
risk of surging 喘振危险
rivalry 敌对状态
rivet 铆钉；铆，铆接
rivet gun 铆钉枪
rivet set 铆钉用具
rivet squeezer 铆钉压缩机
riveted construction 铆接结构，铆合结构
riveting joint 铆接
riveting machine 铆接机，铆合机
riveting mechanism 铆接机械
RMI（radio magnetic indicator）无线电磁指示器
RMS（Remote Monitoring System）远端监视系统；远端测控系统；远程监控系统
RMS concentrator 远端监视系统集中器
RNAV（required navigation）所需导航
RNAV flight plan 所需航行飞行计划
RNP（required navigation performance）所需导航功能;所需导航性能
robot bomb 自动炸弹；自动操纵的飞弹；自动操纵的炸弹
rocker arm 摇臂；气门摇臂
rocker-joint chain 摇力链

rocket artillery 火箭炮

rocket attacks 火箭榴弹炮轰击

rocket ejector 引射火箭

rocket launcher 火箭发射架；火箭筒；火箭发射装置

rocket pod 火箭发射器

rocket projectile 火箭弹

rocket-assisted takeoff 火箭助推起飞

rocket-propelled grenade（RPG）火箭推动榴弹；肩射反坦克火箭弹；火箭弹

rockwell hardness 洛氏硬度，洛克威尔硬度

rod screen 杆筛

rod-cone vision 杆状与锥状视觉

roger 收到（陆空通话用语）

roll 倾斜，（飞机）横滚；横滚飞行；横摇；横倾侧（指舰船或水上飞机绕纵轴转动）；滑跑

roll angle 滚转角；倾侧角

roll attitude 横摇姿态，滚转姿态，坡度

roll attitude beep rate 滚转姿态“滴”音速度

roll axis 横摇轴；滚转轴；翻滚轴

roll baler 滚卷式打包机；卷包机

roll cloud 飑云，滚轴云

roll control 滚转控制；横侧操纵，滚转操纵；横侧操纵机构

roll control input module 横滚操纵输入组件

roll cumulus 滚轴积云；卷轴积云

roll hesitation 滚转停顿

roll indicator 滚转指示器；滚转指示灯

roll linear actuator 滚转线性作动筒；滚转线性作舵机

roll mode 滚转模式；滚转状态

roll moment 滚转力矩

roll off 滚离

roll out 改出坡度；向一侧脱离（编队）

roll pointer 滚动指针；回滚指针

roll reversal 滚转操纵反效；反效滚转

roll trim actuator 滚转配平舵机；滚转配平作动器；滚转配平致动器

roller 滚柱；飞机水平状态着陆；【口】起落架机轮；滚棒；滚柱；滚子

roller bearing 滚棒轴承；滚柱轴承；滚子轴承

roller chain 滚子链

roller conveyor 滚轮运送机；滚柱式输送机

roller pump 滚子泵；滚轮泵

rolling（running）takeoff（直升机或垂直起飞飞机的）滑跑起飞，滑行起飞；直接起飞（不在跑道头停留的起飞）；连续起飞

rolling angle 横滚角；滚转角

rolling bearing 滚动轴承

rolling body 滚动体

rolling cone 滚锥体

rolling contact 滚动接触

rolling eight 并列八字；双横 8 字飞行

rolling friction 滚动摩擦

rolling landing 滚转着陆；滚动落地；滑跑着陆

rolling moment 滚转力矩

rolling moment coefficient 滚转力矩系数

rolling performance 滚转性能

rolling speed 滑跑速率

rolling takeoff 滚转起飞；连续起飞；（直升机或垂直起飞飞机的）滑跑起飞；直接起飞

rolling victory 全面胜利

rolling-up of vortex sheet 涡面卷起

rollout 滑行减速；改出坡度；着陆滑跑，冲出跑道，滑出跑道；（一种新型飞机或原型机的）首次公开展出；机动飞行结束后使飞机处于完成下一个动作的有利位置

rollout（stop）end of the runway 着陆跑道停止端

rollover 翻滚；（飞机地面）侧滚翻转；换用机型

roof panel 舱顶板

roof transparent panel 顶部透明面板

room temperature 室温；常温

room temperature vulcanizing material 室温硫化材料

root constraint 根部约束

root locus diagram 根轨迹；根轨迹图

root mean square 均方根

root sum square 和的平方根

root-locus method 根轨迹法

root-mean-square 均方根值

rotary antenna 旋转式天线

rotary balloon 圆球形气球
rotary disc 转动盘
rotary induction system 旋转式进气系统
rotary joint 旋转接合器，旋转接合；旋转关节
rotary knob 回转旋钮；旋转开关
rotary motion 回转运动；旋转运动
rotary plate 转动圆盘；动环；动盘；转盘
rotary pump 旋转泵；转动真空泵
rotary resistance derivative 转动阻力导数
rotary screen 旋转筛；回转筛
rotary selector switch thermometer 转旋电门温度计
rotary shear 回转剪切；旋转剪切
rotary stabilizer 旋转安定器
rotary switch 旋转开关
rotary trim actuator 旋转配平作动筒；旋转调整作动筒
rotary variable differential transducer 转动可变差动传感器
rotary variable displacement transducer 旋转式可变位移传感器
rotary wind 旋转风
rotary wing 水平螺旋桨；旋翼；旋转翼
rotary wing aircraft 旋翼机，旋转翼飞机
rotary-wing aircraft 旋翼机，旋转翼飞机
rotatable-loop radio compass 可转环形天线无线电罗盘；装有旋转环形天线的无线电罗盘
rotate speed 滚行速度；旋转速度
rotating assembly 转动部件；转子组件
rotating beacon 旋转灯标；旋转信标
rotating beam ceilometer 旋光云幕仪；旋转波束云高计
rotating bezel 旋转表圈
rotating combustion engine 旋转燃烧发动机；转缸式发动机
rotating face 旋转面
rotating load 转动负载；旋转负荷
rotating nozzle 转动喷嘴；旋转式喷嘴
rotating pin shaft 转动销轴；旋转销轴
rotating platform 转台；转动平台
rotating ring 动环；旋转环
rotating shaft 回转轴，旋转轴

rotating speed 转速

rotating stall 旋转失速

rotating surface 旋转面

rotating swash plate 旋转斜板凸轮（动盘）；旋转倾转盘；旋转斜盘；斜盘

rotation 旋转；轮换；转动

rotation anemometer 旋转风速计

rotation axis 旋转轴

rotation construction plane 旋转构造平面

rotation deviating from the center of gravity 不绕重心的转动；旋转偏离重心

rotation shaft 转动轴

rotation speed 转速

rotational flow 旋流

rotational motion 有旋运动；旋转运动

rotational speed 转速

rotative compressor 旋转压气机；循环压缩机

rotative diaphragm 旋转薄膜

rotative surface 旋转面

rotochute 自动旋转伞；旋转降落伞；自转旋翼减速器

rotodome 旋转天线罩，旋转雷达天线罩，旋转雷达罩

rotor 旋翼；转子，叶轮

rotor angular momentum 转子角动量

rotor bearing 转子轴承

rotor blade 旋翼桨叶；转子叶片；转动叶片

rotor blade tip 桨尖；转子叶尖

rotor brake 旋翼刹车

rotor brake handle 旋翼刹车手柄

rotor brake indicator 旋翼刹车指示灯；旋翼刹车显示器

rotor brake switch 旋翼刹车开关

rotor bursting 转子破裂

rotor cloud 滚轴云

rotor collective pitch 旋翼总距

rotor concentricity 转子同心

rotor cone 旋翼锥体

rotor diameter 桨盘直径；旋翼直径

rotor disc 桨盘；旋翼桨盘；转子盘

rotor disc load 旋翼桨盘载荷

rotor downwash 旋翼下洗

rotor drive 旋翼传动装置

rotor dynamics 转子动力学

rotor flight control 旋翼飞行控制；

转子飞行控制
rotor head 旋翼头
rotor head drag 旋翼头阻力
rotor hub 旋翼桨毂；主桨毂
rotor mast 旋翼转轴；旋翼主轴
rotor moment of inertia 转子惯性矩；转子转动惯量
rotor pump 转子泵
rotor radius 桨盘半径；转子半径
rotor rotation detector 转子侦转器
rotor shaft 旋翼转轴；旋翼轴；转子轴
rotor solidity 旋翼实度
rotor speed 转子转速，旋翼转速
rotor speed sensitivity 转子速率灵敏度；转子速度敏感性
rotor structure plane of rotation 旋翼构造旋转平面
rotor tip vortex 翼梢漩涡；转子叶尖涡
rotor torque 旋翼扭矩
rotor track 旋翼锥体
rotor trim tab 旋翼调整片；旋翼配平片
rotor type 旋翼型式；转子类型
rotor vane 转子叶片
rotor wing helicopter 旋转翼直升机
rotorcraft 旋翼飞行器，旋翼机；直升机
rotorcraft flight manual（RFM）旋翼飞机飞行手册
rotor-tip speed 转子叶尖速率
rough balance 均衡不良；初步平衡
rough engine 发动机动作不良
roughness drag 表面粗糙阻力；粗糙度阻力
round counter 射弹计数器
round figure 整数
round head rivet 圆头铆钉
round head screw 圆头螺钉
round it out 拉平飘
round mouth plier 圆嘴钳
round nosed airfoil 圆头翼形
round number 整数
round trip flight 往返飞行
round trip fuel 往返燃油
round-the-clock observing and sighting system 昼夜观瞄系统
route 航线，航路，路，路程，道路；按规定路线发送
route chart 航路航图，航线图
route component 航路分风

route data 航路数据，航线数据；航线资料

route flight 航路飞行，航线飞行；按规定航线空运的任务飞行

route forecast 航线天气预报

route formation 沿途队形；航线飞行队形，疏开队形

route manual 航路手册

route segment 航线段

routine maintenance 日常维护；例行维护，例行维修；定期检修，定期维护

routing indicator 电路标识；位置标志；路线标志

RPG（rocket-propelled grenade）火箭推动榴弹；肩射反坦克火箭弹；火箭弹

RPH（remotely piloted helicopter）遥控直升机

RPM（revolutions per minute）每分钟转数

RPM 365 selector RPM365 选择器

RPM switch 转速开关

RST（reset）重置；复位键

RTG（radiotelegraph）无线电报

RTP（radiotelephone）无线电话

RTP（radiotelephony）无线通话，无线电话学

rubber case 橡皮套；橡胶套

rubber disc strut 橡皮减震支柱

rubber jacket 橡皮套

rubber removal 清除橡胶

rubber wheel 橡胶轮

rubber-base propellant 橡胶基推进剂

rudder 舵；方向舵

rudder bar 方向舵连杆；方向舵脚蹬

rudder control cables 方向舵操纵索

rudder control tab 方向舵操纵片

rudder effective speed 方向舵有效速率

rudder horn 方向舵角柄；方向舵杆

rudder pedal 舵磴，方向舵踏板，方向舵脚蹬板

rudder post 方向舵柱，舵柱

rudder roll 舵滚，蹬舵滚转

rudder torque 方向舵转矩，方向舵扭矩

rudder travel limiter 方向舵偏转角限动器；方向舵行程限制器，方向舵行程限位器

rudder travel limiter actuator 方向舵偏转角作动筒；方向舵行程限制器作动筒，方向舵行程限位器作动筒

rudder trim tab angle 方向舵配平片角

ruddervator（V 形尾翼的）方向升降舵；（空中加油伸缩套管上的）V 形稳定器

rugged 崎岖的，高低不平的，粗糙的

rugged terrain 崎岖地形

rule of the air and ATS 空中规则和空中交通服务

rumble【口】猛推油门；隆隆声；隆隆声（指火箭发动机不稳定燃烧产生的低音轰鸣声），发动机震动声

run away propeller 失控螺旋桨

run in 进入（飞行）；运转配合（机械）；试车

run off the R/W 冲出跑道（滑跑冲出跑道）

run out of the R/W 冲出跑道（滑跑冲出跑道）

run speed 运行速度

runaway failure mode（某种装置的）失控故障模式

runaway speed 飞逸速度；失控速度

run-down check 停车检查，停转检查

run-down time 停转时间；余转时间

run-down time 停转时间，余转时间

run-free 空转

running fix 移动定位，航行定位

running of navigation station 导航台工作情况，导航站工作情况

running of radio station 无线电台工作情况

running off the edge of the R/W 偏离跑道（侧风影响飞机偏出轨道）

running out 流出，用尽，清出；（燃料）耗尽；放下（起落架、襟翼等）

running/rolling takeoff 滚转起飞；连续起飞；（直升机或垂直起飞飞机的）滑跑起飞；直接起飞

running-away 用尽（燃料）

run-off 流出量，溢放口；越出，流出；冲出跑道

run-up area 试车坪，试车区

run-up time 起转时间

runway 跑道

runway alignment factor 跑道方位因数；跑道对准因子

runway alignment indication 跑道方位指示；跑到对准指示

runway alignment indicator 跑道方位指示灯；跑道对准指示灯

runway alignment indicator light 跑道方向指示灯

runway approach threshold 跑道进场端

runway arresting gear 跑道拦截索；跑道拦阻装置

runway barrier 跑道拦截索（网）；跑道拦阻装置

runway basic length 跑道基本长度

runway blast pad 跑道喷流防护垫

runway capacity 跑道承载能力，跑道通行能力：跑道容量（用于起飞、着陆的最大容许频率）

runway center line 跑道中心线

runway center line lights 跑道中心线灯

runway centerline lighting system（RCLS）跑道中心线灯光系统

runway condition reading 跑道状况读数；跑道情况读数

runway control van 跑道管制车辆

runway controller 跑道管制员；起飞线管制员

runway direction number 跑道方位号；跑道中心方向标号

runway end 跑道头，跑道尖，跑道端

runway end identification 跑道头识别，跑道终端识别

runway end identification lights 跑道头辨识灯，跑道终端识别灯

runway end identifier light（REIL）跑道末端标识灯

runway end safety area 跑道头安全区，跑道端安全区

runway floodlight 跑道泛光灯

runway gradient 跑道梯度，跑道坡度

runway incursion 跑道入侵

runway lead-in lighting system 跑道引进灯光系统

runway length limited weight 跑道长度限制重量

runway light 跑道灯

runway localizer 跑道定位器；着陆航向信标台

runway localizing beacon 跑道定

位信标

runway observation 跑道观测

runway occupancy 跑道占用时间，跑道的占用，跑道不空

runway separation 跑道隔离，跑道使用间隔

runway shoulder 跑道肩；跑道的侧安全道

runway snow blower 跑道吹雪车，跑道除雪机

runway spacing 平行跑道间隔

runway strip 跑道带，跑道

runway surface condition 跑道道面状况，跑道道面情况

runway temperature 跑道温度，跑道气温

runway threshold 跑道入口，跑道头

runway threshold speed 跑道头上空速率

runway touchdown zone lights 跑道着陆区灯，跑道接地区灯

runway turning bay 跑道转弯

runway turnoff lights 跑道滑出口标志灯

runway visibility 跑道能见度

runway visibility value 跑道能见度值

runway visual range（RVR）跑道视程

runway-edge light 跑道边界灯

runway-end identification light 跑道头辨识灯，跑道终端识别灯

runway-end identifier light 跑道末端指示灯

rupture 横向撕裂；炸裂；破裂，拆裂；断裂；绝缘击穿

rust remover 除锈剂

rustless steel 不锈钢

rust-proof oil 防锈油

RVR（runway visual range）跑道视程

RVSM（reduced vertical separation minimum）缩小最小垂直间隔

S

S turn S 形转弯

SA（selective availability）选择可用性

SA（situational awareness）态势感知

sabot【炮】炮弹软壳（包在炮弹外面，使次口径炮弹能在大口径炮内发射）；次口径炮弹；超速穿甲弹弹体（穿甲弹钢心外面包的一层铝壳）；火箭支撑环；（风洞内）模型支撑环

safe conduct 军用护照，（军事当局发给的）通行许可证；（尤指战时的）安全通行权，通行证

safe conduct pass 安全通行证

safe flight condition 安全飞行条件

safe gap 安全间隙

safe life structure 寿限内安全结构

safe load 安全负载

safe observer 安全观测员

safe-flying speed 安全飞行速率，安全飞行速度

safety alert 安全警戒

safety altitude 安全高度

safety arming device 安全备炸装置

safety barrier 安全网，安全屏障，安全拦阻装置

safety belt（座椅）安全带，保险带，座椅吊带

safety clutch 安全离合器

safety factor 安全系数，安全因数，保险系数

safety fuel 安全燃料；防爆燃料

safety glass 安全玻璃

safety harness 安全套带，安全吊带，安全挂钩具

safety height 安全高度

safety lever 保险杆

safety load factor 安全负载因数，安全载荷因数

safety means 保险方式

safety mode and fault isolation 安

全状态与失误隔离，安全模式与故障隔离

safety not assured 无安全保障

safety of flight 飞行安全

safety of flight addendum 飞安修正通告

safety operation area 安全工作区

safety pin（炸弹、地雷等爆破器材上的）保险针，保险销（以防引信意外触发），安全销

safety release clutch 安全释放离合器

safety relief device 安全泄压装置，安全减压装置

safety review 安全审查

safety service 安全服务，安全勤务

safety speed 安全速率，安全速度

safety stop 安全停车，安全停止

safety stripe chart 安全图线

safety unit device 保险器；安全装置设备

safety valve 安全活门，安全阀

safety window 飞安视窗，安全窗口

safety wire 保险丝

Saint Elmo's fire 翼梢电火；（暴风雨中桅顶、塔尖上出现的）电极发光，天电光球；圣埃尔莫火球

sally 突围；出击

salt laden atmosphere 含盐大气

salted weapon 高放射性核武器；放射性增强的核武器；加料核武器

salvage 营救；（对船只、货物等的）海上救助；海上打捞；（火灾、洪水、船舶失事时的）财产抢救；抢救出的财物（或船舶、船员等）；废物收集（或利用）；可利用的废物；（从危险的飞行动作中）挽救出来

salvo（火炮的）齐射；（礼炮的）齐鸣；（炸弹的）齐投；（火箭的）齐发；齐射的炮火；齐鸣的礼炮；齐投的炸弹；齐发的火箭翼次射

salvo arm 齐投备炸装置

salvo switch 齐投开关

SAM（surface-to-air missile）地对空导弹

SAM intercept missile 反防空飞弹

sample signal 抽样信号

sampling【地理】采样；取样；

抽样

sand laden atmosphere 带沙子大气，含沙大气

sand mist 沙雾

sandstorm 大风沙，沙暴，沙尘暴

sandwich maneuver 三明治机动，夹心机动（指防御双机中有一架脱离编队转到外侧并加速，另一架则进入敌机后方，形成夹击）

sandwich plate 三夹板，夹心板，夹层板

sandwich structure 夹层结构；三明治结构；叠层结构

sanitary airport 合于卫生条件的机场

sanitation 公共卫生，环境卫生；改善环境卫生；卫生设备，卫生清洁设施；盥洗设备

sap one's will to fight 使其士气低落

SAR（search and rescue）搜索与救援；搜救

SAS（hands-on stability augmentation system）手控增稳系统

SAS（stability augmentation system）增稳系统

SAS release switch 增稳系统解除开关

sastruga 雪脊，雪面波纹

satellite 卫星

satellite communication 卫星通信

satellite control data unit 卫星控制数据单元

satellite data unit 卫星数据单元

satellite interceptor missile 卫星拦截导弹；卫星截击导弹；卫星防空导弹

satellite lifetime 卫星寿命

satellite operational implementation team 卫星运行实施小组

satellite speed 卫星速率

satellite-and-laser-guided bombs 卫星及激光导航导弹

satellite-based landing system 星基着陆系统

satellite-guided missile 巡航卫星导弹；卫星制导导弹

satellite-tracking camera 人造卫星追踪照相机

satelloid 动力卫星；准卫星（指飞行在行星大气层的轨道中，需做连续或间断机动的一种卫星）

saturated adiabatic lapse rate 饱和绝热直减率

saturated air 饱和空气

saturated water 饱和水

saturation 饱和

saturation adiabat 饱和绝热线

saturation adiabatic process 饱和绝热过程

saturation area 饱和区

saturation bombing 饱和轰炸

saturation condition 饱和状态

saturation deficit 饱和差

saturation mixing rate 饱和混合比

saturation point 饱和点

saturation pressure 饱和压力

saturation specific humidity 饱和比湿

saturation temperature 饱和温度

saturation-vapor pressure 饱和蒸气压

savanna climate 热带草原气候

sawtooth 锯齿，锯齿形

sawtooth climb 锯齿形爬升

saybolt viscometer 赛氏黏度计

S-band S 波段

scale error（graduation error）刻度误差

scale height 均质大气高度；太阳大气标高

scale minus key 标度减少键

scale model 缩尺模型，成比例模型

scale plate 标度盘，刻度板

scale plus key 标度增加键

scan（波束）搜索（目标）；扫描，扫掠（电磁波或声波搜索中传感器旋转一周）；目视搜索，搜索，（肉眼对仪表或前方外界景象）扫视一周；（空中截击代语）“搜索指令扇区，遇到情况及时报告”

scan conversion 扫描变换，扫描交换

scan type 扫描方式

scanning antenna 扫描天线

scanning electronic microscope 扫描式电子显微镜

scanning sonobuoy 扫描声呐浮标，扫探声呐

scanning speed 扫描速率，扫描速度

scarf cloud 巾状云，云冠，围巾云

scatter 散布，分散，扩散，散射；（目标）弹着点分布；（炮火的）散射面；（数据）分散度

scattered 散射的

scattered cloud 疏云

scattered light 散射光

scattered showers 零零散散的阵雨，零星雨

scattering 散射

scattering coefficient 散射系数

scattering cross-section 散射截面

scattering factor 散射因数

scattering wave 散射波

scavenge 回油；回油油路，回油管

scavenge circuit 回油路

scavenge filter 回油滤

scavenge line 回油路

scavenge pump 回油泵

scavenge strainer 回油滤

scavenging 回油；回油油路，回油管

scavenging efficiency 回油效率

scavenging pump 回油泵

scavenging stroke 扫气冲程；回油行程

scenario 演习方案，演习计划，演习预案；想定，方案；景象，图幅，像幅，帧

schedule perturbation 排程扰动

scheduled check 定期检查

scheduled data 计划数据

scheduled flight 定期飞行；定期航班；按飞行计划表飞行

scheduled inspection 定期检验，定期检查

scheduled maintenance 定期保养维修，定期维修

scheduled maintenance check 定期维护检查

scheduled service 定期飞航，定期航班，定期航班服务

schematic diagram 原理图，简图，示意图

schlieren 纹影术

schlieren method 纹影法

schlieren photograph 纹影摄影

schmidt number 施密特数

scintillation counter 闪光计数器，闪烁计数器

scissoring 剪形动作，作剪式移动

scissors（起落架的）扭力臂，防扭臂;（主起落架支柱上的）安全电门；剪刀机动（指两个或数个双机组交叉飞行类似剪刀的动作）

SCMS（software configuration

management system）软件配置管理系统

scoop 戽斗形进气口，进风斗，戽斗;（滑油）收油池

scoop wheel 扬水轮；斗式水车；斗式挖掘车轮

scope 射程；视界，视野，范围；示波器；显示器;（箭、导弹等的）射程

scopodromic 朝向目标，航向对准目标的，自导引的

scorching 部分燃烧，高温害

score（压缩机叶片）刻痕；划痕，伤痕，划线；弹着记录，测定弹着偏差，射击效果；击毁（击伤）敌机数

scotopic vision 暗视力，暗视觉

scout 侦察者；侦察机；侦察件；侦察员；侦察部队；侦察；搜索

scouting 侦察，搜索

scouting map front 搜索正面

scouting speed 搜索速率，搜索速度

scramble 紧急起飞（航空母舰上或地面上飞机的紧急起飞）；驾驶员的紧急起飞动作；截击；拦截目标;（电子信号）加扰；加密；用改变的频率通信

scramble order 紧急起飞命令

scramble ramp 警戒坪

scramjet 超音速燃烧冲压（式）喷气发动机

scrap 报废；停止使用；废料；残物；碎片；切屑;【口】将战舰退役并作废钢铁出售;（装备的）报废处理

scraper conveyor 刮板运送机，刮板输送机

scratch 划伤，擦伤，抓痕

scratchpad【计】便条；便笺式存储器；高速暂存存储器

screened pan 网罩蒸发皿

screw 螺钉；螺丝钉

screw conveyor 螺旋运送机，螺旋输送机

screw gearing 螺旋齿轮传动，螺旋齿轮传动装置

screw motion 螺旋运动

screw pump 螺旋泵

screw stud 双头螺栓，螺桩，螺柱

screw thread 螺纹

screwdriver 螺丝起子，螺丝刀

screw-thread 螺纹

scroll 滚动；滚动条；翻卷，滚

动；上卷

scroll collector 螺旋集气管

scroll up 滚动条向上滚动；向上滚屏；上卷

scrolling 滚动显示

scrolling information 滚动信息

scrub 低矮丛林，低矮丛林地，密灌丛；擦掉（雷达标图）；取消计划；（飞行人员等的）淘汰

scrub and maintenance 擦洗与保养

Scud 碎雨云，飞云；“飞毛腿”（战区或战术弹道导弹）

SD（storm detection）风暴探测

SDP（standard datum plane） 标准基准面

sea breeze（尤指白天吹向内陆的）海风

sea clutter 海面杂波，海面回波

sea fog 海雾，海面蒸汽雾

sea level 潮位；海拔；海平面

sea level altitude 海平面高度

sea level density 海平面密度

sea level power 海平面动力

sea level pressure 海平面气压

sea level speed 海平面速率，海平面速度

sea level standard 海平面标准

sea level standard day 海平面标准日

sea level thrust 海平面推力

sea marker 海面标志

sea mine 水雷

sea return 海面回波，海面反射信号

sea scout 侦海火箭

sea state 波浪；海况，海浪情况

sea survival 海上求生

sea water thermometer 海水温度表

sea wave 海浪

seal 密封，封严，封口；焊接；焊封；密封装置，密封圈，密封件，密封垫；密封胶；封口物

seal carrier plate 封严衬套

seal head 动环，封头

sealant 密封胶，密封剂（胶），密封材料

sealed cabin 密封舱室，贵重货舱

sealed silo 密闭圆筒仓

sealing bush 密封衬套

sealing of drives 传动密封

sealing ring 封严圈，密封环，垫圈

seam welding 缝焊接，缝焊

search and rescue（SAR）搜索与

救援；搜救
search and rescue and homing 搜索救援和归航
search and rescue region 搜索救援区
search and rescue service 搜索救援服务
search and strike missions 搜索和攻击任务
search antenna 搜救天线
search jammer 自动干扰器；搜索干扰发射机
search light 探照灯，搜索灯
search lighting 探照灯搜寻
search radar 搜索雷达
searchlight 探照灯
seasat 海洋卫星
season shift check 换季检查
seat allocation 机位分派
seat arm rest 座椅臂靠，座椅扶手
seat belt 座椅安全带
seat capacity 座位容量
seat configuration（机舱）座椅布局
seat cushion 座椅缓冲垫，坐垫
seat electronics unit 座椅电子组件
seat head rest 座椅头靠，坐椅头枕
seat load factor 座位酬载率，实有客座率
seat of war 战场
seat pack parachute 座包式降落伞
seat pitch 座椅间距；座椅排距（指座椅前缘到前排座椅前缘之间的距离）
seating arrangement 座位配置，安排座次
seating chart 座次图，座位表
seating standard 座位配置标准；座椅布置标准
seat-type parachute 座伞，坐垫式降落伞
second（first-officer）pilot 副驾驶员，副机长
second（secondary）harmonic 二次谐波
second cosmic speed 第二宇宙速率
second freedom 第二航权
second hand 间接的，用过的
second law of motion 运动第二定律
second law of thermodynamics 热力学第二定律
second level maintenance 第二级维护

second passing the navigation station 二次通过导航台

second phase of expansion 第二阶段膨胀

second pressure rise 第二次加压

second radar altimeter 第二雷达高度表

secondary action 副作用；助攻行动

secondary air flow 二次气流

secondary air system 二股空气系统

secondary attack 助攻，次要作战

secondary axis 副轴

secondary bending 次弯曲

secondary bonding 二级胶合，二次胶接

secondary cause 次因

secondary circulation 次环流，二级环流

secondary control 辅助操纵，辅助控制

secondary cyclone 副气旋，次生气旋

secondary depression 副低气压，次生低压

secondary flow 第二股气流，次要气流

secondary frequency 次要频率，次频；备用频率

secondary fringe 次级条纹

secondary front 副锋

secondary part 从动部件

secondary probe 辅助探头；辅助探测器

secondary radar 二次雷达（应答机），次级雷达；辅助雷达

secondary rainbow 副虹，霓

secondary stall 二次失速

secondary stop 二次停止

secondary structure 辅助结构，次要结构（指飞机机体结构中破坏后不致随即影响安全的部分）

secondary surveillance radar（SSR）二次监视雷达

secondary surveillance radar code（SSR code）二次监视雷达代码

secondary surveillance radar with selective addressing 二次监视雷达 S 模式（选址模式）

secondary target 次要目标

secondary winding 次级线圈，二次绕组

second-in-command 副司令；副指挥官；副机长

section 切面；小队；区域
section properties 断面特性
section shape 截面形状
sectional aeronautical chart 区域航空地图
sectional boiler 分节式锅炉
sectional chart 分区航图（指五十万分之一航图，低速飞机目视领航用图）
sectional core 分部心型
sectionalized vertical antenna 分段式垂直天线
sector 地段；航线段；（无线电指向标的）扇形空间；空域区分，分区
sector arm 扇形摇臂
sector scan 扇形扫描
sector wind 航路分段风，区段风
secular perturbation 长期摄动
secular variation 长期变化
secure 占领；巩固；掩护；保卫；系紧；弄紧；【令】演习完毕；操练完毕；停止操作；解除任务；安全的；保密的；夺取及保持空中优势
secure flight indication system 安全飞行指引系统
securing bolt 自锁螺栓，紧固螺栓
securing clamp 紧固卡箍，紧固夹子
securing nut 锁紧螺母，自锁螺母，紧固螺母
securing screw 自锁螺钉，紧固螺钉
security 警戒；安全，保卫；保密；（旅客登机前的）安全检查
security classification 安全保密等级
security van 保安运货车
security zone 安全区，安全地带
sedentary fitness 静止
see-and-be-seen system 目视系统
Seebeck effect 塞贝克效应
seek 寻找，搜索；自导引，自寻的
seeker 寻标器；（导弹的）导引头，自动寻的头；探导器，探索者，探测器；传感器
seesaw hinge 跷跷板铰链
seismograph 地震仪
SELCAL（selective call）选择呼叫
selectable 可选择的
selected pressure altitude value 选

择性压力高度值

selected weather mode 选择天气模式

selective availability（SA）选择可用性

selective call（SELCAL）选择呼叫

selective calling system 选择呼叫系统

selective identification facility 选择性识别器，选择识别装置

selective inattention 选择性不在意，选择性忽视

selective jamming 选择性干扰，瞄准式干扰

selective pitch propeller 自动螺旋桨，选择性变距螺旋桨

selective reinforcement 选择性强化

selective transmission 选择变速传动

selectivity 选择性

selector 选择器

selenocentric constant 月心常数

selenocentric coordinates 月心坐标系

selenographic coordinates 月面坐标系

self alignment 自动校准，自主式对准

self sustaining speed 自助转速，自持转速，自持速度

self synchronous 自动同步

self-adaptive control-system 自适控制系统

self-bias 自动偏压；自动偏移，自偏移；自偏流

self-centering chuck 自定中心夹头，自动定心卡盘

self-destroying fuze 自毁引信

self-destruction equipment 自毁装备

self-energy 自有能量，自身能量

self-excited oscillation 自激振荡

self-extinguishing 自熄灭

self-guided missile 自导导弹，自控导弹

self-ignition 自燃，自发火（枪弹在进膛前自行发火的现象），膛外早发火，自动点火

self-ignition temperature 自燃温度

self-imposed stress 自加压力

self-inductance 自感系数

self-induction electro-dynamic force 自感电动势

self-initiated antiaircraft missile（自动完成敌我识别和截击目标的）自主式防空导弹
self-loading wagon 自装式搬运车
self-piloting ignition 自引点火，自动驾驶点火
self-priming pump 自吸式泵，自吸泵，自引泵
self-propelled speed 不用助推器之飞行速率，自航速度
self-sealing fuel line 自封油管
self-sealing housing 自封壳体
self-sealing tank 自封油箱
self-stall 自发性失速
self-sustaining speed 自持转速，自持速度
self-tapping screw 自攻螺钉
semi-active guidance 半主动制导
semi-active homing 半主动归航，半主动寻的
semi-active homing guidance 半主动归向导引，半主动寻的制导
semi-active landing gear 半主动起落架
semi-arid climate 半干燥气候
semi-armor-piercing projectile 半穿甲弹，半穿甲弹丸
semi-circular canals 半规管
semi-circular cruising level 半圆制巡航空层
semiconductor 半导体
semiflush inlet 半埋入式进气口
semi-latus rectum 半正焦弦
semi-monocoque 半硬壳式
semi-potential flow 半无旋流动
semi-retractable landing gear 半收放式起落架
semi-rigid airship 半硬式飞船，半硬式飞艇
semi-rigid rotor 半刚性旋翼
semirigid rotor system 半刚性旋翼系统
semi-span 半翼展
semi-stall 半失速
sense indicator 感测指示器
sense-antenna 感测天线
sensible atmosphere 可感觉天气
sensible enthalpy 可感热焓，显焓
sensible horizon 感觉地平，视地平
sensible-temperature 感觉温度
sensitivity time control 灵敏度时间控制
sensor 传感器，感受器，遥感

器；探测装置，雷达；接收机，接收器

sensor sun 太阳角感测器

sensory perception 感官知觉

sentry 哨兵，警卫；站岗，放哨

separated flow 气流分离

separating calorimeter 分离式蒸气干度计

separation 隔离；（气流）分离；（空中交通管制的）间隔；高度层次；（火箭级的）脱离，分离；脱离时间；退役；离开军队；（截击机与目标之间的）距离

separation angle 分离角

separation distance 分开距离，分隔距离

separation flow 分离流

separation line 分离线，分隔线

separation manoeuver 分离动作，分离机动

separation minima 最低隔离标准；最小间隔，最小距离或高度差

separation minimum 最小隔离，最小间隔

separation of variables 分离变量

separation point（气流）分离点，分流点；（火箭的）脱离点；脱落点

separation rocket 分离式火箭

separation stress 分离应力

separation surface 分离面

separator cloth 分离布

sepcial-use airspace delay 专用空域延迟

sequence flashing lights 顺序闪光灯

sequence of events 事件顺序

sequence report 顺序报告

sequential control 时序控制，程序控制

sequential correlation ranging system 程序相关测距系统

sequential logic circuit 时序逻辑电路

serein 晴天雨，晴空雨

sergeant 士官；军士（非委任军官）

serial data link 系列数据链接，串行数据链路

serial link 串行链路

serial number 系列号，序号，编号

serial rudders 串联方向舵，分段方向舵

series actuators 串联制动器

series double actuating cylinder 串联双作动筒
series-fed antenna 串馈天线
series-fed vertical antenna 串馈式垂直天线
serious accident 一等事故（机毁人亡事故），重大事故
serious injury 重伤害，严重损害
service 服务，服役，使用；勤务；维护，检修，保养；军种；武装部队的；服现役时用的
service apron 修护停机坪
service bulletin 服务通报，勤务通报，使用情况通告
service ceiling 实用升限（使用升限）
service difficult report 使用困难报告
service helicopter 勤务直升机
service life 使用寿命，耐用年限，使用期限；工作期限
service load 实用负载，使用载荷
service manual 维护手册，使用手册
service module 服务舱，勤务舱
service platform 工作台，操作平台
service request 维护要求（申请），服务申请
service stock 勤务备用器材
service type parachute 军用伞
servicing inspection 例行检查，维修检查
servicing placard 使用标牌
servicing procedure 维修程序
servicing truck 地面特种车辆，服务车
servo 伺服（随动）系统产，伺服（随动）机构；助力器；舵机
servo actuator 伺服传动机构；伺服作动器，伺服执行机构；舵机，助力器
servo control 伺服控制
servo control amplifier 伺服控制放大器
servo electronic box 随动电子箱
servo loop 伺服环路；伺服回圈；伺服回路，伺服系统，助力操纵系统
servo mechanism 伺服机构
servo motor 伺服马达，伺服电动机
servo mounting tray 伺服机构安装架（盘）

servo operation 伺服操作
servo piston 伺服活塞
servo position（SP）伺服位置
servo rudder 伺服方向舵；方向舵操纵片
servo system（SES）伺服（随动）系统
servo tab 随动操纵片（指由飞行员直接操纵的舵面上的调整片）
servo unit 伺服器，伺服机构
servo valve 伺服阀，伺服活门
servo-actuator 伺服作动筒；伺服电机；伺服执行机构
servo-aileron 伺服副翼
servo-valve 伺服阀，伺服活门
SES（servo system）伺服（随动）系统
set bolt 固定螺栓
SET inner knob 内部调整旋钮
SET outer knob 外部调整旋钮
set pressure 调定压力，调制压力，设定压力；整定压力，标定压力
set screw 固定螺钉
set the altimeter 高度表拨正
set the engine to idle 慢车位
setting 调整，调定；装定（在某一个位置）；调定值，设定值，调定位置；底座，底脚
setting angle 安装角
setting into operation 调定进入工作
setting pressure 调定压力，调制压力，设定压力；整定压力，标定压力
setting with power 功率设定
settle 调整，调度；固定；下沉，下降
severe injury 严重伤害，重度伤
severe storm 剧烈风暴，猛烈风暴
severe thunderstorm 剧烈雷雨
severe turbulence 严重颠簸，强紊流
severe weather 剧烈天气，恶劣天气
severe windshear 严重风切变
severe-cold 严寒
sextant 六分仪
sextant altitude 六分仪高度
SFC（specific fuel consumption）燃油消耗率，耗油率
shadowgraph 阴影显像
shaft 轴
shaft coupling 轴联结器，联轴节

shaft extension power 出轴功率
shaft horse power 轴马力
shaft liner 轴套，轴衬
shaft power 轴功率
shaft sleeve 轴套，轴套筒
shaft speed 轴速度
shaft thrust 轴推力
shaft turbine（涡轮轴发动机的）动力涡轮
shaft-turbine engine 涡轮轴发动机；涡轴发动机
shakedown flight 初次试飞；试航
shaking conveyor 摇动运送机，振动输送机
shallow 小角度；浅滩，浅水处
shallow approach 小坡度进近；小角度进近；小角度下滑进场着陆
shallow fog 地面雾，浅雾
shallow glide 小角度下滑
shallow turn 缓转弯，大转弯
shape effect 形状效应
shape function 形状函数
shaped beam antenna 成形波束天线
shaped rotor blade 特型桨尖
shaped-beam（phase-shaped）antenna 成形波束天线（相位定形）
shaped-charge 锥形装药
sharp bend 突弯，急弯
sharp nose pliers 尖嘴钳
sharp turn 急转弯，小转弯
sharp-edged gust 突变阵风，突发阵风
sharp-lip inlet 尖唇进气道
shear 切变
shear beam 剪力梁
shear lag 剪力滞后，剪滞
shear line 风切线，切变波，横波，切变线
shear load 剪切载荷
shear nut 抗剪螺帽
shear plane 剪力平面，剪切面，剪断面
shear section 剪切段
shear slide 剪力推压
shear strength 剪力强度，剪切强度
shear stress 剪应力，剪切应力
shear turbulent flow 剪力紊流
shearing effects of blast 爆破切剪效应
sheath 外壳；护套；鞘；（柔性桨叶螺旋桨的）金属包尖
sheathing（propeller blade）包端（螺旋桨叶）

sheet glass 薄片玻璃
sheet lightning 片状闪电，大闪电
sheet metal 金属片
sheet metal gage 金属片号规
sheet of vortex 涡旋片
sheet of vorticity 涡旋面
sheet-stringer construction 板桁结构
shelf life 库存寿命；货架寿命；储藏寿命；保存期限
shell 筒体；炮弹；航炮炮弹；弹药筒，烟火弹；猎枪子弹；炮击
shell crater 弹坑
shell theory 壳体理论
shield 防护罩，护板，盾；防护，遮蔽；屏幕，罩
shielded cable 屏蔽电缆
shielding 防护，屏蔽；防护的，屏蔽的；屏蔽物，放射性防护装置，核效应屏蔽物，辐射防护材料；（空气动力）遮蔽，阴影
shielding layer 掩护层，屏蔽层
shift joint 移动接头
shifting 偏移
shifting yoke 拨叉
shim 垫片
shimmy 摆振
shimmy damper 减摆器
shimmy hinge 摆振铰链；摆振铰
shimmy soft type 摆振柔软式
shining 燃亮
shipboard launching 舰上发射，舰载发射
ship-launched missile 从舰艇上发射的导弹；舰射导弹
shipment 装运
ship-to-air missile 舰对空导弹
ship-to-ship missile 舰对舰导弹
ship-to-underwater missile 舰对潜（艇）导弹；舰对水下目标导弹
shock absorber 缓冲支柱，减震器
shock absorber ability 减震能力
shock absorber leg 缓冲支柱
shock and awe air campaign 震慑空袭行动
shock cord 减震索，减震网；减震橡筋绳，弹力索，缓冲绳
shock damper 减震器
shock formation 震波形成
shock free airfoil 无震波翼形
shock front 震波前，激波波前；激波阵面，冲击波阵面；激波锋
shock isolator 减震器，减震装置
shock mount 减震架

shock polar 震波极图

shock pressure 震波压力，冲击压力

shock reducer 减震器

shock spectrum 冲击谱

shock speed 轴转速

shock spring 减震弹簧

shock stall 震波失速；冲击波气流分离

shock strut 减震支柱

shock tunnel 震波风洞，激波管风洞

shock wave 激波；震波；冲击波

shock wave intensity 震波强度

shock wave structure 震波结构

shock wave thickness 震波厚度

shock-absorbent mounts 减震架；减震装置

shock-expansion interference 震波膨胀干扰

shock-expansion theory 震波膨胀理论

shock-stalling speed 激波速率

shooting range 靶场

shop replaceable unit（SRU） 工厂更换装置；车间可替换件；内场可换（部）件

shop visit rate 维修频率，进厂率

shoreline effect 海岸效应

short（range）distance aid 短程助航设备，短距助航设备

short beam shear test 短梁剪力试验

short circuit 短路

short count 短数，短时计数

short field takeoff 短机场起飞

short final 短五边（指近距导航台以内的进场边）；进近至着陆的最后阶段

short haul 短程运输；短距离，短程；短程运输机（指航程 1 000 英里或 1 610 千米瓣支线飞机）

short landing 短距滑跑着陆（目测低的着陆）

short landing speed 短距滑跑着陆（目测低的着陆）速度

short of R/W（undershoot run）场外接地（提前接地）；没有落进跑道；目测低着陆

short range attack missile 短程攻击导弹，近程攻击导弹

short range ballistic missile 短程弹道导弹

short range forecast 短期预报

short range missile 短程导弹

short range navigation 短程导航，近程导航系统

short range navigation aid 短程助航设施，短程无线电导航仪

short takeoff and landing 短距起降；短距起飞与着陆；短距起落飞机（一般指 50 英尺越障高度的起飞或着陆距离不越过 1 500 英尺，即 457 米的飞机）

short takeoff and vertical landing 短场起飞和垂直降落，短距起飞和垂直降落

short takeoff speed 短距起飞速度

short term conflict alert 短期冲突告警，短期冲突警报

short term storage 短期贮存

short video film 录像片，视频短片

short wave radiation 短波辐射

short — circuit 短路

short-field landing（short approach）短跑道着陆（小航线进场着陆），短距离进场

short-haul aircraft 短程飞机

short-period oscillation 短周期振动

short-range 短距的，短航程的；活动半径小的；射程短的

short-range fighter 短射程战斗机，近程战斗机

short-range navigation vehicle 短程导航载具，近程导航飞行器

short-range omnidirectional beacon 短程全向信标，近距全向指向标

short-range weather forecast 短期天气预报

short-term memory 短期记忆

short-wave radio set 短波电台

shoulder 跑道肩；路肩，山肩；侧安全通道（指跑道两侧的无道区）

shoulder harness 肩套带;（座椅上的，或降落伞上的）肩带

shoulder straps 肩带

shoulder-fired missile 肩射导弹

show line 显示线

shower 阵性降水，阵雨

shrapnel（炮弹、水雷等的）弹片，炮弹破片，炮弹碎片；榴霰弹，子母弹

shrimpboat 标志牌（为空中交通管制员所用，在上面写上航班号、飞行高度等，放在显示器尖头信号附近）

shroud 散热罩；罩环；盖板；套

筒；覆盖物，护罩
shroud line 吊索，伞绳
shunt 分流管，分流器，分路器
shunt-fed vertical antenna 分路馈电式垂直天线
shut off valve 切断活门
shutdown 关车；（故障造成）空中停车
shutdown in flight 空中停车
shutdown of the engine 发动机停车
shutoff 关闭，切断，断开
shut-off 关闭，切断，断开
shutoff module 断路器
shut-off valve 断流阀，关闭阀
shutter-controlled air inlet 节气门控制进气道
SID（standard instrument departure）标准仪器离场；标准仪表离场
SID（standard instrument departure）chart 标准仪器离场图
side arm controller 座侧驾驶杆
side cabin door 侧舱门
side cutter 斜口钳；切边装置
side direction 侧向
side elevation 侧视图
side force 侧力
side force coefficient 侧力系数
side looking airborne radar 侧视空用雷达，侧视机载雷达
side mounted type 侧承载式，侧悬挂式，旁载式
side of body 机身侧面
side strut 侧支柱
side vertical fin 侧垂直尾翼，侧垂直安定面
side view 侧视图
side-by-side rotor 左右旋翼
side-by-side rotors helicopter 横列式双旋翼直升机
side-facing seat 侧向座
sidereal day 恒星日
sideslip 侧滑
sideslip angle 侧滑角
sideslip landing 带侧滑着陆
sidestep maneuver 侧向动作，侧向机动
sidetone 侧音，背景噪声
sidewall panels 侧墙板
sideward flight 侧飞
sideward speed 侧飞速率，侧飞速度
sidewash angle 侧洗角
sideways movement 侧向运动
Sidewinder “响尾蛇”空空导弹

siege 包围；围攻；围困

sieve sorter 筛选机

sight angle 瞄准角

sight depression 光网下降

sight line 瞄准线

sighting 瞄准，观测；目视发现（目标）

sighting circle 瞄准圈

SIGMET（significant meteorological information）重要气象信息，重要气象报告

sign status matrix 符号状态矩阵

signal 信号

signal area 信号区（指简易机场上邻近塔台的一块空地，用来展示布板等信号，供陆空通信用）

signal collection and identification card 信号收集和识别卡

signal conditioning 信号调节

signal engine aircraft 单发动机航空器

signal flare 信号弹

signal gateway 信号网关，信令网关

signal gun 信号枪

signal lamp 信号灯

signal light 信号灯

signal operation 信号操作

signal panel 信号板

signal pressure differential 压差信号

signal speed 信号速度

signal square 信号区

signal strength 信号强度

signalling speed 信号发送速率，发码速率

signal-to-noise ratio 信噪比

signature 性能，特性;（目标指示）特征；特征标记；签名；署名

significant meteorological information（SIGMET）重要气象信息，重要气象报告

significant weather 重要天气

silent chain 无声链，无声传动链

silicon carbide fiber 碳化硅纤维

silk cloth 绸布

similarity parameters 相似参数

simple beam 简支梁

simple doublet antenna 偶极天线

simple harmonic motion 简谐运动

simple helipad 直升机简易起降场地，简易直升机坪

simple injector 简单压力喷嘴

simplex 单路通信;【通】单工，

单向，单向信道（一方只能发，而另一方只能收的通信工作方式）

simplex circuit 单式电路，单工电路

simply connected region 单连通区域

simulated drill 模拟演习

simulated flight 模拟飞行

simulated forced landing 模拟迫降

simulated missile 模拟导弹

simulating obstruction 模拟障碍

simulation 模拟，仿真

simulator 模拟机，模拟器，模拟装置；练习器

simulator for air-to-air combat 空战模拟机，空战模拟器

simultaneous engagement 协同攻击；（拦截飞机和地对空导弹）同时攻击空中目标

simultaneous ILS approach 同时仪器降落进场；仪表着陆系统同时进场

simultaneous ranges 两用导航台

single blade propeller 单叶螺旋桨

single bottom plow 单底板犁

single channel digital electronic control system 单通道数字式电子控制系统

single channel per carrier 单路单载波

single chime 单谐音

single crystal alloy 单晶合金

single crystal blade 单晶叶片

single crystal material 单晶材料

single curvature 单曲率

single cylinder engine 单缸引擎，单缸发动机

single disk brake 单圆盘制动器，单盘式制动器

single disk clutch 单片式离合器，单盘离合器

single drift correction course 单一偏航修正航线

single engine（operations）单发（工作）

single engine landing 单发着陆，单发动机落地

single engine mode 单发模式

single engine rating 单发动机飞行检定

single entry compressor 单进气道压缩器，单面进气压气机

single front wheel tractor 单前轮牵引机，三轮牵引机

single heading course 单向航线
single isolated wheel load 单轮负重
single lever 单杆
single line drawing 单线图
single main rotor 单旋翼，单一主旋翼
single main rotor system 单旋翼系统
single phase three-wire connection 单相三线接法，单相三线连接
single pilot operations 单个飞行员作业，单个飞行员操作
single piloted aircraft 单人驾驶飞机
single point sensor 单点传感器
single pointer key 单指针键
single pole double throw 单刀双掷式
single precision 单倍精度
single rotor 单旋翼
single seat aircraft 单座机
single shot probability 单发炮弹命中概率
single side band（SSB）单边带
single sideband suppressed carrier 单边带抑制载波
single spar wing 单梁翼
single stage compression 单级压缩
single stage rocket 单节火箭
single stage sounding rocket 单节探空火箭
single stage to orbit 单节进轨道，单级入轨
single station analysis 单站分析
single strut landing gear 单支柱起落架
single theodolite observation 单经纬仪观测
single tow 单拖
single wedge airfoil 单楔翼形
single wheel microswitch 单轮微动开关
single-acting pump 单动泵
single-axis gyro 单轴陀螺仪
single-axis integrated gyro 单轴积分陀螺
single-phase rectification circuit 单相整流电路
single-pilot IFR 单个飞行员仪表飞行
single-point filler neck 单点式加油孔，单点式加油口颈
single-return wind tunnel 单次回流风洞
single-rotor helicopter 单旋翼直升

机，单桨直升机

single-rotor helicopter（with tail rotor）单旋翼带尾桨式直升机

single-row cascade 单列叶栅

single-spar construction 单梁结构

single-stage turbine 单级涡轮

single-wheel main landing gear 单轮主起落架

singular matrix 奇异矩阵

singularity 奇异性

singularity function 奇异函数

sink rate 下沉率；下滑速度

sinking speed 下沉速度，下沉速率

SINS（strap-down inertial navigation system）捷联式惯性导航系统

sinusoidal wave 正弦波

siphon chamber 虹吸压力计，虹吸室

situational awareness 环境观察；局势了解；形势感知，态势感知

situational awareness（SA）态势感知

SIU（system interface unit）系统接口单元

six-degree-of-freedom 六自由度

six-force elements of propeller hub 桨毂六力素

sixth freedom 第六自由

size of the cabin door 舱门尺寸

sketch map 示意图，草图，略图；（空中侦察时）手绘地图

skew aileron 斜副翼

skew bevel gear 歪斜齿轮，斜伞齿轮

skew T-log diagram 斜温图

skew-bevel gearing 歪斜齿轮传动

skid 滑板；滑橇；尾橇

skid landing gear 橇形起落架，滑撬式起落架；起落架防翻撬

skid strip 降落滑道

skid turn 侧滑转弯

skid type gear 滑橇式起落架

skidding turn 外侧滑转弯

skid-fin 侧滑直翅，翼上垂直面

skill or technique lack 缺乏技巧

skimp 节约使用

skin 蒙皮

skin buckle 表皮皱摺

skin drag 蒙皮表面的摩擦阻力

skin friction 表皮摩擦阻力（飞机移动时，空气与飞机表面摩擦），表面摩擦阻力

skin rivet 表皮铆钉

skin stress 表皮应力
skin temperature 表皮温度，表面温度
skin track 反射跟踪，雷达追踪；表皮追踪，表皮回波跟踪，蒙皮回波跟踪；回波循迹
skin tracking 反射跟踪，雷达追踪；表皮追踪，表皮回波跟踪，蒙皮回波跟踪；回波循迹
skin-bonded construction 蒙皮胶合结构
skinny 机密情报；内部消息；小道消息
skin-stringer construction 蒙皮桁条结构
skip bombing 跳跃轰炸，超低空投弹，跳弹轰炸
skip distance 跳跃距离，跳距
skip effect 跳跃效应
skip glide bomber 跳跃滑翔轰炸机
skip re-entry 跳跃式重返
skip vehicle 跳跃式飞行器，跳跃式航天器
skip zone 盲区，死角，静区，跳跃区
skirt dipole antenna 裙偶极天线
skirt fog 裙雾（指导弹从水冷装置发射时喷口产生的蒸气云）
sky balloon 高空气球
sky clear 天气晴朗，晴空；晴好；碧空
sky condition 天空情况，天空状态
sky cover 天空遮蔽，天空覆盖；空中掩护
sky diving 高空跳伞
sky force observer 天空力量观测者
sky force observer mark 天空力量观测点
sky obscured 天空不明，朦胧天空覆盖
sky radiation 天空辐射
sky synchronized long range navigation 天波同步长距导航
sky wave 天波（电离层反射的无线电波）
slab tail 全动尾翼
slab tailplane 全动水平尾翼（平尾）
slab trailing edge 平后缘
slab-tail plane 全动水平尾翼飞机（平尾）
slam acceleration 猛然加速，急剧加速，猛推油门
slant 歪斜的

slant contact 倾斜接触，斜面接触
slant distance 斜距
slant range 斜距（指飞机到地面某点之间的倾斜距离）
slant visibility 倾斜能见度
slash 劈刺；斜线
slat 前缘缝翼；固定式前缘缝翼
slat conveyor 板条输送机
slat plow 板条犁
slave ground radio station 地面无线电台
slave gyro 驱动陀螺
slave mode 从动方式，从属方式
slave part 从动部件
slave search（SS）从动搜索
slave station 副电台，从属电台
slaved gyro 驱动陀螺；磁修正陀螺仪
slaved gyro magnetic compass 陀螺磁罗盘
sledge 滑橇式
sleet 雹，冰珠；雨夹雪，融雪；冻雨，霰
sleeve 套筒；衬筒；衬套，（气缸的）中心衬套；风向袋；（外面用颜色表示用途的）电线塑料套
sleeve-dipole antenna 套筒偶极子天线
slender body theory 细长体理论
slenderness ratio 细长比
slew（猛地）转向，旋转，扭转；滑向一侧，侧滑
slew mode 定速驱动型
slide coupling sleeve 滑动连接套筒
slide rule 计算尺
slide shaft 滑动轴
sliding blade compressor 滑片式压缩机
sliding contact 滑动接触，滑动触点
sliding door 滑动门
sliding door ASSY 滑门组件
sliding friction 滑动摩擦
sliding motion 滑动运动
sliding speed 滑动速率
slight breeze 轻风（人面感觉有风，树叶有微响，风向标能转动）
slinger ring 甩油盘；离心喷液环
slip boundary condition 滑动边界条件
slip coefficient 滑流系数
slip flow 滑流
slip indicator 侧滑仪，侧滑指示器

slip joint 滑动接头，伸缩性连接
slip speed 滑速
slipping turn 侧滑转弯（盘旋）
skidding turn 侧滑转弯（盘旋）；滑动转
slipstream 后向气流；（飞机或螺旋桨的）滑流；沿流线方向运动；随另一架飞机的尾迹前进
sliver loss 裂片损失
slop line lighting 倾斜线灯光
slop of lift curve 升力曲线斜率
slope angle（下滑道）倾角，（对水平线的）倾斜角
slope coefficient 斜率系数
slope operations 斜坡作战；斜坡作业；斜坡行动；斜坡运行
slope resistor value 斜率电阻值
slope source internal resistance 斜率电源内阻
sloping ground 下滑降落
sloping shaft 斜轴
slot antenna 缝隙天线；开槽天线，槽式天线
slot time 槽时，时隙
slot type screwdriver 一字解刀；一字改锥；一字螺丝刀
slow down 减慢（速度）；减速，延迟
slow roll 慢滚
slow speed 低速
sluggish 迟滞
sluggish aileron response 副翼反应迟钝
slump pad 松软地表起降块
slush 半融化的雪（或冰）；烂泥，稀泥浆
small aeroplane 小型飞机
small aircraft 小型飞机
small arm 小型武器；（随身携带的）轻武器；随身武器；个人武器；枪械
small computer system interface 小型计算机系统接口
small disturbance theory 小扰动理论
small flow 小股气流
small helicopter 小型直升机
small pitch 小距
small pointer 小指针
small tension bar 小拉杆
smart bomb 灵巧炸弹，激光制导炸弹
smart weapon 智能武器，（激光）制导武器；灵巧武器

smear 污迹，污渍
SMMC（system maintenance monitor console）系统维护监视台；系统维修监控台
smog（常指工业地区排出的烟与雾混合而成的）烟雾，烟尘，霾，雾霾；烟幕，屏障；模糊不清
smoke bomb 烟幕航弹；烟幕炸弹；烟幕迫击炮弹；发烟手榴弹
smoke concentration level 烟幕浓度水平
smoke detection 烟雾探测
smoke detection system 烟雾探测系统
smoke elimination procedure 消烟程序
smoke haze 烟尘，烟雾，烟霾
smoke hoods 防烟头罩
smoke horizon 烟尘层顶
smoke particle 烟雾粒子
smoke sensor 烟雾传感器
smoke tunnel 烟风洞
smoking 烟雾
smoldering 无焰燃烧，阴燃
smooth and positive flow 平稳正向流动
smooth roller 光辊
smudge 熏烟；污点，污迹
snake mode 蛇行模式，蛇形飞行
snake motion 蛇行运动
snaking（飞机）摆头；曲折前进，蛇行
snap roll 快滚
snap shoot 扫射，急射，快速射击
snatch 攫取，抓走，绑架
snatch pick up 空拾
sneak 偷袭
sneak attack 偷袭，突然袭击
snipe 狙击，打冷枪；突然袭击
sniper 狙击手
snow blindness 雪盲（症）
snow clearance 雪清除，清除积雪
snow climate 积雪气候
snow cover 雪掩（量）；积雪，雪被，雪层；积雪厚度
snow density 雪密度
snow drift 吹（积成的）雪堆；吹积雪
snow grains 雪粒
snow gust 大雪（突然一阵）
snow line 雪线
snow removal equipment 除雪设备
snow showers 阵雪

snow skid 雪橇；滑橇
snowflake 雪花
snowplough 扫雪机，雪犁
snowstorm 暴雪
snowtam 雪情通告
SOAP（spectrometric oil analysis program）滑油光谱分析；光谱法油料分析计划
socket spanner 套筒扳手
socket wrench 套筒扳手
soft hail 软雹，软雪，霰，颗粒状雪球
soft landing 软着陆（降低飞行体速度而作的缓慢着陆）
soft soldering 软焊
soft terrain pad 松软地表起降块
software configuration management system（SCMS）软件配置管理系统
software modification 软件修改
software requirements and design description 软件要求和设计描述
solar absorptivity 太阳能吸收能力
solar atmospheric tide 日大气潮，太阳大气潮
solar cell 太阳电池组；太阳能电池
solar climate 太阳气候；天文气候
solar constant 太阳常数
solar day 太阳日
solar eclipse 日蚀，日食
solar energy 太阳能
solar panel 太阳电池板
solar parallax 太阳视差
solar polar 太阳极区
solar propulsion 太阳帆推进；太阳能电池驱动的电火箭推进
solar radiation 日射，太阳辐射
solar sail 太阳帆
solar terrestrial physics 日地物理学
solar wind 太阳风，日射微粒流，
solar year 太阳年（天时分 . 秒）
solenoid 电磁线圈
solenoid valve 电磁阀，电磁活门
solid angle 立体角
solid fuel 固体燃料
solid injection 无空气喷射
solid laminate 固体层压板
solid solution temperature 凝液温度
solid state transformer 固态变压器
solid wing 实心翼
solidified blade 结晶叶片
solidity 固性；实度（指螺旋桨

或旋翼桨叶总面积与桨盘面积之比）；稠度（指螺旋桨或桨叶在一个标准的半径处的弦长总和与圆周长之比）
solidity factor 刚性系数
solidity ratio 实度比
solid-state phased-array radar 固态相列雷达，固态相控阵雷达
solid-state power contactor 固态功率接触器，固态电源接触器
solo 单飞
solo flight 单飞
solution-adaptive grid 解调适网格
solution-ceramic 溶压陶质
solvent 溶剂
somatic condition 身体状况
somatogravic illusion 体重力错觉
somatogyral illusion 体旋转错觉
sonar 声呐；声波定位仪；水声测位仪
sonar buoy 声呐浮标
sonic 声音的，有声的
sonic altimeter 声波测高计
sonic barrier 音障，声障
sonic boom 轰鸣声，音爆
sonic boom carpet 音爆地区
sonic boom control 音爆控制
sonic inlet 音速进气道
sonic limit 音速极限
sonic line 音速线
sonic point 音速点，声速点
sonic speed 声速，音速
sonic wave 音波，声波
Sonne 桑尼（相位控制的区域顺序旋转无线电指向标）
SOS【无】国际通用的（船舶、飞机等的）紧急呼救信号;（海上）遇险信号；无线电呼救信号
soul on board 机上人数
sound range altimeter 声学高度表
sound suppressor 灭音器，消音器
sound system 音响系统
sounding airspace 探测空域
source control drawing 源控制图
source destination identifier 源目标识别器
south magnetic pole 南磁极
south pole 南极
Soviet M-1 苏米 -1 直升机
SP（servo position）伺服位置
SPA（standard pressure altitude）标准气压高度
space ring 分隔圈，垫环
space vehicle 空间运载体（卫

星）；航天器
space vehicle number 卫星号
space wave 空间波；天空（电）波
space-based position determination system 星基位置测定系统
space-charge layer 空间电荷区，空间电荷层
spaced antenna 隔开天线
spacer 垫片
spacing dividers 间隔分线规
spall effect 散裂效应
spalling 剥落（特指弹头撞击后装甲脱落）；散裂（特指喷气流使跑道水泥面破裂）；崩落
span 翼展，跨距，全长；一段时间；跨越；幅度取值范围；【海】跨绳
span efficiency factor 翼展效率因数
span length 翼展长，跨距长度
span loading 翼展负载，展向载荷
span of attention 注意广度
span-load distribution 展向载荷分布
spanner 用指距测量的人，测量器；扳手，扳钳，扳紧器
spanwise 展向，沿翼展方向，顺翼展方向
spanwise axis 展向轴
spanwise lift 展向升力
spanwise lift distribution 展向升力分布
spanwise position 展向位置
spanwise width 展向宽度
spar【空】（旋翼的或机翼的）翼梁；梁；桁；桅梁
spar cap 翼梁帽
spar web 翼梁肋
spare part 备份零件，备用部件，备用品
spare part bulletin 备件通告
spare parts catalogue 备件目录
spare parts kit 备份包
spark 火花
spark arrester 火花避雷器；火星熄灭器；火花罩
spark discharge 火花放电
spark plug 火花塞
spatial 空间的
spatial apparatus test 空间仪器检查
spatial disequilibrium 空间失衡
spatial disorientation【口】空间定向障碍
special aeronautical chart 专用航

空地图

special alloy 特种合金

special ATC factor 特别航管因素，特别空中交通管制因素

special detailed inspection 特别精细检测

special electronic installation 特种电子设备携带机；特殊电子装置

special emergency 特殊紧急情况

special flight VFR 特殊目视飞行

special force 特种部队

special forecast 特别天气预报

special meteorological report 特别气象报告

special navigation chart 专用航行图；专用飞机航空图

special observation 特别观测

special operation forces 特种作战部队

special operations 特种作战，特种行动；辅助活动，支援活动

special planning chart 特种计划航图

special rivet 专用铆钉，特殊铆钉

special service 特殊服务

special test permanent 永久特殊试验（改装后飞机的用途符号）

special test temporary 暂时特殊试验

special tool catalogue 专用工具目录

special tools 专用工具

special use airspace 特殊用途空域；禁区空域

special VFR condition 特殊目视飞行天气

specialized forecast 特别天气预报

special-purpose reconnaissance aircraft 特殊目的侦察机；特种任务侦察机

specific attenuation 特殊衰减；比衰减

specific configuration 特殊结构，特殊配置

specific density 比密度，比重

specific discharge 比流量

specific equipment 规定的设备，特定设备

specific excess power 单位剩余功率

specific fuel consumption（SFC）燃油消耗率，耗油率

specific gravity 比重

specific gravity adjustment 比重

调节

specific gravity sorter 比重选别机

specific heat 比热

specific humidity 比湿度

specific lift 比升力

specific mass 比质量

specific range 比航程，单位燃油航程

specific roughness 比粗糙度，比糙率

specific stiffness 比刚度

specific strength 比强度

specific thrust 单位推力，比推力

specific viscosity 比黏度

specific volume 比容

specification 详细说明；说明书；（对新装备的）性能表，规格，技术要求明细表

specification control drawing 规格控制图

specified heading 指定航向；规定航向

specified speed 额定转速

spectra fibers 光谱纤维

spectra resin 光谱树脂

spectral hygrometer 光谱测湿仪，光谱湿度计

spectrapolishing and finishing 光谱复材磨光及修整

spectrographic oil analysis program 滑油光谱分析程序

spectrology 光谱分析学

spectrometric analysis 光谱分析

spectrometric oil analysis 滑油光谱分析

Spectrometric Oil Analysis Program（SOAP）滑油光谱分析

spectropyrheliometer 分光仪

spectroscope 分光镜，光谱仪

spectrum 谱，光谱，频谱，波谱；系列，领域，范围

speed 速率，速度

speed band 速率带

speed brake 减速板

speed change device 变速装置

speed change flight 变速飞行

speed change gear 变速齿轮

speed change gear box 变速齿轮箱

speed characteristic curve 速率特性曲线，速度特性曲线

speed cone 圆锥形速度信号（用以指示军舰昼间在编队航行中的速度）

speed control 速率控制，速度控制

speed control loop 速度控制回路
speed data 速率数据，速度数据
speed down 减速，降速
speed effects 速率效应
speed error 速率误差，速度误差
speed factor 速率因数，速度系数，速率系数
speed fluctuation 速率变动，速度波动，速度忽高忽低
speed increase 速度增加，速度递增
speed landing 高速落地，大速度着陆
speed line 速率线
speed low pressure rotor 低压转子转速
speed measurement 转速测量；速度测量
speed of light 光速
speed of revolution 转速
speed of rotation 旋转速率，转速
speed of sound 音速
speed range 速率范围，速度范围
speed stable attitude 速度稳定姿态
speed threshold 转速极限
speed/attitude control system 速率姿态控制系统，速度姿态控制系统
speed-up engine 增速发动机，加速引擎
sphere of influence 势力范围
spherical antenna 球形天线，球面天线
spherical bearing 球面轴承
spherical coordinate system 球面坐标系
spherical elastic bearing 球面弹性轴承
spherical fly 球面飞行
spherical hinge 球铰链
spider（旋翼或螺旋桨的）星形桨毂，星型架；多脚架，星形轮
spider-web antenna 蛛网式天线
spike nozzle 塞式喷管，含锥喷管
spillage 溢流；溢出，泄漏量
spillage drag 溢流阻力
spillover 激光束偏离；激光束隘出（未照射在目标部位的激光束）；飘来雨；溢出，溢流；进气道超音速溢流；信息漏失
spin 尾旋，螺旋下降，旋冲
spin axis 螺旋轴；旋转轴；自转轴；回转轴
spin brakes 停转刹车装置；停转

制动器

spin chute 抗旋降落伞，反螺旋伞

spin model 螺旋实验模型机

spin tunnel 螺旋风洞，螺旋试验风洞

spindle 主轴，轴，心轴；杆，柱；（硬式飞艇的）锚杆构架

spindle drive 主轴传动机构，主轴驱动

spindle nephoscope 旋转测云器，旋转测云镜

spindle nose 主轴端部型式；主轴头

spinned stabilization 自旋稳定

spinner 螺旋桨整流罩，桨毂盖；（陀螺）转子，叶轮，旋转器；自旋体；旋转稳定的火箭弹；【口】旋涡

spinning tests 螺旋试验

spiral approach 盘旋进场

spiral axis 螺旋轴线

spiral bevel gears 弧齿锥齿轮；螺旋伞齿轮

spiral cone 螺旋锥，旋锥体

spiral dive 盘旋俯冲，急盘旋下降

spiral divergence 盘旋发散

spiral instability 盘旋不稳定性，盘旋下降不安定性，螺旋不安定性

spiral layer 螺旋层

spiral scanning 螺旋形扫描

spiral separator 螺旋分离机，螺旋分离器

spiral stability 盘旋稳定性

spiral tapered gear 锥形螺旋齿轮

spiral vortex 螺旋形涡流

spissatus【象】厚云；密（卷）云

splash lubrication 飞溅式润滑

splasher beacon 闪光灯塔；溅泼式信标

spline 花键，套齿

splinter 弹片

split 分开，分离，分裂，裂开，劈开

split drag 溢散阻力

split spline crankshaft 裂片式曲轴，键槽连接式曲轴

split system breaker 分裂汇流条跳开关，分离系统断路器

split vertical photograph 交像摄影术，错开影像垂直照片

split-axle landing gear 裂轴式起落架

split-surface control 分裂面控制，

分裂式操纵面
splitter plate 分流板,（进气道）隔板
spoiler 紊流板；阻力板；扰流板，扰流器
spoiler and stabilizer control unit 扰流板和安定面操纵装置
spoiler control module 扰流板控制组件
spoiler down 扰流板放下
spoiler electronic control system 扰流板电子控制系统
spoiler electronic control unit 扰流板电子控制组件
spoiler up 扰流板打开
spoilers and stabilizer control module 扰流板和安定面操纵组件
sponson 舷台，船侧突出部,（舷侧）突出部;（水上飞机的）翼梢浮筒,（水上飞机船身两侧的）水上安定面；机侧短翼；机侧翼状结构
spool down 发动机减速
spool up 发动机加速
sporadic but fierce pockets of resistance 零星但猛烈的抵抗
spot bombing 定点轰炸
spot hover 定点悬空，定点悬停
spot jamming 瞄准式干扰，选择性干扰；定点干扰，定频干扰
spot landing 定点落地，定点着陆
spot speed 点速度
spot takeoff 定点起飞
spot turn 原地转弯，原地转身
spot welder 点焊接机
spot welding 点焊接
spot wind 定点风，瞬间风
spouted bed dryer 喷流层干燥机，喷流床干燥器
sprag clutch 制动锁离合器
spray dryer 喷雾式干燥机
spray nozzle 喷雾嘴
spraying jet 喷嘴
spreaded spectrum multiple access 扩频多址
spring（桅杆等的）裂缝；倒缆；弹簧；弹出
spring belt 弹簧带
spring case 弹簧盒，弹簧罩壳
spring damper 弹簧减震器，弹簧缓冲器
spring plate manometer 弹簧流体压力计
spring pressure 弹簧压力

spring strip 弹簧片
spring tab 弹簧配平片，弹簧补偿片
spring tooth harrow 弹簧齿耙
spring-loaded closed ball valve 靠弹簧力关闭的球形活门
sprinkler 洒水车；洒水装置
spur gear 直齿轮；正齿轮
spur gear type pump 直齿型泵
spy plane 侦察机
spy satellite 间谍卫星，侦察卫星
squall 飑（间歇的强风）；飙；狂风
squall line 飑线，不稳定线，气压涌升线，飙线
square grid 方格网
square head bolt 方头螺栓
square search 方形搜索
square wave 矩形波，方波
squareloop antenna 方环天线
square-wing biplane 方形双翼机，等翼展双翼机
squaring shears 切边机，剪压机
squat 重心下移
squeeze effect 挤压效应
squeeze film 挤压油膜
squeeze film bearing 挤压油膜轴承，膜轴承，压膜轴承
squeeze riveting 压缩铆接法，挤压铆接
squeezed course 压挤航线
squelch 静噪；噪声抑制电路
squirrel cage tactics 松鼠笼战术
SRU（shop replaceable unit） 工厂更换装置；车间可替换件；内场可换（部）件
SS（slave search）从动搜索
SSB（single side band）单边带
SSR（secondary surveillance radar）二次监视雷达
SSR code（secondary surveillance radar code）二次监视雷达代码
SST（supersonic transport）超音速运输机
stabilator 全动式水平尾翼，全动平尾
stabilisation time before shut-down 停车前稳定时间
stability 稳定性
stability and control augmentation 稳定与控制加强
stability augmentation system（SAS）增稳系统
stability axis 稳定轴

stability control 稳定性控制
stability derivative 安定性导数，稳定性导数
stability diagram 稳定性图
stability satellite 稳定卫星
stabilization 稳态
stabilized condition 稳态
stabilized gyro 稳定陀螺仪
stabilized idle speed 稳定的慢车转速
stabilized platform 稳定平台
stabilized voltage circuit 稳压电路
stabilized working line 稳定工作线
stabilizer 安定面
stabilizer chord 安定面弦
stabilizer position module 安定面位置组件
stabilizer setting 安定面定位
stabilizer trim control module 安定面配平控制组件
stabilizer trim wheel 水平安定面配平手轮
stabilizer trim/elevator asymmetry module 安定面配平 / 升降舵不对称组件
stabilizer trim/rudder ratio module 安定面配平 / 方向舵比率组件
stabilizing altitude 稳定高度
stabilizing float 安定浮筒；减摇浮筒
stabilizing gear 安定架，稳定架
stabilizing moment 稳定力矩
stable airfoil 稳定翼形；安定翼面
stable break 稳定中断
stable descent 稳定下降
stable equilibrium 稳定平衡
stable hover flight 稳定悬停飞行
stable phugoid 稳定起伏
stable region 稳定区
stable region in the main 基本稳定区
stable system 稳定系统
stack ①【口】分层飞行等待着陆的飞机；(定高盘旋飞行的飞机)机群，编队 ②令飞机分层盘旋飞行等待着陆 ③使飞机作高度层次配置 ④【口】坠毁，撞坏
stadiametric aiming 视距瞄准；视距测量法瞄准
stadiametric ranging 视距测距
stadiametric warning 视距警示
staff pilot 指挥驾驶员，执行特种

任务的资深飞行员
stage efficiency 级效率
stage flight 各段飞行;(多段航程中的)航段飞行
stage fuel 阶段燃油；航段耗油量
stage length 站程；航线段长度
stage of exhaustion 衰竭期，衰竭阶段
stage of resistance 抗拒期，抵抗期
stage pressure 级压力
stage pressure coefficient 级压力系数
stage speed 航段速率，航段速度
stage temperature 级温度
stage time 级时间
stage velocity 级速度
stagger 斜罩，翼阶，翼差；前伸角；飞机维护间隔
stagger angle 前伸角;(双翼机的)斜罩角
stagger tuning 参差调谐
stagger wing 交错式双翼机
staggered grid 交错式网格
staggered joint 错开接缝；相错式接头
staggered seat 交错座位
staging base 中间基地；飞机中间停留基地；前进基地（机群准备在此发动攻击的基地）
staging point 站点
stagnant hypoxia 停滞性缺氧
stagnation 停滞
stagnation enthalpy 停滞热焓；滞止焓
stagnation point 滞（流）点；驻点
stagnation stall 停滞失速，滞止失速
stagnation streamline 停滞流线
stagnation temperature 停滞温度，滞点温度，滞止（流）温度
stain 污迹
stainless steel 不锈钢
stainless steel tape flight recorder 不锈钢带飞行记录器
stainless steel tube 不锈钢管
stairway 扶梯;(乘客)登机梯
stall 失速（飞机）；停车，出故障
stall angle 失速角
stall characteristics 失速特征
stall fence 失速栅；防失速翼刀
stall margin 失速界限
stall point 失速点
stall protection computer 失速保

护计算机
stall protection system 失速保护系统；防失速系统
stall quality 失速品质
stall sequence 失速程序
stall shaker 失速警告器
stall speed 失速速率，失速速度
stall strip 失速条，失速边条
stall tolerance 失速容限；失速裕度（指燃气涡轮发动机受进气畸变、吸入烟气、脉动上升气流、飞鸟、下雹等影响而不致引起风扇或压力机失速的能力）
stall turn 失速转弯，跃升失速倒转
stall warning 失速警示，失速警告
stall warning computer 失速警告计算机
stall warning system 失速警告系统
stall zone 失速区
stalled flight 失速飞行
stalled pressure 失速压力
stalling（stall out）accident 失速事故
stalling angle of attack 失速攻角，失速迎角
stalling characteristic 失速特性
stalling flutter 失速颤振
stalling speed 失速速度
stalling turn 失速转弯
stall-warning speed 失速警告速率
stamping 冲压
stamping die 模锻模；冲压模具
standard 标准（的）
standard accurate channel 标准精度通道
standard air density 标准空气密度
standard altitude 标准高度
standard and recommended practice 标准和建议措施
standard arrival 标准仪器到场航线
standard atmosphere 标准大气
standard atmosphere pressure 标准大气压力
standard atmospheric condition 标准大气状况
standard avionics configuration 标准航电配置
standard barometer 标准气压表
standard base-supply system 标准基地供应系统
standard beam approach 标准波束引导进场
standard body 标准机身

standard civil transport 标准民航交通

standard condition 标准状态，标准条件

standard conditions on ground 地面标准状态

standard configuration 标准配置

standard datum plane（SDP）标准基准面

standard day 标准天气

standard deviation 标准偏差

standard efficiency 标准效率

standard error 标准误差

standard instrument approach 标准仪器进场；标准仪表进场着陆

standard instrument approach procedure 标准仪表进场程序

standard instrument arrival 标准仪器到场；标准仪表进场

standard instrument departure（SID）标准仪器离场；标准仪表离场

standard instrument departure（SID）chart 标准仪器离场图

standard interchange format 标准相互交换格式

standard mean chord 标准平均弦长，平均几何弦

standard normal distribution 标准常态分布

standard of preparation 预备标准

standard operation procedures 标准操作程序，标准作业程序

standard operative temperature 标准作业温度

standard parallel 标准纬线，标准纬圈

standard part 标准件，标准零件

standard pattern 标准模型，标准样式

standard performance data 标准性能数据

standard pitch 标准螺距

standard position service 标准定位服务

standard practice manual 标准规程手册，标准练习手册，标准实践手册

standard pressure 标准压力

standard pressure altitude（SPA）标准气压高度

standard pressure lapse rate 标准气压直减率

standard propagation 标准大气中

传播

standard radio atmosphere 标准无线电大气层

standard refraction 标准折射率

standard regional route transmitting frequencies 标准地区航路发射频率；标准地区航线发射频率

standard service 标准服务；标准航空客运

standard shape 标准形状，标准形态

standard speed 标准航速；标准速度，标准速率

standard state 标准状态

standard temperature 标准温度

standard temperature and pressure 标准温度和压力

standard terminal arrival chart 标准航站进场图

standard terminal arrival route（STAR）标准抵达终端航线，标准场站到达航线

standard testing 标准试验，标准测试

standard tractor 标准式牵引机

standard turn 标准转弯

standard union 标准联盟

standard unit 标准单位

standard weather 标准天气

standard（-rate）turn 标准（速率）转弯

standardization 标准化，规格化

standards book 标准件手册

standby（STB）①备用的，应急的，待命的；等待（待飞）②备用品，应急设备 ③一等战备状态 ④（呼叫信号）准备发报；准备收报 ⑤（无线电代码）“我要暂停几秒钟”

standby engine indicator 发动机备用指示器；备份发动机仪表

standby frequency 备份频率

standby indicator 备用指示器

standby instrument 备用仪表，备份仪表

standby mode 备用模式；备用状态，等命状态

standing bolt 固定螺栓

standing cloud 停留云，驻云

standing patrol 不间断巡逻，例行巡逻；定位巡逻

standing wave 定波，定振波，波漾，立波，驻波

stand-off 偏离预定航线；在敌防

空区外飞行；（飞向或飞离目标的）巡航飞行；在目标区外
stand-off armour 夹层装甲
stand-off attack 距外攻击
stand-off bomb 远投炸弹
stand-off weapon 距外武器，在防空区外发射的空对地武器；远程武器
STAR（standard terminal arrival route）标准抵达终端航线，标准场站到达航线
star part 星形件
star target 星标
starboard 右舷
starboard navigation light 右航行灯
stardust 星状尘
star-flexible rotor 星形柔性旋翼
start acceleration 起动加速
start accessory 起动附件
start accessory relay 起动附件继电器
start bleed valve 起动放油活门
start C-B（start control button）起动控制按钮
start control has no effect 起动操纵不成功，起动操纵无效果
start drain valve 起动放油活门
start duration 起动持续时间
start electro-valve 起动电磁活门
start envelope 起动包线
start injector 起动喷嘴
start injector ventilation 起动喷嘴吹气
start purge 起动放油
start purge valve 起动放油活门
start selection has no effect 起动步骤不成功，起动步骤无效果
start switches 起动开关
starter 起动机
starter air valve 起动机空气活门
starter circuit breaker 起动断电器，起动断路器，起动断路开关
starter contactor 起动机接触器
starter cut-out 起动机断路，起动机切断
starter exhaust 起动机排气管
starter generator 起动发电机
starter intensity 起动机电流
starter magneto 起动磁电机
starter notch 起动机位置
starter-generator 起动发电机
starter-generator adaptor 起动发电机转接座，起动发电机适配器
starting 起动，开动，开车

starting acceleration 起动加速
starting accessory 起动附件
starting and relight envelope 起动和再点燃包线
starting apparatus 起动装置
starting button 起动按钮
starting chamber 起动燃烧室
starting characteristics 起动特性
starting circuit 起动电路
starting coil 起动线圈
starting cycle 起动循环，起动周期
starting difficulty 起动困难
starting envelope 起动包线
starting fuel flow control 起动燃油流量控制
starting point 起点；发射点；出航点
starting position 起动位置
starting power 起动功率，起动动力，起动电源
starting pump 起动泵
starting rate 起动率
starting sequence 起动程序
starting system 起动系统
starting torque 起动转矩，起动力矩
start-up 起动
start-up circuit 起动电路
state variables 状态变数，状态变量
static 静态的，静止的，不变的；固定的；静电；静电干扰；天电干扰
static air pressure 静空气压力，静压
static air temperature 大气静温，空气静温，静态空气温度
static airspeed indicator 静压空速表
static balance 静平衡，静力平衡
static balance control surface 静力平衡控制面
static balance stand 静力平衡试验架
static cable 固定缆;（降落伞的）引导绳挂钩索
static ceiling 静升限
static conversion 静态转换
static coupling 静力耦合，静电耦合
static curve 静态曲线
static determinacy 静定性
static discharger 静电放电器，静

电释放器

static dispatch 静态调派

static droop 静态下垂

static electricity 静电

static equilibrium 静力平衡，静态平衡

static firing 静态点火

static fracture 静态折断，静态破裂

static friction 静摩擦

static gearing ratio 静态传动比

static ground line 静态地面拉绳，（飞机的）接地线，搭地线

static head 静压头，皮托管的静压受感器

static lateral stability 侧向静安定性

static lift 静升力

static line 固定拉绳（降落伞）；（降落伞用）强制开伞拉绳

static load 静负载，静载荷

static load factor 静负载因素，静载荷因素

static longitudinal stability 纵向静安定性

static margin 静态稳定界限；静稳安定度

static moment 静力矩

static pan 静盘

static performance 静态性能

static port 静压口，静压孔

static pressure 静压力

static pressure port 静压口，静压孔

static pressure source 静压源

static pressure system 静压系统

static pressure system error 静压系统误差

static propeller thrust 静螺旋桨推力

static radar 电子扫描雷达，静态雷达

static rigidity 静态刚度

static source-select valve 静压源选择活门

static stability 静稳定性

static stabilizing moment 静态稳定力矩

static stall 静态失速

static stop 静态停止

static strength 静强度，静力强度

static testing 静力测试，静态测试

static thrust 静推力

static trim 静力配平

static tube 静压管

static vent 静压通气孔
statically determinate structures 静定结构
statically unstable structures 静力不稳定结构
statics 静力学
station 站，台，电台；军队的固定基地；特指航空港；(地球同步卫星的)位置；(飞机或船在编队中的)相对位置
station elevation 测站高度
station keeping(飞机、船)保持编队位置；(卫星)保持位置
station model 测站模式，站位模型
station number 站位标号，站号
station pressure 测站气压，本站气压
stationary components 静止部件；静止件
stationary front 静止锋，滞流锋
stationary jets protecting filters 保护过滤器的静止喷嘴
stationary orbit 静止轨道
stationary part 静止件
stationary part with labyrinth seals 静止件带篦齿封严
stationary satellite 静止卫星，同步卫星
stationary shock 驻震波
stationary shock wave 驻激波
stator 静子
stator blade 静子叶片
stator casing 静子外壳
stator vane 静子叶片
stator vane actuator 静子叶片作动器
status matrix 状态矩阵
statute mile 法定英里
statute miles per hour 每小时里，法定英里每小时
stay bar actuating cylinder 撑杆作动筒
stay time 持久时间，停留时间
STB(standby)①备用的，应急的，待命的；等待(飞行)②备用品，应急设备 ③一等战备状态 ④(呼叫信号)准备发报；准备收报 ⑤(无线电代码)“我要暂停几秒钟”
steady flight 定常飞行，稳定飞行
steady flow 定常流(动)，恒定流，稳定流
steady flow coefficient 稳流系数
steady flow system 稳流系统

steady speed 稳定速率，稳定速度
steady state 稳定状态
steady state system 稳态系统，稳定态系统
steady wind 稳定风，恒风，定常风
steady/unsteady 定常 / 非定常
steady-state flight 稳态飞行，定常飞行
stealth bomber 隐形轰炸机
stealth fighter 隐身战斗机；隐形战机；隐形战斗机
stealth technology 隐秘技术，隐形技术
steam condenser 水蒸气凝结器，蒸汽冷凝器
steam engine 蒸汽机
steam generator 水蒸气发生器；蒸汽发生器；蒸汽锅炉
steam rate 耗汽率
steel alloy 合金钢，钢合金
steel cable 钢索，铁索
steel holder tungsten 钨钢保持架
steel pipe beam 钢管梁
steel reinforced aluminum cable 钢芯铝线
steep（sharp）turn back 大坡度转弯（坡度 45° ~ 70° ）
steep angle of bank 大坡度角
steep approach 大下滑角进场着陆，大角度进场
steep bank 大坡度
steep climbing turn 上升小转弯
steep dive 大角度俯冲，急剧俯冲
steep initial 大角度开始
steep spin 陡螺旋
steep takeoff 大坡度起飞
steep turn 急转弯；大坡度转弯
steeple wave 尖顶波
steer 操纵，驾驶，转向;【通】用指定的磁航向我（或向……）靠拢；驾驶；操纵；引导；控制
steer column 驾驶杆
steer dot 导向光点
steer signal 导向信号
steer torque 方向操纵力矩
steerable antenna 可操纵方向性天线，可转向天线
steerable tail wheel 可操纵尾轮
steer-by-wire 线控转向
steering 操纵，掌控
steering actuator 操纵致动器，转向作动筒

steering and navigation instruments 驾驶航行仪表

steering brake 转向闸，转向制动器

steering engine 舵机

steering gear 驾驶机构；转向装置

steering surface 操纵面

steering wheel 方向盘，驾驶盘，舵轮

Stefan-Boltzmann constant 斯蒂芬 - 波兹曼常数

Stefan-Boltzmann law 斯蒂芬 - 波滋曼定律

stellar inertial guidance 星体惯性导引，恒星惯性制导，天文惯性导航

stellar lightning 星状闪电

stellar magnitude 恒星等级

stellar map matching 星图匹配导航

step bearing 端轴承，阶式止推轴承

step climb 步进爬升，逐步爬升，阶梯爬升

step climb profile 阶梯爬升剖面

step function 阶梯函数

step response 阶跃响应，阶跃应答

steppe climate 草原气候

stepped down 下阶梯队

stepped down descent 阶梯下降

stepped gear 塔齿轮，阶式齿轮

stepped leader 梯级先导（闪电）

stepped up 上阶梯队

stepped up climb 阶梯爬升

stepper motor 步进电机，步进马达

stereogram 立体图

stereographic projection 立体投影；赤平投影（法），球面投影，平射投影，等角方位投影

stereopsis 立体视觉，实体映像，立体观测

stereoscopic parallax 立体视差

stereoscopic vision 立体视野，立体视觉

sterilize 消毒；冻结，封存；（水雷）定期失效；销毁；销毁痕迹

stern attack 尾追攻击，后方攻击，尾后攻击

stern heaviness 尾重

stern-droop 尾部下垂

stern-drop 尾坠

stern-drop stall 尾坠失速

stick 驾驶杆，操纵杆；连续投下

的一批炸弹；连续发射的一批炮弹或导弹

stick force 杆力

stick grip 驾驶杆握柄

stick knocker 自动敲杆器（失速告警），驾驶感失速警告器，抖杆声响器

stick movement per g 单位 g 值杆移量

stick pusher 自动推杆器（失速告警）

stick shaker 振杆器，抖杆器

stick spacing 连投间隔，连续投弹间隔距离

stick travel 驾驶杆行程

stick travel per g 单位 g 值杆移量

stick-fixed 定杆的，握杆的

stick-fixed static stability 静态稳定界限；握杆静稳定性

stick-free 自由驾驶杆；松杆操纵；松开

stick-shaker speed 自动震杆器速率；振杆速度

stiff nut 紧锁螺帽，刚性螺帽；防松螺帽

stiff pavement 刚性道面

stiff wing 固定翼飞机

stiffness 刚度

stiffness criterion 刚度准则

still-air range 静风航程，无风航程

still-well 静凹舱

sting 尾支杆（风洞模型），（风洞内用于固定模型的）尾支臂；探针

stinger soldier 毒刺导弹

stipulation 规定

stock 库存品，备用物，供应物

stock in trade 存货，库存

STOL port 短距起降机场

STOL strip 短距起降跑道

stop 光圈级数；停止，阻止，中断；间断；制动机；制动器，栓；击倒，击败，击毙；停泊，停泊处

stop distance 停止距离

stop electro-valve 停车电磁活门

stop fuel drain valve 停车放油活门

stop purge valve 停车放油活门

stop squawk 停止发送讯号；关闭应答机

stop valve 停止阀

stop/idle/flight selector 停车 / 慢车 / 飞行选择器

stop-and-go 连续停住并起飞

stopover 中途停留
stopper 制动器；限动块
stopper components 限动组件
stopper pawl 限动爪
stopper plate 限动板，止动板
stopway 缓冲区
stopway lights 缓冲区灯
storable propellants 可贮存推进剂
storage 贮存，储存，贮藏；保管；储存量；仓库;【计】存储器；存储
storage battery 蓄电池
storage battery vehicle 电瓶车
storage cell 蓄电池
storage element 储能元件，存储元件
storage limit 库存时限；储备限额
storage procedure 贮存程序
store load 外负载
store separation 外载脱离
storehouse 仓库
storm 风暴，沙暴;（军事上的）猛攻，强击，直捣，强袭
storm cell 风暴中心
storm detection（SD）风暴探测
storm light switch 风灯开关，风暴灯电门
storm model 风暴模式
storm path 风暴中心经过的地区，风暴路径
storm surge 风暴激浪；风暴（大）潮，风暴大浪；风暴增水，风暴海啸；暴风浪涌，风暴涌水
storm tide 风暴涌潮，风暴潮，暴风雨潮；风暴涌潮潮位；风暴波浪
storm track 风暴中心经过的地区，风暴路径，风暴路线，风暴轨迹
storm warning 暴风警报，风暴信号，大风警报
storm wave 风暴（引起的）大浪，风暴波，暴风雨波浪；风暴涌潮；风暴潮
storminess 风暴度；磁暴度
storming party 强击队
storm-warning signal 暴风警报信号
storm-warning tower 暴风警报塔
stow position 天顶位置;（大型雷达天线的）运输状态
stowage bins 存物箱
strafe（从低空）扫射，低空攻击;（猛烈）炮轰，轰炸；猛击，

痛打

straight and level flight 水平直线飞行

straight axle landing gear 直轴式起落架

straight blade 直式刀，直叶片

straight in approach 直接进近，直线进近

straight screwdriver 一字解刀；一字改锥；一字螺丝刀

straight stall 直前失速；无坡度失速；正飞机头急附失速，反尾冲

straight trailing edge 直线后缘

straight wing 直翼

straight-in approach 直接进场；直接进场着陆

straight-line attack 直线攻击

straight-pass attack 直接攻击；直线飞行攻击，直线进入攻击（指飞机不经过目视搜索和跟踪目标，用机载武器瞄准系统以低空高速直线飞行攻击地面点状目标）

strain 应变，（应力下的）变形

strain energy 应变能

strain hardening 应变硬化

strainer 滤网；过滤器；滤波器

strainers sampling 滤网取样

strake 轮铁；包铁；轮箍

strake vortex 辅助翼涡流，肩翼涡流

strap brake 带式制动器带闸；带刹车

strap-down inertial navigation 捷联式惯性导航

strap-down inertial navigation and integrated navigation system 捷联式惯性导航与综合导航系统

strap-down inertial navigation system（SINS）捷联式惯性导航系统

strapdown INS 捷联式惯性导航系统

strapdown platform 捷联式平台

strapdown system 捷联系统

strapon booster 捆绑式加力器，捆绑式助推器

strapping 搭接；捆扎

strategic air command【美】战略空军司令部

strategic reconnaissance 战略侦察

strategic studies institute 战略研究所

strategic support 战略支援，战略保障

stratiform cloud 层状云，成层云
stratocirrus 层卷云
stratocumulus 层积云
stratopause 平流层顶
stratosphere 平流层，同温层；恒温层
stratospheric chamber 平流层实验室
stratus 层云
streak 翼条板；（荧光屏上的）水平白色条纹；条痕；（空气吸气发动机因燃烧不正常而产生的）火舌
streak lightning 枝状闪电
stream function 流函数，流量函数
stream landing 密集接续降落，鱼贯着陆
stream line coordinate system 流线坐标系
stream pattern 流型
stream takeoff 密集接续起飞，跟进起飞
stream tube 流线管
streamline 流线；流线型
streamline airfoil 流线型翼形
streamline covering 流线型罩
streamline flow 流线型流动，流线流
streamline form 流线型
streamline function 流线函数
streamline motion 流线型运动
streamline tube 流线型管
streamlined fuselage 流线型机身
streamlined landing gear strut 流线型起落架支柱
streamlining 流线型
streamtube 流管
streamwise vortices 流向涡
strength data 强度数据
stress 应力
stress analysis 应力分析
stress and balance 受力与平衡
stress axial 轴向应力
stress concentration 应力集中
stress corrosion 应力腐蚀
stress cracking 应力破裂，应力开裂
stress curve 应力曲线
stress cycle 应力周期，应力循环
stress deformation 应力变形
stress derating 应力减额
stress flow 应力流
stress intensity 应力强度
stress raiser 应力集中器

stress skin construction 应力蒙皮构造
stress tensor 应力张量
stressed skin 应力蒙皮
stressed-skin construction 承（应）力蒙皮结构
stretch forming 拉伸成形
stretch receptor 伸张感受器
stretcher 担架
strike control and reconnaissance 打击控制与侦察，突击控制与侦察
strike eagle aircraft 战鹰式战斗机
strike mission 攻击任务，突击任务
striking 对地攻击，冲击，突击
striking velocity 冲击速度，冲撞速度；着速（指射弹在弹着时的速度）
stringer 桁条
strip 简易机场，简易跑道；狭条，窄条，带，片，带状物；拆卸，分解；条幅式侦察照片，航摄照片条
strip chart recorder 纸带记录器；长图记录仪
strip line 带线
strip map 地带地图，航路地图；条幅式航线地图
striped zone 条纹区域
strobe/stroboscope 防撞闪光灯；防撞旋转灯标；选通；选通标志；选通脉冲；频闪观测仪，闪光仪
stroke 冲程；行程；闪击
stroke density 闪击密度
stroke speed 行速
strong wind 强风
strong windshear 强风切变
strope 吊带
structural assembly 结构组合件
structural bond 结构胶合，结构胶结件
structural coefficient 结构系数
structural collapse 结构瓦解
structural damping 结构阻尼
structural design 结构设计
structural durability 结构耐久性
structural element 结构元件，结构组件
structural factor 结构因子
structural failure 结构失效
structural fastener 结构紧固件
structural icing 结构积冰
structural inspection 结构检测；

结构检查

structural integrity 结构完整性

structural limit 结构限制

structural measurement 结构度量

structural member 结构构件；结构部件

structural merit index 结构价值指数

structural placard 结构标牌

structural repair publication 结构修理手册

structural ring 结构圈

structural skin panel 结构蒙皮（板）

structural-mode control system 结构模态控制系统

structure 结构

structure inspection manual 结构检查手册

structure repair manual 结构修理手册

structure significant item 重要结构项目

structure test model 结构测试原型；结构试验模型

structure weight 结构重量

strut 支柱，支撑杆，吊架；支撑；（支承发动机吊舱的）吊杆

strutted wing 有支柱翼

stub antenna 刀形天线；短柱天线

stub exhaust 排气短管

stub runway 短跑道（指跑道与滑行道交叉外伸出的一段短跑道）

stub shaft 短轴

studdle 支撑杆

stunt 特技飞行

S-turn 蛇行转弯；S 形转弯

stylus 触针

subassembly 分组件，子组件；半组件；半装配件，分装件；局部装配，组件装配

subcarrier 副载波

subchassis 辅助底盘

subcommutation 副转接

subcooled liquid 过冷液体；再冷却液

subcooler 再冷却器

subcritical airflow 次临界气流

subcritical Mach number 次临界马赫数

subcritical mass 次临界质量（核物理）

subcritical rotor 次临界旋翼，亚临界旋翼

subcritical speed 次临界速度
subdisk 副反射盘
subgravity 次重力；亚重力
subhumid climate 次湿气候
subhunter【口】反潜飞机
sub-idle 次空转（发动机）；亚慢车转速
sublimation 凝华，升华（作用）
submachine gun 冲锋枪，轻型自动（或半自动）枪
submarine 潜水艇
submenu 子菜单
subscript 下标
substitution 代换；替换；（密码中的）代替
substratosphere 副平流层；次同温层
substratum 胶层
substructure analysis 次结构分析
substructure unit 子结构单元；亚结构单元
subsystem 分（支）系统，子系统，辅助系统
subterranean command centre 地下指挥中心
subthreshold 阈值限以下；亚阈值
sub-transmission 副变速器
subtropical cyclone 副热带气旋；亚热带气旋
subtropics 副热带；亚热带
successive carriage 连续运送
successor activities 后续活动
suck in door 吸进门
suction 吸入，吸除；真空度
suction cup 吸盘；吸杯
suction effect 吸力效应
suction gyro 吸式陀螺；真空驱动陀螺仪
suction head 吸引落差
suction inlet 吸气口；吸入口
suction lift 吸引升力
suction mode 抽吸模式；吸入模式
suction peak 吸力峰值
suction specific speed 抽吸比速
suction to the tank 往油箱吸油
sudden acceleration 突然加速
suicide attack 自杀式袭击；自杀性攻击
suicide bomb attack 自杀式炸弹袭击；自杀性爆炸物袭击
suicide bomber 发动自杀式炸弹袭击的人，人体炸弹；自杀性轰炸机

suicide bombing 自杀式炸弹袭击；自杀性爆炸

sulfidation 硫化；硫化作用

sulfide 硫化物

sump oil vapors 滑油蒸气室；污油蒸气；沉淀池油蒸气

sun glasses 太阳镜

sun line 太阳线

sun orientation 太阳定向

sun pillar 日柱，太阳光柱

sun seeker 太阳追踪器

sun sensor 太阳感测仪；太阳能传感器

sunk screw 埋头螺钉

sunrise to sunset 日出至日没

sunset to sunrise 日没至日出

sunshine autograph 日照计；日照仪

sunshine recorder 日照计

sunspot 太阳黑子，日斑

sun-synchronous orbit 太阳同步轨道

sun-synchronous satellite 太阳同步卫星

Super Cobra Attack Helicopter 超级眼镜蛇攻击型直升机

super emergency valve 超应急活门

Super Frelon 超黄蜂（直升机）

super high frequency 超高频

Super Hornet 超级大黄蜂（直升机）

super low recoil rifle 超低后坐力线膛砲

super search mode 超广搜索模式

super speed 超高速

superadiabatic lapse rate 超绝热直减率

superalloy 超合金

superboom 超音爆

supercharge 增压

supercharger 增压器

supercharger blast gate 增压器排气门

supercharging 增压（作用）

supercirculation 超环流

supercompression 超压缩

superconductive gyro 超导陀螺仪

superconductivity 超导电性

supercritical airflow 超临界气流

supercritical airfoil 超临界翼

supercritical flow 超临界流

supercritical Mach number 超临界马赫数

supercritical pressure 超临界压力

supercritical range 超临界范围

supercritical speed 超临界速度

supercritical velocity 超临界速度

superheated vapor 过热（水）蒸气

superheterodyne 超外差接收机

superior 上级；优势的

superlaminar speed 超层流速率速度

supernumerary rainbows 复虹

superplastic forming and diffusion bond 超塑性成形及扩散结合

superplasticity 超塑性

superposition principle 叠加原理

superpressure 超压

superrefraction 超折射

supersaturation 过饱和度

supersaturation ratio 过饱和比

superscript 上标

supersonic jets 超音速飞机；超音速喷流

supersonic transport（SST）超音速运输机

superstandard propagation 超标准传播

superstratosphere 超平流层

superthermal ion detector 过热离子侦测器

super-wide angle 超广角

supplemental carrier 附属式航空载具

supplemental control unit 辅助控制组件

supplemental performance information 补充性能信息

supplementary acceleration 补充加速度

supplementary aerodrome 辅助机场；备用机场

supplementary flight plan message 补充飞行计划书

supplementary type certification 补充型号认证

supply line 补给线

supply ship 补给舰

supply terminal 电源接线端

supply voltage 电源电压

support 支架；支援；支援部队；支援任务

support bearing 支承轴承

support casing 支承套管

support ferrule 支承套圈

support helicopter force 直升机保障部队

support lug 支耳

support pivot 支撑支点；支撑枢轴
support platform 支撑平台
support strut 支柱
supportability 后勤保障性
supporting aircraft 支援飞机
supporting arm 支臂
supporting surface 升力面
suppressant 抑制剂
suppressed antenna 隐蔽式天线
suppressive attack 压制性攻击
suppressor 抑制器；消除器；消声器
surface 面；表面；翼面
surface acoustic wave 表面声波
surface actuator 翼面致动器
surface boundary layer 表面边界层
surface chart 地面（天气）图
surface condenser 表面冷凝器
surface cooler 表面冷却器
surface corrosion 表面腐蚀
surface drag 表面阻力
surface energy 表面能
surface fire 表面起火；地表火
surface fitting 曲面配合；曲面拟合
surface force 地面部队；水面部队；表面力
surface friction 地面摩擦
surface gauge 平面规
surface inversion 地面逆温
surface movement control 场面活动管制
surface movement guidance and control system 场面活动引导和控制系统
surface movement radar 地面活动雷达
surface of discontinuity 不连续面
surface of equal potential 等位面
surface of revolution 旋转曲面
surface position digitizer 表面位置数字转换器
surface pressure 表面压力；地面气压
surface roughness 表面粗糙度
surface smoothness 表面平滑度
surface strength 表面强度
surface temperature 地面温度
surface tension 表面张力
surface treatment 表面处理
surface turbulence 表面扰动
surface visibility 地面能见度
surface wave 表面波
surface weather observation 地面

气象观测
surface wind 地面风
surface wind speed 地面风速
surface-launched interceptor missile 地射拦截导弹
surface-to-air ballistic-missile interception system 地对空弹道导弹拦截系统
surface-to-air gunsight 地空射击瞄准具
surface-to-air missile（SAM） 地对空导弹
surfusion 过冷却；过冷现象
surge diverter 避雷器
surge line 喘振线；风速骤增线，气压波线
surge point 喘振点
surgical strikes 外科手术式打击
surveillance 监视
surveillance inspection 一般检测
surveillance satellite 监视卫星
survivability of impact 撞击存活率
survival（survivable）accident 人员幸存事故
survival equipment 救生装备
survival kit 救生包
survival radio equipment 救生无线电装备
suspended beam of weapons 武器挂梁
suspension line（气球的）吊索；（降落伞的）伞绳
suspension ring 吊环
sustainable airspeed 可持续空速
sustained flight 稳定持久飞行
sustained G 持续 G 值
swap icon 精确换人；定位换人
swarm tactics 密袭式战术
swash plate 旋转斜盘
swashplate 旋转斜盘
sweat cooling 发汗冷却，蒸发冷却
sweep 搜索；掠过；扫射
sweep angle 后掠角；扫描角
sweep jamming 扫频干扰
sweepback 后掠
sweepback tapering 后掠尖削
sweeper 清扫车；清扫人员
swept back 后掠
swept volume 容积排量
swirl error 旋转误差
swirl plate 涡流片
swirler 旋流器
switch 电门；开关

switch connector 开关接头
switch key 转换键
switch volume control 音量控制电门
switchable transponder 可交换转频器
switcher 调车机车；交换机
switching characteristics 开关特性
switching lobe 交换波瓣
switching unit 开关装置
swivel tail wheel 转向尾轮
swivelling engine 旋转发动机
swivelling nozzle 可转向喷管
symbol generator 字符产生器
symbology（雷达显示器荧光屏上的）目标符号
symbology indication（雷达显示器荧光屏上的）目标符号标识
symmetric airfoil 对称翼型
symmetric axis 对称轴
symmetric flight 对称飞行
symmetric laminate 对称层板
symmetric motion 对称运动
symmetric stability 对称稳定性
symmetrical airfoil 对称翼型
symmetrical configuration 对称构形；对称外形
symmetrical flow 对称流
symmetrical flutter 对称颤振
symmetrical section 对称剖面；对称翼型
symmetry 对称
symptom 征兆
sync flow restrictor 同步限流器
synchro 同步器；同步的
synchro resolver 同步解算器
synchro transmitter 自动同步发送机；自动同步传感器
synchron piston 同步活塞
synchronization 同步；协调
synchronous corridor 同步轨道带
synchronous orbit 同步轨道
synchronous satellite 同步卫星
synchropter 同步交叉式双旋翼直升机
synchroscope 同步指示仪；同步示波器
synergic ascent（沿）最佳（上升曲线）上升
synergic ascent curve 最佳上升曲线
synodic coordinate system 旋转坐标系
synoptic chart 天气形势图

synoptic display 综观显示器
synoptic format 综观模式
synoptic meteorology 天气学
synthetic fiber 合成纤维
synthetic fuel 合成燃料
synthetic leather 合成皮
synthetic oil 合成油
synthetic resin 合成树脂
synthetic training 模拟飞行训练
synthetic vision system 综合视景系统
system accuracy criteria 系统精度准则
system altitude 系统高度
system analysis 系统分析
system analysis recorder 系统分析记录器
system control 系统控制
system control modules 系统控制器
system definition manual 系统定义手册
system diagram 系统图
system dynamics 系统动力学
system embedded training 系统附带培训
system engineering 系统工程
system engineering and integration 系统工程与集成
system flow chart 系统流程图
system identification 系统识别
system integration 系统集成
system interface unit（SIU）系统接口单元
system maintenance monitor console（SMMC）系统维护监视台；系统维修监控台
system management and communication 系统管理和通信
system monitoring 系统监控
system of coordinates 坐标系
system performance 系统性能
system performance degradation 系统性能降低
system reliability 系统可靠性
system requirements document 系统需求文件
system safety 系统安全
system schematics manual 系统图解手册
system simulation 系统模拟
system support facility 系统支持设施
system support laboratory 系统保

障实验室

systematic drift rate 系统漂移率

systematic error 系统误差

systematic maintenance 系统性维护

systematic preventive maintenance 系统预防性维护

systematic test 系统测试

systems integration and display 系统集成与显示

systems monitoring instrument 系统监控仪表

T

T（termination of war）-Day 战争结束日

T/O power 起飞功率

tab area 调整片面积

tab chord 调整片弦

TACAN（tactical air navigation system）"塔康"（战术空中导航系统）

tachometer 转速表；流速表

tachometer indicator 转速指示器

tachometer transmitter 转速计发送器

tachometer unit 转速表组

tack-free 不剥落的；无黏性的

tactical air navigation system（TACAN）战术空中导航系统（"塔康"）

tactical basic weight 战术基本重量

tactical bombing 战术轰炸

tactical chart 战术目标图

tactical decision 战术决心

tactical flight 战术飞行

tactical intention 战术意图

tactical jamming system 战术干扰系统

tactical map 战术地图

tactical missile 战术导弹

tactical mobility 战术机动能力

tactical navigation system 战术导航系统

tactical operation 战术行动

tactical operations center 战术作战中心

tactical radar 战术雷达

tactical radar and navigation 战术雷达与导航

tactical radius 战术活动半径

tactical reconnaissance 战术侦察

tactical surprise 战术突袭

tactical target 战术目标

tactical training center 战术训练中心

tactics 战术；战术学

tactics range 战术靶场

TAF（terminal aerodrome forecast）终点机场气象预报

TAF（total aerodynamic force）总空气动力

taiga climate 副极地气候；泰加林气候

tail alignment 尾桨校准；尾桨校正

tail area 尾翼面积

tail boom 尾梁

tail buffeting 机尾抖震

tail chute 尾伞

tail cone 尾椎；（机身的）尾端整流锥

tail configuration 尾翼构型

tail control 尾翼操纵机构；尾舵

tail design 尾翼设计

tail down 机尾下沉

tail drag 尾翼阻力；尾部擦地

tail droop 尾垂

tail effect 尾端效果；尾端效应

tail efficiency 尾翼效率

tail end 尾端

tail fin 尾翅；垂直安定面；垂直尾翼

tail first landing 尾部着陆；单点着陆

tail float 尾部浮筒

tail flutter 尾翼颤震

tail force coefficient 尾力系数

tail gear-box（TGB）尾桨减速器

tail heavy 尾重

tail light 机尾灯

tail load 尾面负载

tail moment 尾面力矩

tail navigation light 尾部航行灯

tail outrigger 尾桁

tail parachute 尾（减速）伞

tail position light 尾翼闪光灯

tail reducing gear 尾减速器

tail rotor 尾桨

tail rotor blade 尾桨叶

tail rotor drive 尾桨传动装置

tail rotor drive shaft 尾桨（传动）轴

tail rotor drive shaft cover 尾桨（传动）轴盖

tail rotor drive shaft cowling 尾桨传动轴蒙皮

tail rotor drive system 尾传动系统

tail rotor drive train 尾桨传动系

tail rotor fairing 尾桨毂整流罩

tail rotor gear-box 尾桨减速器

tail rotor head 尾桨转子毂

tail rotor hub 尾桨毂

tail rotor pedal 尾桨脚蹬

tail rotor pitch change 尾桨变距

tail rotor pitch change mechanism 尾桨变距机构

tail rotor pitching control link rod 尾桨变距操纵联杆

tail rotor shut-off valve 尾桨液压油切断阀

tail rotor speed 尾桨转速

tail rotor vortex ring state 尾桨涡环状态

tail screw 尾螺旋桨

tail skid 尾撬

tail strike 尾桨触地

tail strobe light 尾部照明灯

tail structure 尾部结构

tail surface 尾面

tail unit 尾翼组

tail volume method 机尾容积法

tail volume ratio 尾翼体积比

tail warning radar 机尾警戒雷达

tail wedge angle 尾楔角

tail wheel 尾轮

tail wheel landing gear 尾轮起落架

tail wheel position indicator 尾轮位置指示器

tail wind takeoff 顺风起飞

tailplane 水平安定面；全动水平尾翼

tailwheel landing gear 后三点式起落架

tailwheel-type aircraft 后三点式飞机

tailwind 顺风

tail-wind landing 顺风着陆

take a sight at 观测

take it around 复飞

takeoff and landing area 起降区

takeoff gross weight 起飞总重量

take post 各就各位（口号）

take the air 夺取空中优势；起飞

takedown 卸下，拆卸

takeoff 起飞

takeoff acceleration 起飞加速度

takeoff accident 起飞事故

takeoff airspeed 起飞空速

takeoff assist 起飞助推；起飞助推器

takeoff boost 起飞助推；起飞加速

takeoff characteristics 起飞特性

takeoff clearance 起飞许可

takeoff climb speed 起飞爬高速度

takeoff climb velocity 起飞爬升速度

takeoff decision point（TDP）起飞决策点

takeoff decision speed 起飞决断速度

takeoff distance 起飞距离

takeoff end of the runway 跑道起飞端

takeoff ground speed 起飞滑跑速率

takeoff jams 起飞拥挤；起飞堵塞

takeoff leg 第一边；起飞边

takeoff mass 起飞质量

takeoff over an obstruction 跨越障碍物起飞

takeoff performance 起飞性能

takeoff phase 起飞阶段

takeoff power 起飞功率

takeoff rating 起飞额定功率；起飞额定推力

takeoff roll 起飞滑跑

takeoff rotor speed 旋翼起飞转速

takeoff run 起飞滑跑；起飞滑跑距离

takeoff safety speed 起飞安全速度

takeoff shutters 起飞节气门

takeoff speed 起飞速度

takeoff strip 起飞跑道

takeoff threshold 跑道的起飞端

takeoff thrust 起飞推力

takeoff time 起飞时间

takeoff weight 起飞重量

taking over 转为手操纵

tally ho 目视发现目标

TAM（technical acknowledgement message）技术认可文电

tandem airplane 串翼（飞）机；串座式飞机

tandem cascade 串列叶栅

tandem drive 串联传动

tandem rotor 纵列式双旋翼

tandem staging 级的串列配置

tandem tractor 串联式拖拉机

tandem-rotor helicopter 纵列式（双旋翼）直升机；串座式直升机

tangent arcs 切弧

tangent cone method 切锥法

tangent control box 切线控制盒

tangent modulus 切线模数；切线模量

tangent ogive 切面尖拱

tangential acceleration 切线加速度

tangential ellipse 正切椭圆轨道

tangential landing 切向着陆

tank unit 油箱计量组；坦克部队
tanker 加油飞机；加油车；坦克手
tanker aircraft 加油机
tank-selector valve 油箱选择阀
tap bolt 自攻螺栓
tape control 带式自动控制
tape laying 窄带叠层
tape measure 卷尺
tape recorder 磁带录音机
taper（机翼的）梯形度
taper aperture 减度口径
taper roller bearing 圆锥滚柱轴承
tapered airfoil 梯形（机）翼
tapered beam 变截面梁
tapered chord 渐缩弦
tapered planform 锥形俯视图，锥形平面图
tapered trailing edge 渐缩后缘
tapered wing 尖（端）翼
tapping 电话窃听
tapping hole 出铁口；螺纹孔
tapping point 开孔点
tare effect 支承效应
tare weight 皮重
target acquisition 目标获得
target airplane 目标飞机
target altitude 目标高度
target analysis 目标分析
target approach chart 进入目标图
target approach point 目标接近点
target area 目标区域
target azimuth 目标方位
target board 目标指示板
target calibration 靶标标定
target chart（轰炸）目标图
target complex 目标群；目标体系
target concentration 目标集中区
target crossing speed 目标接近速率
target data 目标数据
target designation 目标指示
target designator controller 目标选定控制器
target destruction 目标摧毁
target detection device 目标探测装置
target director post 目标引导哨
target discrimination 目标鉴别
target display unit 目标显示装置
target diving speed 目标俯冲速率
target drone 靶机；无人靶机
target elevation 目标仰角；目标高度

target fade 目标信号减弱
target fixation 盯住目标
target helicopter 靶机
target history 目标航迹
target identification and acquisition system 目标识别与捕获系统
target identification set laser 激光目标识别仪
target image generator 目标影像产生器
target indicator 目标指示器
target intelligence 目标情报
target level of safety 安全目标等级
target of opportunity 临时目标
target pattern 目标航线；靶机航线
target priority 目标优先顺序
target profile area 目标剖面面积
target recognition 目标识别
target recognition chart 目标识别图
target return 目标回波
target reverser 推力转向器
target run 进入目标航路
target scintillation 目标回波起伏
target search and acquisition system 目标寻获系统
target selection 目标选择
target shooting 目标射击
target signal 目标信号
target span 目标翼展
target speed 目标（运动）速度
target support 目标支援
target system 目标体系
target threshold speed 跑道头上空规定速率
target timing 目标定时
target tracking radar 目标跟踪雷达
target tray of tape 目标盘式带
target vulnerability 目标弱点；目标要害
target warker 目标标示烟火
target-acquisition weapon delivering system 目标获得武器投射系统
targets of military importance 重要军事目标
tarmac 柏油碎石路面（尤指飞机跑道）
TAS（true air speed）实际空速；真空速
task air force 空军特遣部队
task component 特混组成部队
task element 特混分队；特遣分队
task fleet 特混舰队

task force 特遣部队
task group 特混大队
task numbered shop manual 按任务编号的车间手册
task organization 特混编组
TAT（true air temperature）绝对大气温度
TAWS（terrain awareness and warning system）地形提示和警告系统
taxi 滑行
taxi accident 滑行事故
taxi channel 水上机场滑行地带
taxi channel light 滑行通道灯
taxi collision accident 滑行相撞事故
taxi into position and hold 进入跑道等待
taxi landing 滑行着陆
taxi light 滑行灯
taxi phase 滑行阶段
taxi track 滑行道
taxi-back landing 滑回着陆
taxiing 滑行
taxiing drag 滑行阻力
taxiing guidance system 滑行引导系统
taxiing instruction 滑行须知
taxiing speed 滑行速度
taxiing turn 滑行转弯
taxiway 滑行道
taxiway light 滑行道灯
taxiway marking 滑行道标志
taxiway shoulder 滑行道的侧安全道
Taylor diagram 泰勒图
TB（true bearing）真方位
TBO（time between overhauls）大修间隔时间
TC（true course）真航向
TCA（terminal control area）末端控制区
TCAS（traffic alert and collision avoidance system）空中交通报警与防撞系统
TCAS（traffic collision avoidance system）交通防撞系统
TDP（takeoff decision point）起飞决策点
TDWR（terminal Doppler weather radar）航空港多普勒气象雷达
TDZL（touchdown zone light）着陆地区灯
teaching aid 教具

team of deminers 扫雷小组
teamwork 协同行动；配合
tear down deficiency report 拆卸缺失报告
tear gas 催泪性毒气
tear shell 催泪弹
technical bulletin 技术通报
technical coordination meeting 技术协调会
technical error 技术错误；技术误差
technical instruction 技术说明书；工艺规程
technical knowhow 专技知识
technical landing 技术降落
technical log book 技术记事簿
technical manual 技术手册
technical publication 技术出版物
technical reason 技术原因
technical standards order 技术标准法规
technical terms 术语
technical training 技术训练
technology 技术；工艺
tee pipe coupling 三通管接头
tee-drive unit 三路传动器
teetering hinge 跷板式轴承；摇摆轴承
teeter-totter 跷跷板
teeth arm 战斗兵种
Teflon 特氟龙；聚四氟乙烯
teleflex 软套管
telemeteorograph 遥测气象仪
telemeter 测远计
telemeter channel 无线电遥测波道
telemetering 遥测
telemetering FM 调频遥测法
telemetering pick-up 遥测传感器
telemetering pulse duration modulation 脉波波期调变遥测
telemetering pulse width modulation 脉波宽度调变遥测
telemetry 测距术；遥测术
telemotor 遥控传动装置；遥控马达
teleran（television radar air navigation）电视雷达导航
telethermoscope 遥测示温器
teletype control system 电传操纵系统
television microwave link 电视微波设备；电视微波线路
television radar air navigation（teleran）电视雷达导航

temperate climate 温带气候
temperate zone 温带
temperature altitude 温度高度
temperature amplifier 测温（电路）放大器
temperature anomaly 温度距平；温度异常
temperature coefficient 温度系数
temperature combustion 燃烧温度
temperature compensating device 温度补偿装置
temperature contour 等温线
temperature control valve 温度控制活门
temperature distribution 温度分布；温度场
temperature efficiency 温度效率
temperature field 温度场
temperature gradient 温度梯度
temperature lapse rate 温度直减率
temperature limits of flammability 可燃性温限
temperature probe 温度探测仪；温度探头
temperature recovery factor 温度恢复系数
temperature rise 温度上升
temperature scale 温标
temperature sensing element 感温元件
temperature sensitivity coefficient 温度灵敏度系数
temperature sensor 温度传感器
tempered glass 强化玻璃
template 模板
temporary flight restriction（TFR）临时性飞行限制
temporary restricted area 限时限航区
temporary revision 临时修订
temporary storage 暂时存储器；临时储存（库）
tender 供应船
tensile proportional limit 张力比例极限
tensile strength 抗拉强度
tension 张力；拉力
tension clip 张力夹
tension fail 拉伸失效
tension field 张力场
tension spring 拉力弹簧
terminal 候机楼区；终点站
terminal advanced automation 终端高级自动化

terminal aerodrome forecast（TAF）终点机场气象预报
terminal airspace 机场空域
terminal and en-route navigation facilities 场站及航路导航设施
terminal approach control 终端进场管制
terminal area 航站区
terminal area chart 机场区航图；航空港区域航图
terminal area delay 航站区延误
terminal area forecast 终点（降落站）气象预报
terminal area radar 机场区雷达；航站区雷达
terminal area surveillance radar 航站区监视雷达
terminal ballistics 终点弹道学
terminal building 航站大楼
terminal condition 终端条件
terminal configured vehicle 综合布局飞行器
terminal control area 终点站控制区；航站管制区
terminal control area（TCA）末端控制区
terminal control center 终端管制中心；终端控制中心
terminal Doppler weather radar（TDWR）航空港多普勒气象雷达
terminal forecast 着陆区天气预报；航线终点天气预报；航站天气预报
terminal guidance 终点引导
terminal instrument procedure 终端仪表程序
terminal maneuvering area 航站空域；航站空中管制区；航站起落和滑行地带
terminal maneuvering area（TMA）航站空中管制区；终端移动区
terminal meridian 国际日期变更线
terminal nose dive 极限垂直俯冲
terminal phase 末段
terminal procedures 场站程序
terminal radar and control system 终端雷达和控制系统
terminal radar approach control 终端雷达进场控制
terminal radar approach control in the tower cab 在塔台工作间的终端雷达进近控制
terminal radar service area（TRSA）航站雷达勤务区

terminal route 航线
terminal surveillance radar 终端监视雷达
terminal system support facility 终端系统保障设施
terminal velocity 极限速度；末速
terminal VHF omnidirectional radio range 极高频终端全向导航台
terminal voltage 端电压
terminal VOR 终端多向导航台
terminal-homing accuracy demonstrator 归航精度指示仪
terrain 地形；地区；领域
terrain avoidance 地形回避
terrain awareness 地形提示
terrain awareness alerting and display 地形提醒警告与显示
terrain awareness and warning system（TAWS）地形提示和警告系统
terrain clearance altitude 离地高度
terrain comparison 地形匹配；地形比较
terrain masking 地形遮蔽
terrain model 地形模型
terrain restrictions 地形限制
terrain-following 地形跟踪
terrain-following radar 地形跟踪雷达
terrain-hugging cruise missile 贴地式巡航飞弹
terrestrial dynamical time 地球力学时
terrestrial globe 地球仪
terrestrial magnetic force 地磁力
terrestrial meridian 地球子午线
terrestrial pole 地极
terrestrial radiation 地球辐射
terrestrial reference guidance 地球基准制导
terrestrial refraction 地面折射
terrestrial scintillation 地面闪烁
terrestrial space 地球空间；近地太空
terrestrial swing 空对地罗盘校正
terrestrial triangle 地理三角形
terrestrial whirl wind 陆旋风
territorial water 领海
tertiary airflow 环流气流
tertiary circulation 三级环流
tertiary hole 通气孔
tesla 特斯拉（磁感应强度单位）
test and evaluation 试验与鉴定

test bench 试验台
test circuit 测试电路
test control panel 检测控制面板
test flight 试飞
test information sheet 试验资料表
test load 试验负载；试验负荷
test model 试验模型
test pilot 试飞员
test point 测试点
test pressure 试验压力
test procedure 试验程序
test program set 试验项目设备
test replaceable unit 测试更换单元
test report 试验报告
test requirement data 测试需求数据
test station 测试站
test tube 试管；风洞
test valve 试验阀
test voltage 试验电压
test, measurement and diagnostic equipment 测试、计量与诊断设备；检测与诊断设备
testbed 试验台
testing constant 试验常数
testing device 试验装置；测试设备
testing fueltank 试验燃油箱
testing machine 试验机
testing result 试验结果
testing room 试验室
TFR（temporary flight restriction）临时性飞行限制
TGB（tail gear-box）尾桨减速器
TGB assembly 尾减速器组件
TH（true heading）真航向
T-handle T 操纵杆
the cyclic pitch control stick 周期变距操纵杆
theater missile defense（TMD）战区导弹防御（系统）
theatre command 战区
theoretical aerodynamics 理论空气动力学
theoretical air 理论空气量
theoretical best gliding angle 理论上最佳滑翔角
theoretical ceiling 理论升限
theoretical discharge 理论流量
theoretical efficiency 理论效率
theoretical engine cycle 理论发动机循环
theoretical maximum horsepower 理论最大马力
theoretical size 理论尺寸；理论

大小
theoretical static ceiling 理论静升限
theoretical thrust 理论推力
theoretical thrust coefficient 理论推力系数
theory of elasticity 弹性理论
therapeutic oxygen 治疗用氧气
thermal anti-ice 热防冰
thermal anti-icing 加温防冰
thermal barrier 热障
thermal battery 热电池
thermal belt 高温带
thermal boundary layer 热边界层
thermal capacity 热容量
thermal climate 温度气候
thermal conductivity 导热性；导热率
thermal convection 热对流
thermal couple 热电耦
thermal efficiency 热效率
thermal energy 热能
thermal equator 热赤道
thermal equilibrium 热平衡
thermal expansion 热膨胀
thermal gradiometer 温差侦测计
thermal high 热高压
thermal image 热像
thermal imager 热成像器
thermal insulation 热绝缘
thermal jet 热喷射流
thermal jet engine 热喷气发动机
thermal load 热负荷
thermal load limitation 热负荷限制
thermal lock 热锁
thermal low 热低压
thermal photograph 红外线照片；热照像
thermal pile 热电堆
thermal radiation 热辐射
thermal relief valve 热安全活门
thermal sleeve 保温套
thermal strain 热应变
thermal stress 热应力
thermal switch 热电门；热控开关
thermal turbulence 热力湍流
thermal vacuum chamber 热真空室
thermal viscosity 热黏性
thermal wind 热成风
thermic load 热载荷
thermion 热离子
thermionic converter 热离子换能器
thermionic emission 热离子放射
thermistor 热敏电阻

thermo jet 热喷流；热喷气发动机
thermo-baric warhead 温压弹
thermochemical equilibrium 热化学平衡
thermocontact 热接触
thermocouple 热电偶
thermocouple e.m.f. 热电偶电动势
thermocouple harness 热电偶电缆
thermocouple line resistance 热电偶导线电阻
thermocouple plug 热电偶插头
thermocouple probe 热电偶探针
thermocouple system 热电偶系统
thermodynamic cycle 热力循环
thermodynamic diagram 热力学图
thermodynamic efficiency 热力效率
thermodynamic equilibrium 热力平衡
thermodynamic temperature scale 热力学温标
thermodynamical operation 热力学操作
thermodynamics 热力学
thermometer 温度计
thermometry 测温法；测温学
thermonuclear 热核的
thermoplastic material 热塑性材料
thermoplastic resin 热塑性树脂
thermoplastics 热塑性塑料
thermoset material 热固性材料
thermoset plastics 热固性塑料
thermoset resin 热固性树脂
thermosphere 热（成）层
thermostat 恒温器
thermostatic switch 恒温开关
thermostatic valve 恒温活门
Thevenin's theorem 戴维宁定理
thick airfoil 厚翼型
thick cylinder 厚壁圆筒
thick film assembly 厚膜组合
thick fluidity 难流动性
thick fog 浓雾
thickness chart 厚度图
thickness distance 厚距
thickness distribution 厚度分布
thickness effect 厚度效应
thickness line（等）厚度线
thickness ratio 厚度比
thickness-chord ratio（翼）厚（翼）弦比
thimble 套筒；套管；疙瘩形雷达整流罩
thin airfoil 薄翼型
thin boundary layer theory 薄边界

层理论
thin cylinder 薄壁圆筒
thin film 薄膜
thin film circuit 薄膜电路
thin film lubrication 薄膜润滑
thin film transistor 薄膜电晶体
thin fluidity 易流动性
thin nose pliers 扁嘴钳
thin route 轻量航线
thin-aerofoil stall 薄翼失速
thin-airfoil stall 薄翼切面失速
thin-case bomb 薄壳炸弹
thinking order 思路顺序
thiokol 聚硫橡胶
third line maintenance 第三级维护
thread gauge 螺纹规
thread rolling machine 滚丝机
threaded connecting 螺纹连接
threaded fastener 螺纹紧固件
threaded hole 螺纹孔
threat evaluation 威胁判断
threat simulation 威胁模拟
three degree-of-freedom 三自由度
three dimensional boundary layer 三维边界层
three dimensional coordinates 立体坐标
three dimensional flow 三维流
three dimensional motion 三维运动
three greens 起落架已放下并上锁
three impulse orbital transfer 三冲轨道转移
three one tactics 三一战术
three phase stator 三相定子
three phase voltage 三相电压
three position key 三位式键
three position relay 三位继电器
three wheel landing 三轮落地
three-abreast seating 并列三座座椅
three-axis autopilot 三轴自动驾驶仪
three-axis gyro 三轴陀螺仪
three-axis stabilization 三轴稳定
three-axle velocity vector senser 三轴速度矢量传感器
three-body problem 三体问题
three-control aeroplane 三轴操纵飞机
three-dimensional warfare 立体战；三度空间战
three-phase 三相的
three-phase AC 三相交流电流
three-phase alternating current 三相交流电流

three-phase coil 三相线圈
three-phase current 三相电流
three-phase four-wire system 三相四线制
three-phase full wave rectifier 三相全波整流器
three-phase generator 三相发电机
three-phase synchronous motor 三相同步马达
three-phase system 三相制
three-phase three-wire system 三相三线制
three-point landing 三点着陆
three-point mooring 三点系留
three-point suspension 三点悬挂
three-point switch 三点开关
three-pole circuit breaker 三极断电器
three-pole switch 三极开关
three-position switch 三位开关
three-way connection 三通管接头
three-way mono-stable valve 三通单稳态活门
three-way union 三通道接头
three-wheel tractor 三轮式牵引机
three-winding transformer 三绕组变压器
three-wire system 三线制
threshold 跑道入口
threshold contrast 阈值对比
threshold control speed 最小操纵速率
threshold crossing height 入口处穿越高度
threshold crossing waypoint 跑道跨越点
threshold detect 门限探测
threshold frequency 极限频率；临界频率；阈频率
threshold limit 阈限
threshold of hearing 最低可听值
threshold of sound intensity 限界声强
threshold speed 进跑道上空速度；速度阈值；速度临界值
threshold wavelength 限界波长
throttle 油门；节流阀
throttle actuator 油门传动机构
throttle angle 油门杆角度
throttle back the engine 收油门
throttle burst 突然加大油门；猛推油门
throttle chop 收油门；关油门
throttle control 节流控制

throttle curve 节流曲线
throttle displacement 油门杆位移
throttle down 关油门
throttle down the engine 收油门
throttle handle 油门手柄
throttle hold 油门保持
throttle jockey 战斗机飞行员
throttle lever 油门杆
throttle override 节气阀超控
throttle pitch lever 油门桨距杆
throttle quadrant 油门弧座
throttle quadrant assembly 油门杆扇形支架组件
throttle ring lamp 油门环灯
throttle stop 节气阀触止
throttle system 油门系统
throttle up the engine 加油门
throttleable thrust 可调推力
throttling calorimeter 节流式热量计
throttling governer 节流调速器
throttling process 节流过程
through bolt 贯穿螺栓
through deck 全通式甲板
throught flight 过站飞行；过场飞行
thrust 拉力；推力
thrust angle 推力角
thrust augmentation 推力增大
thrust augmented wing 推力增益翼
thrust augmenter 推力增益器
thrust axis 推力轴
thrust balance 推力平衡
thrust bearing 止推轴承
thrust chamber 推力室
thrust chamber efficiency 推力室效率
thrust coefficient 推力系数
thrust construction 推力承力结构
thrust control 推力控制
thrust cutoff 推力关断
thrust decay 推力衰减
thrust deduction 推力减除
thrust drag 推阻力
thrust equalizer 推力平衡器
thrust equation 推力方程式
thrust equivalent horsepower 推进马力；推进功率
thrust face（螺旋桨桨叶的）拉力面
thrust factor 推力因数
thrust force 推力
thrust horsepower 推进马力
thrust lever 推力操纵杆；油门杆
thrust lever angle 油门杆角度

thrust lever resolver angle 油门杆解算器角度
thrust lever travel 油门杆行程
thrust line 推力（作用）线
thrust load 推力负载
thrust lost 推力消失
thrust management computer 推力管理计算机
thrust margin 推力限度
thrust meter 推力计
thrust misalignment 推力偏差
thrust mode select panel 推力方式选择面板
thrust output 推力输出量
thrust power 推进动力
thrust power loading 推进功率负载
thrust range 推力（变化）范围
thrust ratio 推进功率比
thrust required curve 需求推力曲线
thrust requirement 推力需求
thrust reverser 反推装置；反推器；反推手柄
thrust reverser actuation system 反推作动系统
thrust reverser directional valve 反推换向活门
thrust section 发动机舱
thrust specific fuel consumption 单位推力燃油消耗量
thrust stand 推力试验台
thrust structure 发动机（固定）架
thrust surface 推力面
thrust termination 推力终止
thrust termination equipment 推力终止装备
thrust test 推力试验
thrust vector control 推力方向控制
thrust vectoring 推力换向
thrust weight ratio（飞机或发动机的）推重比
thruster 推力器
thrust-to-weight ratio 推力重量比
Thunderbolt Plane（Warthog）雷霆对地攻击机（昵称“疣猪”）
thundershower 雷阵雨
thunderstorm 雷暴
thunderstorm cirrus 雷暴卷云
thyratron 闸流管
tide on 就位
tie bolt 系紧螺栓
tie bus differential protection 连接汇流条差动保护
tie contactor 连接条接触器
tie current transformer assembly

连接条电流互感器组件
tied escort 固定护航队
tied gyro 连接陀螺仪
tiedown 系留
tie-in relay 导入继电器
tie-in system 导入系统
tight formation 密集队形
tight spiral 急速盘旋
tight turn 急转弯
tightening torque 紧固力矩
tilt 倾斜
tilt ducted propeller 变角导管螺旋桨
tilt landing gear 倾斜起落架
tilt rotor 倾斜式旋翼
tilt rotors helicopter 倾转旋翼式飞机
tilt table 倾转台
time approach 计时进场着陆
time before overhaul 大修前时间
time between overhauls（TBO）翻修间隔时间；翻修时限
time change items 定时更换件
time delay 延时
time delay manual relay 人工延时继电器
time diagram 时（间）图（表）
time difference 时差
time dilation 时间膨胀
time dilution of precision 时间精度扩散因子
time division data link 时间划分资料链
time division multiple access 时分多址连接方式
time division multiplex 时分多路复用
time division multiplexing 时分多路复用
time frequency technology 时间频率技术
time from chocks off to chocks on 从轮挡去除到重新放置的时间
time hack 时间校准；预定时间
time in service 服役时间；值勤时间
time lag 时滞
time of arrival 抵达时间
time of day 日时；时刻
time of endurance 续航时间
time of flight 飞行时间
time of operation 工作时间；操作时间
time of perihelion passage 经过近

日点时刻
time of useful consciousness 有效意识时间
time on target（飞机）对预定目标攻击或照相时间
time over steerpoint 转向点时间
time over target 飞临目标时间
time reference navigation 计时航行
time since installation 安装后时间
time since last shop visit 上次进厂以来的使用时间
time since new 从新机以来的使用时间
time since overhaul 翻修后工作时间
time slot 时槽；时隙
time spacing 时间隔离
time tick 对时；报时信号
time to first fix 第一次定位时间
time to go 飞向转弯点的时间；飞向截击点的时间；剩余时间
time to go（TTG）readout 飞向转弯点的时间读取值；飞向截击点的时间读取值；剩余时间读取值；发射时间读取值
time to intercept 截击时间
time to takeoff 起飞（滑跑）时间
time to target 接近目标时间
time zone 时区
time-out 超时，过时，断开时间
time-out function 超时提示，超时警示
timing pulse 定时脉冲
tip 翼尖
tip aileron 翼尖副翼
tip chord 翼尖弦
tip drive（旋翼的）翼尖驱动
tip droop 翼尖下垂
tip effects 末端效应
tip loss 翼梢损失
tip off 前坠
tip speed 桨尖速度
tip speed limitation 翼尖速限
tip stall 翼尖失速
tip tank 翼尖油箱
tip vortex 桨尖涡；翼尖涡流
tip-driven helicopter 喷气旋翼直升机；桨尖喷气式直升机
tip-driven rotors helicopter 叶尖推进式旋翼直升机
tip-path plane 桨尖旋转面
tipping-bucket rain gage 倾斗式雨量器；翻斗式雨量计
tire burst screens 轮胎防爆格栅

tire limit speed 轮胎最大滚动速率
tire pressure 胎压
tire pressure indicating system 胎压指示系统
tire pressure monitor unit 胎压监控组件
tire spinup speed 轮胎预转速率
titanium alloy 钛合金
TMA（terminal maneuvering area）航站空中管制区；终端移动区
TMD（theater missile defense）战区导弹防御（系统）
to be announced 待发布
to be determined 待定
toe brake 脚刹车；机轮刹车
to-from indicator 向背指示器；方向指示器
toggle 拨动电门；拨动式开关
toggle switch 拨动开关
tolerance 公差；偏差；容差
Tomahawk cruise missile“战斧”巡航导弹
Tomcat F-14“雄猫”舰载战斗机
tommy gun 汤姆逊式冲锋枪
tomodromic course 拦截航线
tongue-shaped plate 舌形片
tonnage 吨位
tool designation 工具名称
tool equipment 工具设备
tool keeper 工具管理员
tool kit 工具包；工具箱
tooling catalogue 工具目录
tooling hole 工具孔
tooling resins 工具用树脂
tooth ring 齿圈
top bracket 顶部支架
top cover 高空掩护；空中掩护；空中掩护的飞机
top fuselage inlet 机身顶部进气道
top light 顶灯
top of climb 爬升顶点
top of descent 下降极限；下降起点
top off 加满油箱
top rudder 上舵
top view 俯视图
top-loaded vertical antenna 电容负载垂直天线
topocentric coordinate system 地面中心点坐标系；以观察者为中心的坐标系
topographic map 地形图
topographic troops 测绘部队
topography 地形

torch 手电筒；闪光灯；(飞机油箱中弹后)起火

tornado 龙卷风

tornado fighter jet 旋风式战机

torque 扭矩；转矩

torque amplifier 转矩放大器

torque arm 扭力臂

torque axis 扭矩轴线

torque balance accelerometer 力矩平衡加速计

torque box 扭力盒

torque characteristic 力矩特性

torque command storage 扭矩指令暂存

torque conformation 扭矩匹配

torque conformation box 扭矩构造盒

torque device 扭矩装置

torque dynamometer 扭矩测力计；扭力计

torque effect 扭矩效应

torque energy 转矩能

torque horse power 转矩马力

torque limiter 扭矩限制器

torque limiter switch 扭矩限制器开关

torque link 扭力臂

torque matching 扭矩匹配

torque measurement 扭矩测量

torque motor 力矩电动机；力矩马达；扭矩马达

torque motor current 扭矩马达电流

torque on output shaft 输出轴扭矩

torque pressure 扭力压

torque sensing 扭矩传感

torque stand 转矩架；扭力架

torque tube assembly 扭力管组件

torque wrench 扭力扳手；限力扳手

torquemeter 扭矩计

torquemeter device 测扭装置

torquemeter nominal pressure 扭矩计标称压力

torquemeter pressure tapping 扭矩计压力表接头

torquemeter reference shaft 扭矩计基准轴

torquemeter restrictor 扭矩计节流嘴

torquemeter transmitter 扭矩计传感器

torquing 加扭矩(陀螺仪)

torrential rain 暴雨

torrid zone 热带

torsion 扭转
torsion bar 扭杆
torsion dynamometer 扭转式测力计
torsion failure 扭转失效
torsional angular displacement 扭转角位移
torsional buckling 扭转失稳
torsional displacement 扭转位移
torsional flange 扭转法兰盘
torsional force 扭力
torsional load 扭转负载
torsional mode 扭转模态
torsional modulus of rupture 扭力破坏模数
torsional oscillation 扭转振动
torsional resonance 扭转共振
torsional rigidity 扭转刚度；抗扭刚度
torsional shear flow 扭转剪力流
torsional stiffness 扭转刚度
torsional stress 扭应力
torsional vibration system 扭振系统
total aerodynamic force（TAF）总空气动力
total air temperature 空气总温
total blade area（直升机旋翼的）桨叶总面积
total blade-width ratio 总桨叶宽度比
total drag 总阻力
total drag coefficient 总阻力系数
total driving horsepower 总驱动马力
total failure 完全失效
total field of view 总视场；覆盖区总宽度
total head 总落差
total impulse 总冲量
total lift 总升力
total load 总负载
total pressure 总压力
total pressure loss 总压力损失
total quality control 全面质量管理
total quality management 总体质量管理
total system error 总系统误差
total thrust 总推力
total time 总时间
total weight 总重量
total weight of fuel 总燃油量
totalizer 总加器
total-package procurement 整套采购

touch and go 连续起飞
touch down 降落
touch sensor switch 触摸开关
touch-and-go 连续起飞
touch-and-go landing 着陆后连续起飞
touch-and-go operation 连续起降作业
touchdown 着陆；接地
touchdown and liftoff area 着陆离地区
touchdown autorotation 自动旋转着陆
touchdown end of the runway 跑道着陆接地端
touchdown point 接地点
touchdown runway visual range 落地跑道视距
touchdown shocks 着陆冲击波
touchdown sinking speed 接地下沉速率
touchdown speed 着陆速度
touchdown velocity 接地速度
touchdown zone 着陆区；接地区
touchdown zone light（TDZL）着陆地区灯
toughness 韧度；韧性
tow bar 拖车把；牵引杆
TOW missile（tube-launched, optically-tracked, wire-guided missile）陶式导弹；陶式反坦克导弹
towed flight 拖曳飞行
towel rack 机载劳兰天线
tower cab 塔台工作间
tower communications system 塔台通信系统
tower control 塔台管制
tower control area 塔台管制区
tower controller 塔台管制员
tower data link system 塔台数据链系统
tower launcher 发射塔
tower tape 塔台（录音）磁带
towering cumulonimbus 塔状积雨云
towering cumulus 塔状积云
towering takeoff 垂直起飞
tower-snag recovery 回收塔
towing 拖曳；牵引
towing bar 拖车把；牵引杆
towing eye 挂钩
towing sleeve 筒形拖靶
towing target 拖靶
towing tractor 牵引车

towline model 拖索式滑翔模型机
toxic agent 毒剂
toxic bomb 毒气弹
trace 踪迹；轨迹
trace gas 示踪气体；微量气体
trace impurity 痕量杂质
tracer 曳光弹；曳光剂
track 航迹；轨迹；跟踪
track angle 航迹角
track angle error 航迹角误差
track ball 跟踪球
track carrier roller 履带托带轮
track classification 航迹分类
track crossing angle 航迹交叉角
track definition message 跟踪定义信息
track gauge 轨距
track geometry 轨道几何形位
track intervals 航迹间隔
track roller 履带支重轮
track separation 航路间隔；轨道间隔
track telling 航迹报知
track type tractor 履带式牵引机
tracked armoured vehicle 全履带装甲车
tracked gear 履带式起落架
tracking 跟踪；（旋翼的）同锥度调整；（螺旋桨叶的）同轨迹调整；共面调整
tracking and data-relay satellite 追踪及数据中继卫星
tracking and data-relay satellite system 追踪及数据中继卫星系统
tracking and ranging 追踪与测距
tracking camera 跟踪摄影机
tracking radar 跟踪雷达
tracking range 跟踪距离；跟踪范围
tracking station 跟踪站
tracking system 跟踪系统
traction 牵引
tractor 牵引车；牵引式飞机；拖靶飞机
tractor propeller 拉进式螺旋桨
trade wind 信风
traffic advisory 交通咨询
traffic alert 交通警告
traffic alert and collision avoidance system（TCAS）空中交通报警与防撞系统
traffic circuit 起落航线
traffic collision avoidance system（TCAS）交通防撞系统

traffic congestion 交通拥堵；交通拥挤；信号拥挤
traffic control 交通管制；飞行管制
traffic control projector 航管信号灯
traffic management processor 交通管理处理器
traffic management system 交通管理系统
traffic pattern 起落航线
traffic pattern flight 起落航线飞行
traffic permit 通行证
traffic resolution 交通决断
traffic right 航权
traffic situation display 交通状态显示器
trailer 拖车；跟踪飞机；盯梢飞机
trailer aircraft 跟踪空中目标的飞机；拖靶机
trailing antenna 拖曳天线；下垂天线
trailing edge（翼型的）后缘
trailing edge angle 后缘角
trailing vortex 尾涡
trailing vortex resistance 尾涡阻力
trailing vortex system 尾迹涡系
trailing-edge stall 后缘失速
trailing-edge strip 后缘条
train bombing 连续投弹轰炸
trainer 教练机
trainer aircraft 教练机
trainer equipment design specification 训练器装备设计规范
trainer flight simulator 训练飞行模拟器
trainer subsystem design specification 训练器次系统设计规范
training aid 教具；训练辅助器材
training aircraft 教练机
training device 训练装置
training facilities 训练设施
training flight 训练飞行
training helicopter 教练直升机；教练机
training idle 慢车训练
training idle mode 慢车训练模式
training rating 训练等级
training selector 训练选择器
training sortie 训练（起飞）架次
trajectory of principal stress 主应力轨迹
trajectory phase 弹道段
trajectory shift 弹道偏差
tranquilizing system 增稳系统

trans gear 分动（齿轮）箱（车）

transceiver 无线电收发机

transcribed information briefing service（TIBS）转录信息简报服务

transcribed weather broadcast 抄录的气象广播

transcriber 抄录器；复制装置；信息转换装置

transducer 转换器

transfer 转移；调动

transfer beam 转换梁

transfer button 转换按钮

transfer gearbox 传动齿轮箱

transfer of control 移交控制；指挥交接

transfer of control massage 控制文电传递

transfer orbit 转移轨道

transfer recording 复制录音

transfer trajectory 转移轨道

transferring control responsibility 转换管制责任

transformer 变压器；变量器；变换器

transformer loss 变压器耗损

transformer/rectifier unit 变压 / 整流器

transient 暂时停留的；瞬间的

transient condition 过渡状态；瞬变状态

transient drop 瞬时性下降，短暂下降

transient error 瞬态误差

transient increase 瞬时性升高，短暂升高

transient overtorque 瞬态超转矩

transient variation 瞬时变化

transistor 晶体管；晶体三极管

transistor circuit 晶体管电路

transit time 过渡时间

transition 改装（训练）；过渡

transition altitude 转换高度；过渡高度

transition boundary layer 过渡边界层

transition engineering 过渡工程

transition layer 过渡层

transition level 过渡高度层

transition liner 传接筒

transition phase 过渡阶段

transition route 转移航路，变迁途径

transition speed 过渡飞行速度

transition strip 转换道（机场）

transition training 改装训练

transitional maneuver 过渡策略

transitional surface 转换面

transitional training 改装训练

transitory target 瞬间目标；临时目标

transitron 负互导管

translating nozzle 转向喷口

translational lift 过渡（状态的）升力

translatory resistance derivative 平移阻力导数

translucidus 透光（云）

translusidus stratocumulus 透光层积云

transmission 传动装置；变速器；传递；传动;（直升机）主减速器

transmission（gear）box 变速（齿轮）箱（车）

transmission frame 传动框

transmission oil 变速器油

transmission shaft 传动轴

transmission shaft assembly 传动轴组件

transmission shaft rupture 传动轴损坏

transmission system 传输系统；发送系统

transmission unit 传送器

transmissivity 透过率；透射率

transmissometer（大气）透射仪；视距测量计；能见度仪

transmit 传输；发射

transmit button 传输按钮；发射按钮

transmit switch 传输开关

transmittance 透射比；传递系数

transmitter 发射机；传感器

transmitter receiver limiter 发射接收机限制器

transmitter tune light 发射机调谐灯

transmitter-off light 发射机故障信号灯

transmutation 嬗变；变形；蜕变

transonic barrier 音障

transparent cover 透明罩

transpiration cooling 发汗冷却

transponder（XPDR）应答机

transponder code 应答机代码

transponder code setting 应答机代码设置

transponder decoder 应答机译

码器

transponder modulator 应答机调制器

transponder selector switch 应答机显示电门

transponder transmitter 应答机发射机

transport helicopter 运输直升机

transport services 运输勤务；运输勤务部队

transportation 运输；运输工具

transportation medium 运输工具

transporter 运输机

transtage 变轨级；中间级

transuranic element 超铀元素

transverse acceleration 横向加速度

transverse axis 横（向）轴（线）

transverse balance 横向平衡

transverse corrections 横向修正

transverse displacement 横位移

transverse flow effect 横流效应

transverse force axis 横力轴线

transverse load 横向载荷

transverse motion 横向运动

transverse oscillation 横向振荡

transverse ring structure 横向环结构

transverse stability 横向稳定性

transverse velocity 横向速度

transverse vibration 横向振动

transverse wave 横（向）波

trap circuit 陷波电路

trap gas 空腔气体

trapeze beam 悬吊梁

trapezium distortion 梯形失真；梯形畸变

trapezoidal modulation 梯形调变；梯形调制

trapped fuel 陷滞燃油

trapped gas 圈闭气体；困气

trapped propellent 陷滞推进剂

travelling pan 动盘

traverse error 方向瞄准误差

trench 堑壕；深沟

trench mortar 迫击炮

trial speed 试飞速度

trial-and-error method 试错法

triangle flange 三角法兰边

triangular parachute 三角形降落伞

triangular wave 三角波

triangular wave signal 三角波信号

triangulation balloon 三角测量气球

tribo-electrification 摩擦起电

tribometer 摩擦计

trichlorethylene 三氯乙烯
trick flying 特技飞行
tricycle landing gear 前三点式起落架
tricycle tractor 三轮式牵引机
tridipole antenna 三偶极天线
triform missile 三升力面导弹
trigatron 触发管
trigger（枪的）扳机；启动
trigger action 触发作用
trigger button 起动按钮；击发按钮
trim 配平
trim actuator 配平致动器
trim actuator drive 配平致动器驱动
trim after forming 成形后修边
trim angle of attack 配平迎角
trim balance 配平平衡
trim control 配平操纵
trim cord 调整索
trim die 修整模
trim drag 配平阻力
trim for takeoff 起飞配平
trim lift coefficient 配平升力系数
trim master 修边机
trim motor 配平（装置）电动机
trim run away 配平失控
trim speed 纵倾平衡速度
trim strip 调整条
trim tab 补偿调整片
trim velocity 配平速度
trimetrogon 垂直倾斜混合空中照相；三镜头航空照相
trimmable stabilizer 可调式安定面
trimmed state 配平状态
trimming mechanism 配平机构
trimming moment 配平力矩
trimming tab 配平片
triode 三极管
trioxygen 臭氧
triple point 三重合点；三相点
triple release 三连投
trip-sonic speed 三倍声速
trisonic 三种声速（亚声速、跨声速、超声速）
trisonic tunnel 三种声速风洞
trivalve spring contact 三瓣卡箍式弹簧触点
trolley conveyor 悬吊输送机
trommel sieve 圆筒筛
troop 部队；骑兵连
tropical air 热带空气
tropical atlantic air mass 热带大西洋气团

tropical continental air mass 大陆热带气团
tropical cyclone 热带气旋
tropical easterlies 热带东风带
tropical gulf air mass 热带海湾气团
tropical maritime air mass 热带海洋气团
tropical meteorology 热带气象学
tropical rainforest climate 热带雨林气候
tropical rainy climate 热带多雨气候
tropopause 对流层顶
troposphere 对流层
tropospheric wave 对流层波
trouble shooting 故障探查与排除
trough（低气压）槽
trough line 槽线
TRSA（terminal radar service area）航站雷达勤务区
true air speed（TAS）实际空速；真空速
true air temperature（TAT）绝对大气温度
true altitude 真高度；绝对高度
true anomaly 真近点角
true azimuth 真方位角
true bearing（TB）真方位
true course（TC）真航向
true cycle 实际循环
true flight path 真航迹
true heading（TH）真航向
true heading reconfiguration flag 真航向重构标记
true height（of the flight track）真高
true Mach number 真马赫数
true meridian 真子午圈
true north 真北
trunnion 耳轴
trunnion bearing 耳轴轴承
truss 构架；桁架
trussed type 桁架式
T-shaped wrench T 型扳手
T-tail T 形尾翼
TTG alert 剩余时间警示
tuba 管状（云）
tube anemometer 管形风速表
tube axle 管轴
tube of flow 流管
tube with welding 焊接套管
tube-launched, optically-tracked, wire-guided missile（TOW missile）陶式导弹；陶式反坦克导弹
tubular boiler 火管锅炉；水管式

锅炉

tubular combustor 管状燃烧室，管型燃烧室

tug 拖车

tuned radio frequency 调谐射频

tungsten 钨

tunnel blockage 风洞阻塞

tunnel blockage effect 风洞阻塞效应

tunnel dryer 隧道式干燥机

tunnel turbulence factor 风洞紊流因子

tunnel turbulence level 风洞紊流度

tunnel wall correction 风洞壁修正

tunnel wall effect 风洞壁效应

tunnel wall interference 风洞壁干扰

turbidity 混浊度

turbine 涡轮机；轮机

turbine aerodynamics 涡轮空气动力学

turbine bearing 涡轮轴承

turbine blade 涡轮叶片

turbine blade and vane cooling 涡轮转子和静子叶片冷却

turbine blank 轮盘

turbine bypass valve 涡轮旁通活门

turbine case cooling 涡轮机匣冷却

turbine cooling 涡轮冷却

turbine cooling control 涡轮冷却控制

turbine creep 涡轮蠕变

turbine disc 涡轮盘

turbine engine 涡轮发动机

turbine engine modular performance 涡轮发动机模块化性能

turbine engine modular performance estimating routine 涡轮发动机模块化性能评估程序

turbine entry temperature 涡轮进口温度

turbine inlet 涡轮进口

turbine inlet temperature 涡轮进口温度

turbine nozzle guide vane 涡轮喷嘴导向叶片

turbine power assembly 涡轮动力组件

turbine pump 涡轮泵

turbine rating 涡轮定级，涡轮定类

turbine rear frame 涡轮后框架

turbine shroud 涡轮叶冠

turbine stage 涡轮级

turbine stator 涡轮静子
turbine total-to-static efficiency 涡轮全静效率
turbine total-to-total efficiency 涡轮全全效率
turbine vane cooling 涡轮叶片冷却
turbine windmilling speed 涡轮风转转速
turbine-airscrew unit 涡轮螺旋桨组
turbo compressor 涡轮压缩器，涡轮压缩机
turbo pump 涡轮泵
turbo-annular chamber 涡轮环形燃烧室
turbocharger 涡轮增压器
turbocopter 涡轮发动机直升机
turbofan 涡轮风扇
turbofan engine 涡轮风扇发动机
turbojet aircraft 涡轮喷气式飞机
turbojet engine 涡轮式喷气发动机
turbomachine 涡轮机
Turbomeca 透博梅卡公司
turboprop 涡轮螺旋桨发动机；涡轮螺旋桨（式）飞机
turboprop aircraft 涡轮螺旋桨飞机
turboprop engine 涡桨发动机；涡轮螺旋桨式发动机
turboramjet 涡轮冲压喷气发动机
turborocket 涡轮火箭发动机
turboshaft engine 涡轮轴发动机
turbostarter 涡轮起动机
turbulator 紊流器
turbulence 紊流，湍流；紊流率，湍流度
turbulence accident 湍流事故，涡流（飞行）事故
turbulence cloud 湍流云
turbulence factor 紊流因子
turbulence intensity 紊流强度
turbulence level 紊流度
turbulence penetration speed 穿越紊区速率
turbulence shedding 紊流尾迹
turbulent boundary layer 湍流边界层
turbulent boundary layer separation 湍流边界层分离
turbulent flame speed 紊流火焰速率
turbulent flow 紊流；扰流
turbulent mixing 紊动混合；湍流混合
turbulent motion 湍动
turbulent region 紊流区

turbulent resistance 扰流阻力，紊流阻力

turbulent separation 紊流分离

turbulent stress 紊流应力

turbulent wake 扰流尾流；紊流尾流

turbulent-wake state 紊流尾流状态，湍流尾流状态

turn altitude 转弯高度

turn and bank（slip）indicator 转弯倾斜（侧滑）仪；转弯倾斜（侧滑）指示器

turn error 转弯误差

turn height 转弯高

turn in 转入

turn indicator 转弯指示器

turn initiation point 转弯起始点

turn to base（turning base）三转弯

turn to final（turning final/final turn）四转弯

turnaround time 回程起飞准备时间；往返（飞行）时间；再次出动准备时间，停放时间

turnback 转 180° ；转到反航向；中途返航

turnbuckle 紧缆器；松紧螺旋扣

turning angle 转折角

turning bank 转弯坡度

turning error 转弯误差

turning moment 转弯力矩

turning point（bank/time）转弯点（坡度 / 时间）

turning speed 转弯速度

turning time 转弯时间

turnoff 转弯滑出跑道；脱离跑道点，跑道出滑行道的连接点；脱离跑道用的滑出道

turnoff taxiway 岔道滑行道

turnover structure 翻倒防护结构

turns around a point 绕点转弯

turns ratio 匝数比

turnstile antenna 正交叉天线

turret 回转装置；（军舰、坦克、飞机等的）回转炮塔；旋转枪架

TV recording 电视录像

TVOR（terminal VHF omni-range）机场多向导航台；终端甚高频全向无线电信标

twin engine adaptation 双发协调

twin engine rating stop 双发动机额定停车

twin engine stop 双发停止

twin helicopter single engine stalling 双发直升机单发停车

twin machine gun 双联机枪，双管机枪
twin tail 双尾翼
twin turbo-shaft engines 双涡轮轴发动机
twin-ball bearing 双滚珠轴承
twin-chip joint 双耳片接头
twin-disk clutch 双片式离合器
twin-engine configuration 双发动机式飞机布局
twin-engine operation alignment 双发工作协调
twinjet 双喷气发动机飞机
twinkle roll 编队横滚
twin-rotor helicopter 双旋翼直升机
twin-spool compressor 双转子压缩机
twin-wheel nose gear 双轮前起落架
twist 弯曲，扭转；缠绕，盘旋，螺旋运动；螺旋状；缠度；横滚，半滚，滚转
twist bevel gear 扭转斜齿轮
twist drill 螺旋钻
twist joint 扭接
twisted gear 扭齿轮
twisted pair 双绞线
twisted rope 绞绳
twisted spur gear 扭转正齿轮，正扭齿轮
twisted wing 扭转机翼
twist-grip throttle 转把式油门
twist-grip throttle lever 转把式油门杆
twisting couple 扭转力偶
twisting force 扭力
twisting moment 扭转力矩
twizzle 旋转机动
two cavity hydropneumatic type 双腔油气式，双腔液压气动式
two position valve 双位活门
two segment combination approach 两段混合进场
two way exchanger 双路交换器
two-axis gyro 双轴陀螺仪
two-axis homing head 双轴导引头
two-bladed wooden propeller 双木质螺旋桨
two-colour pyrometer 两色高温计
two-control aircraft 双操纵面飞机（用升降舵与副翼操纵而无方向舵的飞机）
two-course radio range 双航道无线电指向标；双正航向信号导航台；双正航向信号电塔

two-dimension 二维
two-dimensional flow 二维流动
two-dimensional inlet 二维进气道
two-dimensional nozzle 二维喷管
two-dimensional radar 二坐标雷达
two-ear piece 双耳件
two-engine landing 双发着陆
two-engine speed 双发飞行速率
two-helicopter formation 双机编队
two-phase flow 二相流
two-phase motor 二相马达
two-phase selsyn 二相自动同步器
two-phase servomotor 二相伺服马达
two-phase system 二相制
two-spar wing 双梁翼
two-stage compression 二级压缩
two-stage ignitor 二级点火器
two-stage supercharger 双级增压器
two-terminal network 二端网络
two-terminal trajectory 双端点弹道
two-way communications 双向通信
two-way data link 双向数据链路
two-way radio 双向通信无线电台
two-way switch 双向开关；双路开关
type certificate 型号生产许可证；型式证明；型号合格证
type frequency 类型频率
typhoon 台风
typical anticyclone 标准反气旋
typical cyclone 标准气旋
typical helicopter DC power supply system 典型直升机直流供电系统
tyre 轮胎
tyre barometer 轮胎气压计

U

U tube manometer U 形管流体压力计
U-bomb 铀原子弹
UHF（ultrahigh frequency）超高频
UHF communication system 超高频通信系统
UHF radio 超高频电台
UIR（upper flight information region）高空飞行情报区
ULB（underwater locator beacon）水下定位信标
ultimate fly-up limit 最终起飞限制
ultra short wave 超短波
ultra short waves transceiver 超短波电台
ultraharmonic 高次谐波
ultra-high bypass-ratio 超高涵道比
ultrahigh frequency（UHF）超高频
ultra-high frequency navigation system 超高频导航系统
ultrahigh speed 超高速
ultra-high vacuum 超高真空
ultralight vehicle 超轻型载具
ultra-long range 超远程
ultramicroscope 超倍显微镜
ultraradar 超远程警戒雷达
ultra-short wave radio set 超短波无线电台
ultrasonic wave 超声波
ultraviolet light 紫外线；紫外光
ultraviolet radiation 紫外辐射
ultrawide angle 超宽视角
umbilical tower 脐带式管线塔架
umbrella type vibration 伞形振动
unbalanced weight 不平衡重量
unburnt fuel 未燃烧的燃油
uncertainty phase 情况不明阶段
uncommanded deceleration 非预定减速
uncontrolled spin 失控螺旋
unconventional warfare 非常规战争；非正规战争

uncouple 解除连接；使解耦联；使非耦合

uncovering of the covering cloth 揭开蒙布

under motivation 动机不足

undercarriage 飞机起落架

undercarriage control test 起落架收放试验

undercarriage track 起落架轮距；主轮距

underexpanded flow 膨胀不足流

underexpanded nozzle 欠膨胀喷管

underexpansion 不完全膨胀，不足膨胀

underfrequency 频率过低

underground bunker 地下掩蔽部，地下掩体

underground launcher 地下发射装置

underground storage facilities 地下储存设施

undermanned（舰、炮等）人员配备不足；缺额

undermine 爆破；挖坑道

undershoot 射击过近，投弹过近（炸弹或射弹落在目标前方的投弹或射击）；目测过低的着陆

undervoltage 欠压，电压不足

underwater locator beacon（ULB）水下定位信标

undetermined connection 不确定的连接

undulatus 波状云

uneven ground 不平地

unexpected bleed valve closing 放气活门意外关闭

unexpected illumination 意外灯亮

unexpected Ng deceleration 燃气涡轮转速（Ng）意外减速

unexplained accident 不明事故

unexploded ordnance 非爆炸武器；未爆弹药

unexploded shell 未爆弹

unguided rocket launcher 非制导火箭发射器

unhurried landing 从容着陆

unidentified liquid and powder 不明液体及粉末

unidirectional aerial 单向天线

unidirectional composite 单向复合材料

unidirectional couplet antenna 单向成对天线

unidirectional solidification 定向

（单向）凝固

uniform acceleration 匀加速度

uniform acceleration motion 匀加速运动，等速运动

uniform annual value 等年值

uniform circular motion 等速圆周运动

uniform flight 等速飞行

uniform flow 均匀流；等速流

uniform grid 均匀网格

uniform illuminated aperture 均匀照射口径

uniform line 均匀流线；等粗线

uniform load 均匀负载；均布荷载

uniform motion 等速运动

uniform planform 统一俯视图，统一（从上向下看物体的）平面图

uniform retarded motion 等减速运动，匀减速运动

uniform varying load 等变负载

unilateral conductivity 单向导电性

unilateral glass fiber 单向玻璃纤维

unilateral war 单边战争

unimproved landing areas 未加改善的着陆区域

uninterruptable power supply 不间断供电，不间断电源

uninterruptible power system 不间断供电系统

union 管节（管接头）

union for compressor washing 压气机清洗接头

unipole 单极（天线）

unipole antenna 单极天线

unit 部队；单元

unit area 单位面积

unit deformation 单位变形

unit interval 单位时间间隔

unit load method 单位负载法

unit of measure 计量单位

unit of time 时间单位

unit pole 单位极

unit price 单价

unit pulse 单位脉冲

unit time 单位时间

unit under test 被测部件

unit weight 单位重量

unity coupling 全耦合

universal asynchronous receiver 万用非同步接收器（机）；通用异步接收器（机）

universal asynchronous receiver/transmitter 通用异步收发机（器）；万用非同步收发机（器）

universal asynchronous transmitter 通用异步发射机（器）；万用非同步发射机（器）
universal chuck 万能卡盘，自动定心卡盘
universal flight data recorder 通用型飞行参数记录仪
universal gas constant 通用气体常数
universal gravitation 万有引力
universal hanging rack 通用挂架
universal head rivet 通用头铆钉
universal inclinometer 通用倾斜仪
universal joint 万向接头，十字接头
universal meter 通用电表
universal method 通用方法
universal mounting tray 通用安装架
universal pylon 通用挂架
universal shift joint 通用接头
universal time 世界时
unloading gate 卸载门
unloading point 卸载点
unloading unit 卸载机构
unlock 打开；解锁
unmanned helicopter 无人驾驶直升机
unmanned reconnaissance aircraft 无人驾驶的侦察飞行器
unmanned shutdown 意外停车
unmanned spy plane 无人驾驶侦测机
unpowered flight 无动力飞行
unpowered rotor speed 自转旋翼的转速
unprepared airfield 无道面机场
unreadable 听不清；没听明白
unrefueled range 不进行空中加油的航程
unreliable brake 刹车不可靠，不可靠制动
unrestricted propellant 多面燃烧推进剂
unrestricted visibility 无障能见度；无限制能见度
unsaturated cloud 未饱和云
unscheduled engine removal 非计划更换发动机
unscheduled landing 非计划内着陆
unscheduled trim 非预定配平
unsealed strip 无防水铺筑道面的跑道
unserviceability 使用不可靠性；

运转不安全性
unserviceable 不适用的；不适航的
unshock 消除激波
unstabilized 不稳定的
unstable air 不稳定空气
unstable airfoil 不稳定翼型
unstable combustion 不稳定燃烧
unstable control 不稳定控制
unstable equilibrium 不稳定平衡
unstable flow 不稳定流
unstable motion 不稳定运动
unstable oscillation 不稳定振荡
unstable region 不稳定区
unstable state 不稳定状态
unstable vibration 不稳定振动
unstalling 终止气流分离
unstart 发动机不起动
unsteady flight 不稳定飞行
unsteady flow 不稳定流
unstick 起飞；离地瞬间
unstick ratio 离地速率比
unstick run 离地前滚行
unstick speed 离地速度
unsymmetrical configuration 不对称构型
unsymmetrical flight 不对称飞行
unsymmetrical flow 不对称流
unsymmetrical frame 不对称构架
unsymmetrical loading 不对称负载
unsymmetrical propeller thrust 不对称螺旋桨拉力
unsymmetrical vortex 不对称涡旋
unusable 不能使用的
unusable fuel 无用燃油
unusable power 无用功率
up speed 增速
up to the standard 符合规定
up-and-down current 升降流
up-chaff 无干扰拦截
upcurrent 上升气流
update 更新；修正
up-dating documentation 最新有效文件
up-datings 最新资料
updraft 上升气流
updraught 上升气流
updraught carburetor 上吸式化油器
upflow angle 上升流角
upfront controller 上前方控制器
upkeep 保养
up-leg（弹道、航迹等的）上升段
uplift rate 上升率
uplink 上传链路；上行链路

upload 向上负载；供弹；加负荷，加载
uplock 上位锁
upper air route 高空航线
upper atmosphere 高层大气
upper branch 上时圈；上半子午线
upper camber 上弧线
upper cloud 高云
upper control area 高空管制区域
upper culmination 上中天
upper deck door 上舱门
upper deck fairing 平台上蒙皮
upper deviation 上偏差，上尺寸差
upper flight information region（UIR）高空飞行情报区
upper inversion 高空逆温
upper level anticyclone 高空反气旋
upper level chart 高空图
upper level cyclone 高空气旋
upper level trough 高空槽
upper limb 上缘
upper limit 上限
upper limit of flammability 可燃性上限
upper limit of stability 稳定性上限
upper mark 上标
upper mixing layer 高空混合层
upper rudder 方向舵上段
upper side band 上边带
upper sideband data 上边带数据
upper sideband voice 上边带音频
upper transit 上半径圈通过
upper vertical tail 上垂直尾翼
upper wing 上翼
upper-surface aileron 上翼面副翼
upper-surface blowing 上翼面吹气
uprate 提高功率；提高性能
uprush 强上升气流；垂直气流
upset in pitch 自动俯仰
upset in roll 自动滚转
upside down loop 倒飞筋斗
upside-down 机腹朝上
upside-down flight 倒飞
upside-down formation 倒飞编队
upslope fog 上坡雾
upstairs 在空中；爬高；高空
upstream flow 上游流
upstream pressure 上游压力
uptime 在用时间
upward air current 向上气流
upward force 向上力
upward identification light 机背识别灯；上部标志灯

upward pressure 向上压力

upward roll 上升滚转，上升横滚；上升侧滚

upward rotation 上仰转动

upward spin 上升螺旋

upward voltage 升压

upwash 上洗流

upwind 顶风；逆风

upwind effect 上风效应

upwind end of the runway 跑道逆风端

upwind leg 第一边

upwind takeoff 逆风起飞

urgency 紧急情况

usable ceiling 可用升限

usable fuel 可用燃料

usable power 有用功率

useful lift 有用升力

useful load 有用负载

user differential range error 用户差分距离误差

user equivalent range error 用户等效距离误差

user range accuracy 用户测距精度

user range error 用户测距误差

UTC（Coordinated Universal Time）协调通用时；协调世界时

utility aircraft 通用机

utility airplane 通用飞机

utility airport 通用机场

utility circuit shut-off valve 通用线路截止阀

utility helicopter 通用直升机

utility light 通用工作灯

utilization and maintenance 使用和维护

V formation V 形编队，V 字飞行队形

vacuum speed 真空飞行速率

vacuum trajectory 真空弹道

vacuum tube voltmeter 真空管电压表

vacuum tunnel 真空式风洞

valid 有效的

valid external power 有效外部力量

validate 确认

validity 有效性

value engineering 价值工程

valve 阀；活门

valve arrangements 活门装置

valve blocked 活门阻滞

valve closing 活门关闭

valve overlap 活门重叠

valve position logic card 活门位置逻辑卡

valve rigging 阀门系统调整

valve support 活门座

Van Allen radiation belt 范艾伦辐射带

vane anemomete 导片风速表，旋板式风速表

vane pump 旋板泵；叶轮泵

vane sensor 翼片式传感器

vane-type pump 叶片型泵，轮叶式泵

vapor concentration（absolute humidity）水蒸气浓度

vapor cycle system 蒸发循环系统

vapor trail 凝结尾迹

vaporizer 蒸发器

vaporizing combustor 汽化燃烧室

vapor-phase inhibitor 气相抑制剂

variable 变量

variable acceleration 变加速度

variable area 可变面积

variable area exhaust nozzle 可变排气喷口

variable bleed valve 可变（调）

放气活门

variable bypass valve 可调旁通活门

variable camber wing 可变弧翼

variable capacitance 可变电容

variable ceiling 变动云幕

variable chord wing 变弦机翼

variable density tunnel 变密度风洞

variable descent gradient 不定下降梯度

variable exhaust nozzle 可变喷嘴

variable geometry 可变几何形状

variable geometry diffuser 形态（状）可变扩散器

variable incidence 可变倾角

variable inlet 可变进气口

variable inlet guide vane 可调进气导流片

variable leak 可变泄漏量

variable lift device 可变升力装置

variable mass system 变质量系

variable pitch propeller 变距螺旋桨

variable pressure 变压

variable resistance 可变电阻

variable separation 分离变量

variable signal 可变信号

variable span wing 变展机翼

variable speed 变速

variable speed constant frequency generator 定频变速发电机

variable stator 装有可调叶片的静子；叶片安装角可变的整流环

variable stator vane 可变（调）静子叶片

variable stroke 可变冲程

variable variation 可变的变化

variable visibility 变动能见度

variable wind 不定风；无定风

variable wing 可变翼

variable wing area 可变翼面积

variable wing section 可变翼剖面

variable-datum boost control 变基准加力控制

variable-delivering pump 变流量泵

variable-depth sonar 可变深度声呐

variable-discharge turbine 变流量涡轮

variable-displacement pump 变量泵；可变容积泵；可变排量泵

variable-geometry inlet 变几何形状进气道；可调进气道

variator 可变电阻器；变阻器

variocoupler 可变耦合器

variolosser 可变损耗器

variometer 磁变仪，磁力偏差计

varistor 可变电阻器
VASIS（visual approach slope indicator system）目视进场下滑道指示系统
VDC bus-bar 直流汇流条
vection illusion 相对运动错觉
vector 矢量
vector analysis 矢量分析
vector approach 引导进场
vector computer 矢量计算器
vector flight control 向量飞控；矢量飞行控制
vector hard port 向左急转
vector hard starboard 向右急转
vector nozzle 向量喷嘴
vector sight 定向瞄准；矢量瞄准器
vector steering 矢量控制；矢量操纵
vectored attack 引导攻击
vectored thrust 矢量（可调的）推力；转向推力
vectoring nozzle 向量喷嘴，变向喷嘴
veeder counter 计程仪
veer 改变方向;（风的）顺转
veil cloud 卷层云
velocimeter 速度计；测速计
velocity 速度
velocity at burnout 主动段终点速度
velocity attitude heading scale 速度姿态方向刻度
velocity boundary layer 速度边界层
velocity envelope 速度包线
velocity gradient 速度梯度
velocity head 流速水头；速度头
velocity hold mode 速度保持模式
velocity image 速度像
velocity minimum control（VMC）最低速度控制
velocity never to exceed（VNE）最高限速；速度上限
velocity of maximum operating 最大可操作速度
velocity of separation 分离速度
velocity pressure 速度压力
velocity search 速度搜索
velocity signature 速度标示
velocity stability derivatives 速度安定性导数
velocity storage 速度贮存
velocity vector 速度矢量
velocity-altitude map 速度高度图

vendor component maintenance manual 厂商部件维修手册
vent 出口；排出；通气孔
vent line 放气管；通风系统
ventilate membrane 呼吸膜
ventilate the start injector 给起动喷嘴供气
ventilated storage 通风储藏
ventilated temperature sensor 通气（风）温度传感器
ventilation 通风
ventilation air intake 通风进气
ventilation device 通风设备
ventilation garment 通气服
ventilation outlet 通风出口
ventilation position 通气状态
ventilation switch 通风开关
ventilation system 通风系统
ventilator 通风器；送风机
venting nozzle 排气喷嘴
ventral fin（飞机的）腹鳍
ventral inlet 机腹进气口
ventral turret 机腹炮塔
venturi 文丘利管（文氏管）
venturi effect 文丘利效应（文氏效应）
verge ring 边环
verification 验证；确认；核实
verification and validation 验证及确认
vernier engine 微调发动机
vernier rocket 微调火箭
versatility 多功能性；通用性
vertebratus 脊状云
vertex 顶点
vertical 垂直的；垂直线
vertical axis 垂直轴
vertical bank 垂直坡度
vertical boiler 立式锅炉
vertical circle 垂直圈；地平经圈
vertical climb 垂直上升
vertical cross hair 垂直交叉瞄准线
vertical deviation scale 垂直偏移刻度
vertical dilution of precision 垂直精度扩散因子
vertical dive 垂直俯冲
vertical downward 垂直下降
vertical drive shaft 垂直驱动轴
vertical ejector rack 垂直弹射架
vertical envelopment 垂直包围
vertical fin 垂直安定面；直尾翅
vertical flight 垂直飞行
vertical glide path（VGP）垂直

下滑道
vertical gust 垂直阵风
vertical gyro 垂直陀螺仪
vertical gyro unit 垂直陀螺仪装置
vertical hinge 垂直铰；摆振铰
vertical interval 垂直间隔
vertical landing 垂直着陆
vertical launch 垂直发射
vertical lever 纵杆
vertical line 垂直线
vertical navigation 垂直面导航
vertical ordinate 纵坐标
vertical plane 垂直面
vertical pressure gradient 垂直压力梯度
vertical pump 垂直泵
vertical ramp 2-D inlet 立置斜板二维进气道
vertical reverse turn 垂直倒转
vertical scale 垂直刻度
vertical separation 垂直间隔
vertical separation minimum 垂直间隔最低标准
vertical shaft 立轴
vertical situation display 垂直情况综合显示器
vertical speed 垂直速度；升降速度
vertical speed indicator（VSI）垂直速度表
vertical speed pointer 垂直速度指针
vertical speed readout 垂直速度读取值
vertical speed selected value FDC 垂直速度选择值 FDC
vertical speed value 垂直速度值
vertical spin tunnel 立式螺旋风洞
vertical split 垂散开花，垂直分裂
vertical stabilizer 垂直安定面；垂直尾桨；垂尾
vertical stabilizer station 垂直安定面站位
vertical tail 垂直尾翼
vertical tail area 垂直尾翼面积
vertical tail efficiency 垂直尾效率
vertical tail fin 垂直尾桨；垂尾；垂直安定面
vertical tail surface 垂直尾翼面
vertical tail unit 垂直尾翼组
vertical tail volume coefficient 垂直尾容积系数
vertical takeoff（旋翼机的）垂直起飞
vertical takeoff and landing 垂直

起降

vertical target speed 目标垂直分速

vertical temperature gradient 垂直温度梯度

vertical track 垂直航迹

vertical track alert caption 垂直航迹警告说明

vertical translation 垂直位移

vertical turn 垂直转弯

vertical uplift 垂直上升

vertical uplift performance 垂直上升性能

vertical velocity 垂直速度

vertical velocity indicator 垂直速度表

vertical vibration 垂直振动

vertical visibility 垂直能见度

vertical/short takeoff and landing 垂直 / 短距起降

vertical-attitude takeoff and landing 立姿起飞和着陆

vertifuge 错觉模拟机

vertigo 晕眩；头晕

vertigon 错觉训练机

vertipad 垂直起降点

vertiplane 垂直起落飞机

very high frequency（VHF） 甚高频

very high frequency direction-finding 甚高频测向

very high frequency omnidirectional radio range 甚高频全向无线电信标

very high frequency omnidirectional radio range navigation system（VOR）特高频全向无线电导航台

very high frequency receiver 甚高频收信机

very low frequency 甚低频

VFR flight 目视飞行

VFR flight following 目视飞航指引

VFR freeway 目视非管制航路

VFR on top 云上目视飞行

VFR radar advisory 目视飞行雷达咨询

VFR separation standard 目视飞行隔离标准

VFR terminal area chart 航站区目力飞行航图

VFR terminal minimums 目视飞行终点最低限

VFR weather 目视飞行天气

VGP（vertical glide path）垂直下

滑道
VHF（very high frequency）甚高频
VHF data radio 甚高频数据无线电
VHF datalink 甚高频数据链
VHF navigation receiver 甚高频导航接收机
VHF omnidirectional navigation beacon 甚高频全向导航信标
VHF omnidirectional radio range（navigation beacon）甚高频全向无线电导航台；甚高频全向无线电指向标
VHF radio 甚高频电台
vibration differential equation 振动微分方程
vibration frequency 振动频率
vibration frequency meter 振动频率计
vibration galvanometer 振动电流表
vibration isolator 隔振器
vibration level 振动水平
vibration level check 振动水平检查
vibration plane 振动面
vibration reducer 减震器
vibration test 振动试验
vibration unit 振动单元
vibrational energy 振动能
vibratory frequency 振频
vibrex balancing kit 振动平衡套具
vibro-impact 撞振
vibrometer 振动计
vibropendulous error 振动误差
vice 老虎钳
video frequency 视频；图像频率
video mapping 视频地图装置；频谱扫描指示；扫调指示
video recorder 录像机
video recording 录像
video signal 视频信号
video tape recorder 磁带录像机
video tape recording 磁带录像
vidicon 光导摄像管
vigilance 警惕性
vinyl chloride 氯乙烯
violet 紫外线
virga 雨幡；幡状云
virtual displacement 虚位移
virtual force 虚力
virtual mass 虚质量
viscoelastic material 黏弹性材料
viscosity 黏性
viscosity coefficient 黏性系数
viscosity index 黏性指数

viscous aquaplaning 黏滞水飘
viscous damping 黏滞阻尼；黏性阻尼
viscous flow 黏滞流动；黏性滞（动）
viscous fluid 黏滞流体
viscous force 黏性力
viscous friction 黏性摩擦
viscous hydroplaning 黏性水漂
viscous stress 黏应力
visibility 能见度
visibility meter 能见度表
visible horizon 可见地平
visible spectrum 可见光谱
visual approach 目视进近
visual approach and departure protection area 目视进离场保护区
visual approach and landing chart 目视进场及降落图
visual approach chart 目视进近图
visual approach slope indicator（VASI）目视进场坡度指示器
visual approach slope indicator system 目视进近坡度指示系统
visual axis 视轴
visual bombing 目视轰炸
visual check 目视检查
visual check list 目视检验表
visual contact 目视联络
visual contact approach 目视进场着陆
visual cue 目视信号；目视提示
visual descent point 目视下降点
visual display 直观显示；可见显示
visual display unit 目视显示单元；可见显示装置；视觉显示装置；视频显示器
visual envelope 目视范围
visual flight 目视飞行
visual flight rules 目视飞行规则
visual glideslope indicator 目视下滑道指示器
visual head-up image 目视抬头影像
visual illusion 视觉性错觉；目视错觉
visual inspection 目视检查
visual measurement landing 目测着陆
visual measurement stilling 目测消速
visual meteorological avoidance system 目视气象回避系统
visual meterological conditions 目

视飞行气象条件
visual omnirange 目视全向信标
visual orientation 目测方位；目测定向；目视定向
visual photometry 目视测光；目视光度测量
visual range 目视距离；视距；能见距离
visual reference 目视基准；目视参考物
visual separation 目视间隔；目视分离
vital area 重要地区
vital target 重要目标
vitreous 玻璃体
VMC（velocity minimum control）最低速度控制
VNE（velocity never to exceed）最高限速；速度上限
voice coder 话音编码器
voice keying 语音键控
voice recorder 话音记录器；录音机
voice recorder system 录音系统
voice response system 话音响应系统
voice rotating beacon 语音旋转信标
voice speech synthesizer 语音综合器
voice switching and control system 话音交换和控制系统
voice-grade channel 音频级信道
voiceless homing 非语音归航，无声归航
volatility 挥发性
VOLMET（meteorological information for aircraft in flight）对空天气广播
VOLMET broadcast 飞行气象资料广播
voltage 电压
voltage comparator 电压比较器
voltage controlled oscillator 电压控制振荡器
voltage division 分压
voltage regulator 调压器
voltage transformation 变压
voltmeter 电压表
volts alternating current 交流电压
volts direct current 直流电压
volume 体积
volume adjustment 音量调节
volute pump 涡卷泵
VOR（VHF omnidirectional radio

range）甚高频全向（无线电）信标

vortex core 涡心

vortex cylinder 涡柱面

vortex drag 涡旋阻力

vortex filament 涡丝，涡旋线

vortex generators 涡轮发生器

vortex hazard 涡旋危害

vortex lift 涡升力

vortex line 涡线

vortex motion 涡动

vortex pair 涡对

vortex path 涡旋轨迹

vortex ring 涡环

vortex ring of tail rotor 尾桨涡环

vortex separation 涡旋分离

vortex strength 涡旋强度

vortex tangential velocity 涡旋切线速度

vortex train 旋涡列

vortex-induced velocities 旋涡诱导速度

vortex-ring state 涡环状态

vortices 旋涡系

vortillon 涡流器；旋涡发生器

V-shaped depression V 形低气压

V-tail V 形尾翼

VTOL port 垂直起降机场

W

WAC（world aeronautical chart）世界航空图
waist 机身中部
waiting position 待命阵地
waiting status 待命状态
wake 尾迹；尾流
wake analysis 尾流分析
wake current 尾流
wake drag 尾流阻力，扰流阻力
wake drag coefficient 尾流阻力系数
wake effects 尾流效应
wake flow 尾流
wake loss 尾迹损失
wake shape 尾流形状
wake stream 尾流
wake turbulence 尾流紊流度
wake vortices 尾迹，尾流涡系
wake-interaction noise 尾迹干涉噪声
walk round check 绕走检查
walk-around inspection 巡视检查
walkround and pre-flight check 绕走与飞行前检查
walkround check 绕走检查
wall panel 壁板
war consumables 战时消耗品
war expenses 军费
war fund 军费
war game 军事演习；兵棋推演
war of annihilation 歼灭战
war of attrition 消耗战
war of conquest 征服战
war of masses 凭兵力优势制胜的战争
war of movement 运动战
war of nerves 神经战
war of position 阵地战
war of propaganda 宣传战
war on two fronts 两线作战
war organization 战时编制
war reserve 战备物资；战争储备

war spoils 战利品
war stores 军需品
war strength 战时定员；战斗力
war theater 战场
war wastage reserves（WWR）战时消耗贮备
war zone 交战地带；作战地带
warehouse 仓库
warhead 弹头；战斗部
warhead booster 弹头助爆药
warhead gain 弹头增益
warhead-blast cluster 爆震集束弹头
warhead-fragmenting 杀伤弹头
warload 战斗载重；战斗装载
warlord 军阀
warm front 暖锋
warm gas thruster 暖气推进器
warm high 暖高压
warm low 暖低压
warm season 暖季
warm sector 暖区
warming switch 加温开关
warming trouble light 加温故障灯
warming-up device 加温设备
warning 警告
warning area 危险区；警告区
warning caution panel 警告板
warning device 警告装置
warning electronics unit 警告电子组件
warning flag 警告旗
warning horn 警报器
warning horn switch 警报器开关
warning indicator 警告信号器
warning lamp 报警灯
warning light 警告灯
warning net 警报网；防空警报网
warning panel（座舱内的）警告板
warning placards 警告标牌
warning range 告警范围
warning signal lamp 警告信号灯
warning streamer 警告飘带；警告布条
warning system 警报系统
WARP（weather and radar processor）天气和雷达处理器
warped wing 挠曲机翼；挠翼
warping function 扭挠函数
warplane 军用飞机；战机
warranty 保用期；保险期
war-readiness spares kit（武器系统）战备成套零件箱

warship 军舰
wartime rate 战时飞行强度；战时飞机利用率；战时飞行率
WAS（water activated switch）水激活开关
wash out circuit 摒弃电路
washer 垫片；垫圈
washin 内洗；（机翼的）正扭转
wastage 损耗量；废品
waste heat 废热
watch item 列管项目
watcher 岗哨；哨兵
water activated switch（WAS）水激活开关
water ballast 水压载
water barrier 水障碍；防水层
water clock 漏壶；水钟；滴漏
water cloud 水云
water collecting sump 集水槽
water contact 水接点
water cooler 水冷却器
water equivalent depth 等量水深
water ingestion 水注入
water injection 注入水
water jacket 水套
water landing speed 接水速率
water level 水平面
water line 水线
water or ice ingestion 吸进水或冰
water pump 水泵
water pump plier 水泵钳
water rudder 水舵
water service truck 供水车
water suit 水夹层抗荷服
water system filler 加水口
water tank 水箱
water tunnel 水洞
water wheel 水车
water-cooled engine 水冷式发动机
water-displacing fluid 防水液
waterproof container 防水包装箱
waterspout 水龙卷
water-tube boiler 水管式锅炉
Watt 瓦特
wave analyzer 析波仪，波形分析器
wave angle 波传播角
wave antenna 行波天线
wave band 波段
wave carrier 载波
wave constant 波常数
wave crest 波峰
wave detection 检波
wave detector 检波器

wave distortion 波形失真
wave disturbance 波扰动
wave drag coefficient 波阻力系数
wave equation 波动方程式
wave filter 滤波器
wave filtering 滤波
wave filtering circuit 滤波电路
wave form 波形
wave frequency 波频；电波频率
wave front 攻击波正面；波前
wave guide 波导（管）
wave length 波长
wave mechanics 波动力学
wave meter 测波表，波长计
wave mirror 电波镜
wave motion 波动
wave number 波号
wave propagation 波传播
wave trough 波谷
waveguide 波导
waveguide coupler 波导耦合器
waveguide elbow 导波管肘
waveguide plumbing 导波管系
waveguide plunger 波导管断路器
waveoff 复飞
waveshape 波形
wave-shape distortion 波形失真
wave-shaping circuit 整波线路
wavetrain frequency 波列频率
wave-trap 陷波器
waving wing 扑翼
way point（WPT）航路点
way point number 导航点号码
waypoint 航路点
WCA（wind correction angle）风力修正角
weak link 可断连接（点）
weapon 武器；兵器
weapon bay 武器舱；军械舱
weapon carriage 武器悬挂架
weapon control system 武器控制系统
weapon delivery 发射武器；武器投射
weapon delivery panel 武器投掷面板
weapon fit 兵器配备；武器配备
weapon insert panel 武器信息输入板
weapon mounting 武器挂装
weapon occupation 武器占位
weapon pit 射击掩体
weapon pylon 武器外挂梁
weapon release 武器发射；武器

投放

weapon state of readiness 武器战备状态

weapon system 武器系统

weapon system specification 武器系统规格

weapon type 武器型类

weapon-aiming system 武器瞄准系统

weapons embargo 武器禁运

weapons inspectors 武器检查人员

weapons of mass destruction（WMD）大规模杀伤性武器

wear 磨损

wear out 磨损

wear out period 磨损期

wear strip 防磨条

wearing piece 摩擦片

weasel 履带式水陆两用人员运输车

weather abort 因天气条件中断任务

weather advisory 天气报告

weather analysis 天气分析

weather and radar processor（WARP）天气和雷达处理器

weather beam 气象侦察波束

weather category 气象类型

weather center 气象中心

weather communications processor 气象通信处理器

weather forecast 天气预报

weather map 天气图

weather minimum 最低天气标准

weather modification 人工影响天气

weather observation 气象观测

weather rad scanner antenna 雷达扫描天线

weather radar（WX）气象雷达

weather radar control panel 天气雷达控制面板

weather radar mode display 天气雷达模式显示

weather radar system 气象雷达系统

weather reconnaissance 气象侦察

weather satellite 气象卫星

weather service radar 气象服务雷达

weather station 气象台

weather symbols 天气符号

weather type 天气类型

weathercock 风标

weathercock stability 风标安定

性，方向稳定性
weathercocking 风标效应
weathervane 风标
weaver（交叉飞行的）护航战斗机
web average burning-surface area 分隔平均燃烧表面积
web joint 腹板接合
web plate 腹板
web splices 腹板接片
wedge 楔形物
wedge angle 楔角
wedge formation 楔形队形
wedge inlet 楔形进气道
wedge slope 锥角，楔形边坡
weighing rain-gauge 衡重雨量计
weight and balance 重量与平衡
weight and balance manual 重量平衡手册
weight and balance sheet（飞机）重量与平衡记录单；载重与平衡表
weight and balance system 装载与配平系统
weight balance 载重平衡
weight barometer 衡重气压表
weight breakdown 重量分配
weight capability 承重能力
weight of displaced air 被排开空气的重量
weight on wheels 着陆重量
weight schedule 载重程序表，重量明细表，重量表
weight sorter 重量选别机
weight-on-wheel（WOW）机轮承重
weight-on-wheel（WOW）micro-switch 机轮承重微动开关
weld joint 焊缝
welded patch 焊接补片
welded seam 焊缝
welded tubular frame 复合材料管状门框
welding goggles 焊接护目镜
welding seam 焊缝
western longitude 西经
Westland Lynx 韦斯特兰“山猫”直升机
wet clutch 湿式离合器
wet drain 湿漏
wet emplacement 发射用冷却水池
wet engine 已加好燃料的发动机
wet float 湿浮筒
wet fuel rocket 湿燃料火箭，液体燃料火箭

wet H-bomb 湿氢弹
wet layup 湿式叠层
wet pad 水冷式发射台
wet start 湿起动
wet strength 湿强度
wet sump 湿机匣；（活塞式发动机的）滑油收油池
wet surface 湿道面
wet takeoff thrust 湿起飞推力
wet thrust 湿推力
wet weight 湿重
wet winding 湿绕法
wet workshop 燃料箱工场
wet-bulb potential temperature 湿球位温
wet-bulb thermometer 湿球温度计
wetout 湿润
wetted area 浸湿面积
wetting agent 润湿剂
wheel 轮；机轮
wheel base 前后轮距
wheel doors 起落架舱门
wheel fairings 轮圈罩
wheel interface unit 机轮接口组件
wheel load capacity 轮承载能力
wheel mode 转动模式（卫星稳定）
wheel pants 机轮减阻罩
wheel track 主轮距
wheel up 收机轮，起落架收起
wheel well 轮舱
wheelbarrow 独轮手推车
wheel-ski 起落轮橇
wheels-up landing 机腹着陆
wheel-type landing gear 轮式起落架
whip stall 尾冲失速
whirl wind 旋风
whirling mode 旋转振态
whirling speed 旋翼转速
whirling thermometer 旋转温度计
whirling wing 旋翼
whistle 啸声
whistling wind 呼啸风
white alert 白色警报；解除空袭警报
whiteout 雪盲
whitworth thread 英制标准螺纹；惠氏螺纹
whole body counter 整体辐射计数器
whole number 正整数
whole range 全程
wide angle 宽视角
wide area augmentation system 广

域增强系统
wide area differential system 广域差分系统
wide band 宽频带；宽带
wide latitude 广泛的行动自由
wide mesh filter 宽网滤
wide open 全开；油门最大；用最大转速；用最大功率
wide turn 小坡度转弯（坡度小于 25°）
wideangle collimating window 广角瞄准窗
wild shot 流弹
winch 绞盘；绞车
wind aloft 高空风
wind aloft report 高空风报告
wind angle 风向角
wind arrow 风向指针
wind axis 风轴
wind bar lights 翼排灯
wind blast 风暴；疾风伤害；风力摧残；气流吹袭；(飞行员弹射出座舱时）迎面气流
wind calm 静风
wind chill 风寒
wind component 风的分量
wind conditions 风（的状）况
wind cone 风袋
wind correction angle（WCA）风力修正角
wind coverage 风频涵盖率
wind direction 风向
wind direction readout 风向读取值
wind direction shaft 风向杆
wind divide 风界
wind drift 风偏流；气流偏差
wind eddies 风涡
wind factor 风因子
wind field 风场
wind fluctuation 风波动
wind gust 阵风
wind indicator 风向指示器
wind load 风力负载
wind milling 风转
wind pattern flying 气流型飞行
wind ram effects 风力冲压效应
wind resistance 风阻力
wind rose method 风频图法；风花图法
wind scale 风级
wind shear 风切变
wind shear indicator 风切指示器
wind shield 风挡；(炮弹）风帽
wind shift 风向转变；风变向

wind shift line 风向转变线
wind sleeve 风向袋
wind sock 风向袋
wind speed readout 风速读取值
wind stream 气流
wind tee T 字布
wind tetrahedron 风标
wind triangle 风速三角形；航行速度三角形
wind tunnel balance 风洞平衡区
wind tunnel model 风洞模型
wind tunnel tests 风洞试验
wind up turn 绕紧转弯
wind vector 风矢量
wind W backing SW 风由西转向西南
wind W veering NW 风由西向西北转
wind-gauge 风速计
winding speed 缠绕速率
winding structure 缠绕结构
windmill 风车；直升机；风转；螺旋桨
windmilling speed 风转转速
window cloud 浪云
window heat control unit 风挡加温控制器，窗户加热控制组件
window panel 窗玻璃
pilot's（captain's）window panels 正驾驶风挡玻璃
windscreen 风挡；挡风玻璃
windscreen panel 挡风玻璃面板
windscreen wiper 风挡雨刷
windshear 风切变
windshield 风挡；挡风玻璃
windshield anti-icer 风挡防冰装置
windshield anti-icing switch 风挡除冰开关
windshield washer button 风挡清洗按钮
windshield wiper 风挡雨刷
windshield wiper speed selector 风挡雨刷速度选择开关
wind-tunnel model 风洞模型
wind-tunnel speed 风洞气流速率
windward 逆风；顶风；上风
wing 机翼；联队
wing airplane 僚机
wing anti-icing 机翼防冰
wing butt 翼根
wing commander 空军中校（英）；联队长
wing fence 机翼导流片，翼面挡流板

wing loading 翼载荷
wing rib 翼肋
wing scan light 机翼扫描灯
wing span 翼展
wing store 翼下外挂物
wing zyglo 机翼荧光渗透检验
wing-illuminating light 机翼照射灯
wings level 无坡度
wingspan 翼展
wingtip 翼尖；翼梢
wipe ring 刮油环；刮油涨圈
wire cutting and throwing mechanism 切线与抛线机构
wire gage 线号规
wire glass 铁丝网玻璃，铅丝玻璃
wire integration unit 导线集成组件
wire marking 线号；电线标识
wire strike 撞电线失事
wiresonde 有线电探空仪
wire-wound 线绕
wiring 线路（系统）；布线；装线；接线
wiring data 线路数据
wiring defect 线路缺陷
wiring diagram 布线图
wiring diagram manual 线路图手册
wiring diagrams and data 线路图及数据
wiring manual 线路图手册
withdrawal 退出战斗
WMD（weapons of mass destruction）大规模杀伤性武器
wobble pump 手摇泵
woolpack cloud 积云
work specification 操作说明书
work-done stroke 做功行程
working energy 工作能源
working piston 工作活塞
working socket 工作插座
workload 工作量
workstation 工作站
world aeronautical chart（WAC）世界航空图
world helicopter speed record 直升机速度世界纪录
worm 螺杆；蜗杆
worm gear 蜗轮；蜗杆与蜗轮
worm gearing 蜗轮传动装置
worn bearing 老化轴承
wound 负伤
WPT（way point）航路点
wreck 遭难；失事；残骸
wreckage 失事；残骸
wrench 旋钳；扳紧器；扳手

writing speed 写入速率

wrong logic 逻辑错误

wrought aluminum alloy 锻造铝合金

WWR（war wastage reserves）战时消耗贮备

WX（weather radar）气象雷达

X target 待判明的目标；待证实目标

x-parallax 左右视差

XPDR（transponder）应答机

X-ray inspection module X 射线检查组件

X-site 露天弹药临时储存所

yaw 偏航；偏转

yaw angle 偏航角

yaw axis 偏转轴（线）；偏航轴

yaw control 方向操纵；偏航操纵

yaw control augmentation system 偏航控制增稳系统

yaw damper 偏航阻尼器

yaw damper module 偏航阻尼器组件

yaw damper system 偏航阻尼系统

yaw force trim 偏航力配平

yaw force trim release switch 偏航力配平释放开关

yaw heading hold 航向保持

yaw heading hold reference 航向保持基准

yaw linear actuator 偏航线性作动器

yaw pedal 偏航踏板，偏航脚蹬

yaw rate damping 偏航角速度阻尼

yaw rod 偏航杆

yaw stiffness 方向稳定性，抗偏刚度

yaw trim actuator 偏航力配平作动器

yaw unstability 偏航失稳

yaw vane 偏航翼

yawing moment 偏航力矩

yawing moment coefficient 偏航力矩系数

yawing moment-due-to roll 滚转偏航力矩系数

yawing moment-due-to sideslip 侧滑偏航力矩系数

yawmeter 偏航表；偏航计

yaw-stability 偏航稳定性

Y-connection Y 形接线

year of service 服役年龄

yellow alert 预备警报（空袭）

yellow caution zone 黄色警戒区

yellow wind 黄风

yield load 降伏负载，屈服载荷

yield point 降伏点，屈服点

yield strength 降伏强度，屈服强度

yield stress 降伏应力，屈服应力

yield tensile stress 降伏抗张应力

yoke 操舵柱；横舵柄；飞机操纵杆；驾驶盘

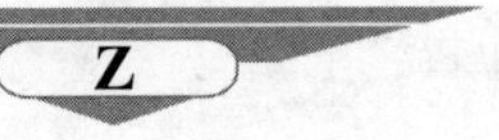

Z-9 helicopter 直 -9 直升机
Zebra time 格林尼治平均时（间）
zenith angle 天顶角
zenith distance 天顶距
zenith-nadir axis 天顶天底轴
zero airspeed 零空速
zero azimuth 零方位（角）
zero ceiling 零高云幕
zero defects 无缺点；无故障；无缺陷；无疵残
zero fuel weight 无燃油重量
zero gravity 失重；零重力
zero ground speed 零地速
zero hour 进攻发起时刻；零时；军事行动开始时间
zero launcher 无导轨发射装置
zero length 零长
zero line 基准线；零位线
zero mark 零位；零刻度；零位标记
zero meridian 零子午线
zero method 零度测量法
zero offset 零点偏移
zero phase shift 零相变换，零相移
zero position 零位
zero potential 零电位
zero rate of climb speed 零上升率速度
zero reader 零度指示器；零位读出器
zero reading 零读数；起点读数
zero resistor value 零点电阻值
zero sequence 零序
zero setting 调零；定零位；归零
zero sight line 零瞄准线
zero signal zone 无信号区；零信号区
zero stability derivatives 零安定性导数
zero stage 零级
zero stick force trim speed 松杆配平速率

zero thrust pitch 无推力螺距
zero visibility 零能见度
zero wave drag 零波阻
zero wind 无风
zero-zone time 零时区时间
zero-crossing voltage comparator 过零电压比较器
zero-entry swirl inlet 无预旋进气装置
zero-gross gradient 零爬升梯度高度
zero-length launching 零长发射
zero-lift angle 零升力角
zero-lift angle of attack 零升力迎角
zero-lift chord 零升力弦
zero-lift drag 零升力阻力
zero-lift drag coefficient 零升力阻力系数
zero-lift line 零升力线
zero-lift moment 零升力力矩
zero-lift trajectory 零升力弹道
zero-torque pitch 零扭矩螺距
zero-zero weather 零零天气
zigzag 锯齿形航线
zigzag antenna 锯齿形天线
zigzag pieces 曲折形件
zigzag resistance unit 锯齿形电阻器
zigzag riveting 差排铆接，交错铆接
zigzag scanning 曲折扫描
zigzag search 曲折搜索
zip fastener 拉链紧固件
zip fuel 高能燃料
Z-marker beacon Z 型指点信标；区域指点信标
zodiacal band 黄道带
zodiacal light 黄道光
zonal flow 纬向气流
zonal guidance facility 区域导航设施
zonal index 纬向指数
zonal wind 纬向风
zone of action 战斗地区，战斗地带；作战区；作用区
zone of interaction 航道交错区
zone of silence 盲区；静音区
zone of trade wind 信风带
zone of variable wind 变风带
zone temperature control 区域温度控制
zone temperature sensor 区域温度传感器
zone time 区时

Z-search 单机 Z 形搜索飞行

Z-time 格林尼治平均时

Zulu time 格林尼治标准时间

zyglo 荧光透视法

各国常见直升机参数

中国

1. Z-8　直 -8 多用途直升机

技术参数 Technical Parameters	
机身长度 Fuselage Length	23.05m
机身高度 Fuselage Height	6.66m
旋翼直径 Rotor Diameter	18.9m
最大起飞重量 Maximum Gross Weight	1,300kg
最大速度 Maximum Speed	315km/h
最大航程 Maximum Range	830km
最大升限 Maximum Ceiling	6,000m

2. Z-9　直 -9 多用途直升机

技术参数 Technical Parameters	
机身长度 Fuselage Length	13.46m
机身高度 Fuselage Height	3.21m
旋翼直径 Rotor Diameter	11.93m
乘员 Crews	2

续表

技术参数 Technical Parameters	
最大起飞重量 Maximum Gross Weight	4,000kg
最大速度 Maximum Speed	304km/h
最大航程 Maximum Range	1,030km
最大升限 Maximum Ceiling	6,780m

3. Z-10　Fiery Thunderbolt　直 -10“霹雳火”攻击直升机

技术参数 Technical Parameters	
机身长度 Fuselage Length	14.15m
机身高度 Fuselage Height	3.85m
旋翼直径 Rotor Diameter	13m
乘员 Crews	2
最大起飞重量 Maximum Gross Weight	7,000kg
最大速度 Maximum Speed	270km/h
最大航程 Maximum Range	800km
最大升限 Maximum Ceiling	6,400m

4. Z-11　直 11 多用途直升机

技术参数 Technical Parameters	
机身长度 Fuselage Length	13.012m
机身高度 Fuselage Height	3.14m
旋翼直径 Rotor Diameter	10.69m

续表

技术参数 Technical Parameters	
乘员 Crews	2
最大起飞重量 Maximum Gross Weight	2,200kg
最大速度 Maximum Speed	278km/h
最大航程 Maximum Range	560km
最大升限 Maximum Ceiling	5,240m

5. Z-19　Black Hurricane 直 -19“黑旋风”武装侦察直升机

技术参数 Technical Parameters	
机身长度 Fuselage Length	12m
机身高度 Fuselage Height	4.01m
旋翼直径 Rotor Diameter	12.01m
乘员 Crews	2
最大起飞重量 Maximum Gross Weight	4,500kg
最大速度 Maximum Speed	305km/h
最大航程 Maximum Range	800km

俄罗斯

1. Mi-4 Hound　Mi-4“猎犬”通用直升机

技术参数 Technical Parameters	
机身长度 Fuselage Length	20.2m
机身高度 Fuselage Height	4.4m
旋翼直径 Rotor Diameter	21m

续表

技术参数 Technical Parameters	
乘员 Crews	2
最大起飞重量 Maximum Gross Weight	7,600kg
最大速度 Maximum Speed	210km/h
最大航程 Maximum Range	520km
最大升限 Maximum Ceiling	5,500m

2. Mi-6 Hook　Mi-6“吊钩”运输直升机

技术参数 Technical Parameters	
机身长度 Fuselage Length	33.18m
机身高度 Fuselage Height	9.86m
旋翼直径 Rotor Diameter	35m
乘员 Crews	5
最大起飞重量 Maximum Gross Weight	42,500kg
最大速度 Maximum Speed	300km/h
最大航程 Maximum Range	620km
最大升限 Maximum Ceiling	4,500m

3. Mi-8 Hip　Mi-8“河马”运输直升机

技术参数 Technical Parameters	
机身长度 Fuselage Length	18.17m
机身高度 Fuselage Height	5.65m

续表

技术参数 Technical Parameters	
旋翼直径 Rotor Diameter	21.29m
乘员 Crews	3
最大起飞重量 Maximum Gross Weight	12,000kg
最大速度 Maximum Speed	260km/h
最大航程 Maximum Range	450km
最大升限 Maximum Ceiling	4,500m

4. Mi-24 Hind　Mi-24“雌鹿”武装直升机

技术参数 Technical Parameters	
机身长度 Fuselage Length	17.5m
机身高度 Fuselage Height	6.5m
旋翼直径 Rotor Diameter	17.3m
乘员 Crews	2
最大起飞重量 Maximum Gross Weight	12,000kg
最大速度 Maximum Speed	335km/h
最大航程 Maximum Range	450km
最大升限 Maximum Ceiling	4,500m

5. Mi-26 Halo　Mi-26“光环”通用直升机

技术参数 Technical Parameters	
机身长度 Fuselage Length	40.03m

续表

技术参数 Technical Parameters	
机身高度 Fuselage Height	8.15m
旋翼直径 Rotor Diameter	32m
乘员 Crews	5
最大起飞重量 Maximum Gross Weight	56,000kg
最大速度 Maximum Speed	295km/h
最大航程 Maximum Range	1,920km
最大升限 Maximum Ceiling	4,600m

6. Mi-28 Havoc　Mi-28“浩劫”武装直升机

技术参数 Technical Parameters	
机身长度 Fuselage Length	17.01m
机身高度 Fuselage Height	3.82m
旋翼直径 Rotor Diameter	17.20m
乘员 Crews	2
最大起飞重量 Maximum Gross Weight	11,500kg
最大速度 Maximum Speed	325km/h
最大航程 Maximum Range	1,100km
最大升限 Maximum Ceiling	5,800m

7. Mi-35 Hind E　Mi-35“雌鹿 E”武装直升机

技术参数 Technical Parameters	
机身长度 Fuselage Length	18.8m

续表

技术参数 Technical Parameters	
机身高度 Fuselage Height	6.5m
旋翼直径 Rotor Diameter	17.1m
乘员 Crews	2
最大起飞重量 Maximum Gross Weight	11,500kg
最大速度 Maximum Speed	330km/h
最大航程 Maximum Range	500km
最大升限 Maximum Ceiling	4,500m

8. Ka-25 Hormone　Ka-25“激素”反潜直升机

技术参数 Technical Parameters	
机身长度 Fuselage Length	9.75m
机身高度 Fuselage Height	5.37m
旋翼直径 Rotor Diameter	15.7m
乘员 Crews	4
最大起飞重量 Maximum Gross Weight	7,500kg
最大速度 Maximum Speed	209km/h
最大航程 Maximum Range	400km
最大升限 Maximum Ceiling	3,350m

9. Ka-27 Helix　Ka-27“蜗牛”反潜直升机

技术参数 Technical Parameters	
机身长度 Fuselage Length	11.3m

续表

技术参数 Technical Parameters	
机身高度 Fuselage Height	5.5m
旋翼直径 Rotor Diameter	15.8m
乘员 Crews	3
最大起飞重量 Maximum Gross Weight	12,000kg
最大速度 Maximum Speed	270km/h
最大航程 Maximum Range	980km
最大升限 Maximum Ceiling	5,000m

10. Ka-50 Black Shark　Ka-50“黑鲨”武装直升机

技术参数 Technical Parameters	
机身长度 Fuselage Length	13.5m
机身高度 Fuselage Height	5.4m
旋翼直径 Rotor Diameter	14.5m
乘员 Crews	1
最大起飞重量 Maximum Gross Weight	10,800kg
最大速度 Maximum Speed	350km/h
最大航程 Maximum Range	1,160km
最大升限 Maximum Ceiling	5,500m

11. Ka-52 Alligator　Ka-52“短吻鳄”武装直升机

技术参数 Technical Parameters	
机身长度 Fuselage Length	15.96m

续表

技术参数 Technical Parameters	
机身高度 Fuselage Height	4.93m
旋翼直径 Rotor Diameter	14.43m
乘员 Crews	2
最大起飞重量 Maximum Gross Weight	10,400kg
最大速度 Maximum Speed	310km/h
最大航程 Maximum Range	1,100km
最大升限 Maximum Ceiling	5,500m

12. Ka-60 Kasatka　Ka-60“逆戟鲸”直升机

技术参数 Technical Parameters	
机身长度 Fuselage Length	15.6m
机身高度 Fuselage Height	4.6m
旋翼直径 Rotor Diameter	13.5m
乘员 Crews	2
最大起飞重量 Maximum Gross Weight	6,500kg
最大速度 Maximum Speed	300km/h
最大航程 Maximum Range	615km
最大升限 Maximum Ceiling	5,150m

美国

1. H-21 Shawnee　H-21“肖尼”通用直升机

技术参数 Technical Parameters	
机身长度 Fuselage Length	16.01m

续表

技术参数 Technical Parameters	
机身高度 Fuselage Height	4.8m
旋翼直径 Rotor Diameter	13.14m
乘员 Crews	5
最大起飞重量 Maximum Gross Weight	6,609kg
最大速度 Maximum Speed	204km/h
最大航程 Maximum Range	644km

2. CH-47D/F CHINOOK　CH-47D/F“支奴干”运输直升机

技术参数 Technical Parameters	
长（桨叶旋转时）Length（rotors turning）	98 feet 10 inches 98 英尺 10 英寸
高（桨叶旋转时）Height（rotors turning）	8 feet 11 inches 18 英尺 11 英寸
宽（桨叶旋转时）Width（rotors turning）	60 feet 60 英尺
机身长度 Fuselage Length	50 feet 9 inches　50 英尺 9 英寸
机身宽度 Fuselage Width	12 feet 5 inches 12 英尺 5 英寸
主旋翼直径 Main Rotor Diameter	60 feet　60 英尺
最大起飞重量 Maximum Gross Weight	50,000 pounds 50 000 磅
巡航速度 Cruise Airspeed	120 to 145 knots120-145 节
作战半径（运载 16 000 磅货物时）Combat Radius（16,000 pounds cargo）	50 nautical miles（90 kilometers）50 海里（90 公里）
作战半径（运输 33 名人员时）Combat radius（33 troops）	120 nautical miles（180 kilometers）120 海里（180 公里）

3. H-43 Huskie　H-43“哈斯基”通用直升机

技术参数 Technical Parameters	
机身长度 Fuselage Length	7.6m
机身高度 Fuselage Height	5.18m
旋翼直径 Rotor Diameter	14.3m
乘员 Crews	2
最大起飞重量 Maximum Gross Weight	4,150kg
最大速度 Maximum Speed	190km/h
最大航程 Maximum Range	298km
最大升限 Maximum Ceiling	7,620m

4. UH-60 L/M Blackhawk　UH-60 L/M“黑鹰”通用直升机

技术参数 Technical Parameters	
长（桨叶旋转时）Length（rotors turning）	64 feet 10 inch 64 英尺 10 英寸
高 Height	12 feet 4 inches at center hub; 17 feet 11 inches at tail rotor（turning） 中部桨叶 12 英尺 4 英寸；尾部桨叶 17 英尺 11 英寸（旋转时）
宽（桨叶旋转时）Width（rotors turning）	53 feet 8 inches 53 英尺 8 英寸
主旋翼和尾旋翼直径 Main Rotor & Tail Rotor Diameter	53 feet 8 inches at main rotor; 11 feet at tail rotor at 20-degree angle 主桨叶 53 英尺 8 英寸；尾部桨叶 20° 角时 11 英尺

续表

技术参数 Technical Parameters	
机身长度 Fuselage Length	41 feet 4 inches with rotors and pylons folded 桨叶和挂架折叠时 41 英尺 4 英寸
机身宽度 Fuselage Width	9 feet 8.6 inches 9 英尺 8.6 英寸
最大起飞重量 Maximum Gross Weight	22,000 pounds 22 000 磅
巡航速度 Cruise Airspeed	120 to 145 knots 120-145 节
作战半径 Combat Radius	225 kilometers 225 公里

5. H-19 Chickasaw H-19“契卡索人”通用直升机

技术参数 Technical Parameters	
机身长度 Fuselage Length	19.1m
机身高度 Fuselage Height	4.07m
旋翼直径 Rotor Diameter	16.16m
乘员 Crews	2
最大起飞重量 Maximum Gross Weight	3,587kg
最大速度 Maximum Speed	163km/h
最大航程 Maximum Range	652km
最大升限 Maximum Ceiling	3,200m

6. UH-1 Iroquois UH-1“伊洛魁”通用直升机

技术参数 Technical Parameters	
机身长度 Fuselage Length	17.4m

续表

技术参数 Technical Parameters	
机身高度 Fuselage Height	4.4m
旋翼直径 Rotor Diameter	14.6m
乘员 Crews	4
最大起飞重量 Maximum Gross Weight	4,310kg
最大速度 Maximum Speed	220km/h
最大航程 Maximum Range	510km
最大升限 Maximum Ceiling	5,910m

7. SH-2 Sea Sprite　SH-2“海妖”搜索直升机

技术参数 Technical Parameters	
机身长度 Fuselage Length	15.9m
机身高度 Fuselage Height	4.5m
旋翼直径 Rotor Diameter	13.4m
乘员 Crews	3
最大起飞重量 Maximum Gross Weight	6,120kg
最大速度 Maximum Speed	256km/h
最大航程 Maximum Range	1,000km
最大升限 Maximum Ceiling	3,000m

8. ARH-70 Arapaho　ARH-70“阿拉帕霍”攻击 / 侦察直升机

技术参数 Technical Parameters	
机身长度 Fuselage Length	10.57m

续表

技术参数 Technical Parameters	
机身高度 Fuselage Height	3.56m
旋翼直径 Rotor Diameter	10.67m
乘员 Crews	2
最大起飞重量 Maximum Gross Weight	2,268kg
最大速度 Maximum Speed	259km/h
最大航程 Maximum Range	362km
最大升限 Maximum Ceiling	6,096m

9. AH-64D/E Apache AH-64D/E“阿帕奇”攻击直升机

技术参数 Technical Parameters	
长（桨叶旋转时）Length（rotors turning）	58 feet 1 inch 58 英尺 1 英寸
高（桨叶旋转时）Height（rotors turning）	17 feet 6 inches 17 英尺 6 英寸
宽（桨叶旋转时）Width（rotors turning）	48 feet 48 英尺
机身长度 Fuselage Length	49 feet 8 inches 49 英尺 8 英寸
机身宽度 Fuselage Width	16 feet 4 inches 16 英尺 4 英寸
主旋翼直径 Main Rotor Diameter	48 feet 48 英尺
最大起飞重量 Maximum Gross Weight	23,000 pounds 23 000 磅
巡航速度 Cruise Airspeed	110 to 120 knots 110-120 节
作战半径 Combat Radius	120 kilometers 120 公里

10. S-97 Raider　S-97“侵袭者”攻击直升机

技术参数 Technical Parameters	
机身长度 Fuselage Length	11m
乘员 Crews	2
最大起飞重量 Maximum Gross Weight	4,990kg
最大速度 Maximum Speed	444km/h
最大航程 Maximum Range	570km
最大升限 Maximum Ceiling	3,048m

11. RAH-66 Comanche　RAH-66“科曼奇”侦察 / 攻击直升机

技术参数 Technical Parameters	
机身长度 Fuselage Length	14.28m
机身高度 Fuselage Height	3.37m
旋翼直径 Rotor Diameter	11.9m
乘员 Crews	2
最大起飞重量 Maximum Gross Weight	7,790kg
最大速度 Maximum Speed	324km/h
最大航程 Maximum Range	485km
最大升限 Maximum Ceiling	4,566m

12. MH-68A Sting Ray　MH-68A“刺鳐”近程武装拦截直升机

技术参数 Technical Parameters	
机身长度 Fuselage Length	13.5m
机身高度 Fuselage Height	3.3m

续表

技术参数 Technical Parameters	
旋翼直径 Rotor Diameter	11m
乘员 Crews	3
最大起飞重量 Maximum Gross Weight	3,000kg
最大速度 Maximum Speed	305km/h
最大航程 Maximum Range	565km

13. CH-34 Choctaw　CH-34“乔克托人”运输直升机

技术参数 Technical Parameters	
机身长度 Fuselage Length	17.28m
机身高度 Fuselage Height	4.85m
旋翼直径 Rotor Diameter	17.07m
乘员 Crews	2
最大起飞重量 Maximum Gross Weight	6,350kg
最大速度 Maximum Speed	198km/h
最大航程 Maximum Range	293km
最大升限 Maximum Ceiling	1,495m

14. AH-1 Cobra　AH-1“眼镜蛇”攻击直升机

技术参数 Technical Parameters	
机身长度 Fuselage Length	13.6m
机身高度 Fuselage Height	4.1m

续表

技术参数 Technical Parameters	
旋翼直径 Rotor Diameter	14.63m
乘员 Crews	2
最大起飞重量 Maximum Gross Weight	4,500kg
最大速度 Maximum Speed	277km/h
最大航程 Maximum Range	510km
最大升限 Maximum Ceiling	3,720m

15. SH-3 Sea King　SH-3“海王”搜索直升机

技术参数 Technical Parameters	
机身长度 Fuselage Length	16.7m
机身高度 Fuselage Height	5.13m
旋翼直径 Rotor Diameter	19m
乘员 Crews	4
最大起飞重量 Maximum Gross Weight	10,000kg
最大速度 Maximum Speed	267km/h
最大航程 Maximum Range	1,000km
最大升限 Maximum Ceiling	4,481m

16. AH-6 Little Bird　AH-6“小鸟”攻击直升机

技术参数 Technical Parameters	
机身长度 Fuselage Length	9.94m

续表

技术参数 Technical Parameters	
机身高度 Fuselage Height	2.48m
旋翼直径 Rotor Diameter	8.3m
乘员 Crews	2
最大起飞重量 Maximum Gross Weight	1,610kg
最大速度 Maximum Speed	282km/h
最大航程 Maximum Range	430km
最大升限 Maximum Ceiling	5,700m

17. CH-46 Sea Knight　CH-46“海骑士”运输直升机

技术参数 Technical Parameters	
机身长度 Fuselage Length	13.66m
机身高度 Fuselage Height	5.09m
旋翼直径 Rotor Diameter	15.24m
乘员 Crews	5
最大起飞重量 Maximum Gross Weight	11,000kg
最大速度 Maximum Speed	267km/h
最大航程 Maximum Range	1,020km
最大升限 Maximum Ceiling	5,180m

18. CH-53 Sea Stallion　CH-53“海上种马”运输直升机

技术参数 Technical Parameters	
机身长度 Fuselage Length	26.97m

续表

技术参数 Technical Parameters	
机身高度 Fuselage Height	7.6m
旋翼直径 Rotor Diameter	22.01m
乘员 Crews	2
最大起飞重量 Maximum Gross Weight	19,100kg
最大速度 Maximum Speed	315km/h
最大航程 Maximum Range	1,000km
最大升限 Maximum Ceiling	5,106m

19. CH-54 Tarhe　CH-54“塔赫”运输直升机

技术参数 Technical Parameters	
机身长度 Fuselage Length	26.97m
机身高度 Fuselage Height	7.75m
旋翼直径 Rotor Diameter	21.95m
乘员 Crews	3
最大起飞重量 Maximum Gross Weight	21,000kg
最大速度 Maximum Speed	240km/h
最大航程 Maximum Range	370km
最大升限 Maximum Ceiling	5,600m

20. OH-58 Kiowa　OH-58“基欧瓦”观察直升机

技术参数 Technical Parameters	
机身长度 Fuselage Length	12.39m

续表

技术参数 Technical Parameters	
机身高度 Fuselage Height	2.29m
旋翼直径 Rotor Diameter	10.67m
乘员 Crews	2
最大起飞重量 Maximum Gross Weight	2,358kg
最大速度 Maximum Speed	222km/h
最大航程 Maximum Range	556km
最大升限 Maximum Ceiling	6,250m

21. SH-60 Seahawk　SH-60“海鹰”搜索直升机

技术参数 Technical Parameters	
机身长度 Fuselage Length	19.75m
机身高度 Fuselage Height	5.2m
旋翼直径 Rotor Diameter	16.35m
乘员 Crews	4
最大起飞重量 Maximum Gross Weight	9,927kg
最大速度 Maximum Speed	333km/h
最大航程 Maximum Range	834km
最大升限 Maximum Ceiling	3,580m

22. UH-72 Lakota　UH-72“勒科塔”通用直升机

技术参数 Technical Parameters	
机身长度 Fuselage Length	13.03m

续表

技术参数 Technical Parameters	
机身高度 Fuselage Height	3.45m
旋翼直径 Rotor Diameter	11m
乘员 Crews	2
最大起飞重量 Maximum Gross Weight	3,585kg
最大速度 Maximum Speed	269km/h
最大航程 Maximum Range	685km
最大升限 Maximum Ceiling	5,791m

23. VH-71 Kestrel　VH-71 "茶隼" 总统直升机

技术参数 Technical Parameters	
机身长度 Fuselage Length	22.81m
机身高度 Fuselage Height	6.65m
旋翼直径 Rotor Diameter	18.59m
乘员 Crews	4
最大起飞重量 Maximum Gross Weight	15,600kg
最大速度 Maximum Speed	309km/h
最大航程 Maximum Range	1,389km
最大升限 Maximum Ceiling	4,575m

英法德意

1. SA341/342 Gazelle　SA341/342“小羚羊”武装直升机

技术参数 Technical Parameters	
机身长度 Fuselage Length	11.97m
机身高度 Fuselage Height	3.19m
旋翼直径 Rotor Diameter	10.5m
乘员 Crews	2
最大起飞重量 Maximum Gross Weight	1,900kg
最大速度 Maximum Speed	260km/h
最大航程 Maximum Range	710km
最大升限 Maximum Ceiling	4,100m

2. Lynx　“山猫”通用直升机

技术参数 Technical Parameters	
机身长度 Fuselage Length	15.16m
机身高度 Fuselage Height	3.66m
旋翼直径 Rotor Diameter	12.8m
乘员 Crews	2
最大起飞重量 Maximum Gross Weight	4,535kg
最大速度 Maximum Speed	289km/h
最大航程 Maximum Range	630km
最大升限 Maximum Ceiling	3,230m

3. Super Lynx　“超级山猫”通用直升机

技术参数 Technical Parameters	
机身长度 Fuselage Length	15.24m
机身高度 Fuselage Height	3.67m
旋翼直径 Rotor Diameter	12.8m
乘员 Crews	2
最大起飞重量 Maximum Gross Weight	5,125kg
最大速度 Maximum Speed	289km/h
最大航程 Maximum Range	630km
最大升限 Maximum Ceiling	3,230m

4. AW159 Wildcat　AW159“野猫”武装直升机

技术参数 Technical Parameters	
机身长度 Fuselage Length	15.24m
机身高度 Fuselage Height	3.73m
旋翼直径 Rotor Diameter	12.8m
乘员 Crews	2
最大起飞重量 Maximum Gross Weight	6,000kg
最大速度 Maximum Speed	291km/h
最大航程 Maximum Range	777km

5. SA 316/319 Alouette Ⅲ　SA 316/319“云雀 Ⅲ”通用直升机

技术参数 Technical Parameters	
机身长度 Fuselage Length	12.84m

续表

技术参数 Technical Parameters	
机身高度 Fuselage Height	3m
旋翼直径 Rotor Diameter	11.02m
乘员 Crews	1
最大起飞重量 Maximum Gross Weight	2,200kg
最大速度 Maximum Speed	220km/h
最大航程 Maximum Range	605km
最大升限 Maximum Ceiling	4,000m

6. SA 321 Super Frelon　SA 321“超黄蜂”通用直升机

技术参数 Technical Parameters	
机身长度 Fuselage Length	23.03m
机身高度 Fuselage Height	6.66m
旋翼直径 Rotor Diameter	18.9m
乘员 Crews	2
最大起飞重量 Maximum Gross Weight	13,000kg
最大速度 Maximum Speed	275km/h
最大航程 Maximum Range	1,020km
最大升限 Maximum Ceiling	3,150m

7. SA 330 Puma　SA 330“美洲豹”通用直升机

技术参数 Technical Parameters	
机身长度 Fuselage Length	19.5m

续表

技术参数 Technical Parameters	
机身高度 Fuselage Height	5.14m
旋翼直径 Rotor Diameter	15m
乘员 Crews	2
最大起飞重量 Maximum Gross Weight	7,500kg
最大速度 Maximum Speed	271km/h
最大航程 Maximum Range	572km
最大升限 Maximum Ceiling	6,000m

8. SA 360/361/365 Dauphin　SA 360/361/365“海豚”通用直升机

技术参数 Technical Parameters	
机身长度 Fuselage Length	13.2m
机身高度 Fuselage Height	3.5m
旋翼直径 Rotor Diameter	11.5m
乘员 Crews	2
最大起飞重量 Maximum Gross Weight	3,000kg
最大速度 Maximum Speed	315km/h
最大航程 Maximum Range	675km
最大升限 Maximum Ceiling	4,600m

9. SA 532 Cougar　SA 532“美洲狮”通用直升机

技术参数 Technical Parameters	
机身长度 Fuselage Length	18.7m

续表

技术参数 Technical Parameters	
机身高度 Fuselage Height	4.92m
旋翼直径 Rotor Diameter	15.6m
乘员 Crews	2
最大起飞重量 Maximum Gross Weight	9,000kg
最大速度 Maximum Speed	278km/h
最大航程 Maximum Range	870km
最大升限 Maximum Ceiling	4,100m

10. SA 565 Panther　SA 565 “黑豹” 通用直升机

技术参数 Technical Parameters	
机身长度 Fuselage Length	13.7m
机身高度 Fuselage Height	4.1m
旋翼直径 Rotor Diameter	11.9m
乘员 Crews	2
最大起飞重量 Maximum Gross Weight	4,250kg
最大速度 Maximum Speed	296km/h
最大航程 Maximum Range	875km

11. Bo　105 武装直升机

技术参数 Technical Parameters	
机身长度 Fuselage Length	11.86m
机身高度 Fuselage Height	3m

续表

技术参数 Technical Parameters	
旋翼直径 Rotor Diameter	9.84m
乘员 Crews	2
最大起飞重量 Maximum Gross Weight	2,500kg
最大速度 Maximum Speed	270km/h
最大航程 Maximum Range	575km
最大升限 Maximum Ceiling	5,180m

12. A129 Mangusta　A129“猫鼬”武装直升机

技术参数 Technical Parameters	
机身长度 Fuselage Length	12.28m
机身高度 Fuselage Height	3.35m
旋翼直径 Rotor Diameter	11.90m
乘员 Crews	2
最大起飞重量 Maximum Gross Weight	4,600kg
最大速度 Maximum Speed	278km/h
最大航程 Maximum Range	1,000km
最大升限 Maximum Ceiling	4,725m

13. AW139　多用途直升机

技术参数 Technical Parameters	
机身长度 Fuselage Length	16.65m
机身高度 Fuselage Height	4.95m

续表

技术参数 Technical Parameters	
旋翼直径 Rotor Diameter	13.80m
乘员 Crews	1
最大起飞重量 Maximum Gross Weight	6,400kg
最大速度 Maximum Speed	310km/h
最大升限 Maximum Ceiling	6,000m

14. AH-64D Apache Longbow　AH-64D“长弓阿帕奇”武装直升机

技术参数 Technical Parameters	
机身长度 Fuselage Length	17.73m
机身高度 Fuselage Height	3.87m
旋翼直径 Rotor Diameter	14.63m
乘员 Crews	2
最大起飞重量 Maximum Gross Weight	10,433kg
最大速度 Maximum Speed	293km/h
最大航程 Maximum Range	1,900km
最大升限 Maximum Ceiling	6,400m

欧洲

1. Tiger　“虎”式武装直升机

技术参数 Technical Parameters	
机身长度 Fuselage Length	14.08m
机身高度 Fuselage Height	3.83m

续表

技术参数 Technical Parameters	
旋翼直径 Rotor Diameter	13m
乘员 Crews	2
最大起飞重量 Maximum Gross Weight	6,000kg
最大速度 Maximum Speed	315km/h
最大航程 Maximum Range	800km
最大升限 Maximum Ceiling	4,000m

2. NH90　通用直升机

技术参数 Technical Parameters	
机身长度 Fuselage Length	19.56m
机身高度 Fuselage Height	5.44m
旋翼直径 Rotor Diameter	16.00m
乘员 Crews	3
最大起飞重量 Maximum Gross Weight	10,000kg
最大速度 Maximum Speed	310km/h
最大航程 Maximum Range	1,204km
最大升限 Maximum Ceiling	6,000m

3. EH101 Merlin　EH101“灰背隼”通用直升机

技术参数 Technical Parameters	
机身长度 Fuselage Length	22.81m
机身高度 Fuselage Height	6.65m

续表

技术参数 Technical Parameters	
旋翼直径 Rotor Diameter	18.59m
乘员 Crews	3-4
最大起飞重量 Maximum Gross Weight	14,600kg
最大速度 Maximum Speed	309km/h
最大航程 Maximum Range	833km
最大升限 Maximum Ceiling	4,575m

4. AS555 Fennec　AS555 “小狐” 轻型直升机

技术参数 Technical Parameters	
机身长度 Fuselage Length	12.94m
机身高度 Fuselage Height	3.34m
旋翼直径 Rotor Diameter	10.69m
乘员 Crews	2
最大起飞重量 Maximum Gross Weight	2,250kg
最大速度 Maximum Speed	246km/h
最大航程 Maximum Range	648km
最大升限 Maximum Ceiling	5,280m

其他国家

1. 印度 LCH　轻型武装直升机

技术参数 Technical Parameters	
机身长度 Fuselage Length	15.8m

续表

技术参数 Technical Parameters	
机身高度 Fuselage Height	4.7m
旋翼直径 Rotor Diameter	13.3m
乘员 Crews	2
最大起飞重量 Maximum Gross Weight	5,800kg
最大速度 Maximum Speed	330km/h
最大航程 Maximum Range	700km
最大升限 Maximum Ceiling	6,500m

2. 印度 Rudra　“楼陀罗”武装直升机

技术参数 Technical Parameters	
机身长度 Fuselage Length	15.87m
机身高度 Fuselage Height	4.98m
旋翼直径 Rotor Diameter	13.2m
乘员 Crews	2
最大起飞重量 Maximum Gross Weight	5,500kg
最大速度 Maximum Speed	290km/h
最大航程 Maximum Range	827km
最大升限 Maximum Ceiling	6,096m

3. 南非 CSH-2 Rooivalk　CSH-2“石茶隼”武装直升机

技术参数 Technical Parameters	
机身长度 Fuselage Length	18.73m

续表

技术参数 Technical Parameters	
机身高度 Fuselage Height	5.19m
旋翼直径 Rotor Diameter	15.58m
乘员 Crews	2
最大起飞重量 Maximum Gross Weight	7,500kg
最大速度 Maximum Speed	309km/h
最大航程 Maximum Range	1,200km
最大升限 Maximum Ceiling	6,100m

4. 伊朗 Toufan “风暴”武装直升机

技术参数 Technical Parameters	
机身长度 Fuselage Length	14m
机身高度 Fuselage Height	4m
旋翼直径 Rotor Diameter	15m
乘员 Crews	2
最大起飞重量 Maximum Gross Weight	4,500kg
最大速度 Maximum Speed	280km/h
最大航程 Maximum Range	550km
最大升限 Maximum Ceiling	3,800m

5. 日本 OH-1 Ninja “忍者”武装直升机

技术参数 Technical Parameters	
机身长度 Fuselage Length	12m

续表

技术参数 Technical Parameters	
机身高度 Fuselage Height	3.8m
旋翼直径 Rotor Diameter	11.6m
乘员 Crews	2
最大起飞重量 Maximum Gross Weight	4,000kg
最大速度 Maximum Speed	278km/h
最大航程 Maximum Range	540km
最大升限 Maximum Ceiling	4,880m

6. 韩国 KUH-1 Surion “雄鹰”通用直升机

技术参数 Technical Parameters	
机身长度 Fuselage Length	19m
机身高度 Fuselage Height	4.5m
旋翼直径 Rotor Diameter	15.8m
乘员 Crews	2
最大起飞重量 Maximum Gross Weight	8,709kg
最大速度 Maximum Speed	259km/h
最大航程 Maximum Range	480km
最大升限 Maximum Ceiling	3,000m

7. AS355NP Ecureuil “松鼠”直升机

技术参数 Technical Parameters	
机身长度 Fuselage Length	12.94m

续表

技术参数 Technical Parameters	
机身高度 Fuselage Height	3.34m
旋翼直径 Rotor Diameter	10.69m
乘员 Crews	1
最大起飞重量 Maximum Gross Weight	2,427kg
最大速度 Maximum Speed	287km/h
最大升限 Maximum Ceiling	6,100m